Landscape Ethnoecology

Environmental Anthropology and Ethnobiology

General Editor: **Roy Ellen**, FBA
Professor of Anthropology, University of Kent at Canterbury

Interest in environmental anthropology has grown steadily in recent years, reflecting national and international concern about the environment and developing research priorities. This major international series is a vehicle for publishing up-to-date monographs and edited works on particular issues, themes, places or peoples which focus on the interrelationship between society, culture and environment.

Landscape Ethnoecology

Concepts of Biotic and Physical Space

Edited by

Leslie Main Johnson & Eugene S. Hunn

Berghahn Books
New York • Oxford

First edition published in 2010 by
Berghahn Books

www.berghahnbooks.com

Library of Congress Cataloging-in-Publication Data

Landscape ethnoecology : concepts of biotic and physical space / edited by
Leslie Main Johnson and Eugene S. Hunn.
    p. cm. — (Environmental anthropology and ethnobiology)
    Includes bibliographical references and index.
    ISBN 978-1-84545-613-9 (hbk.) -- ISBN 978-0-85745-631-1 (pbk.)
    1. Landscape assessment. 2. Landscape ecology. 3. Human geography.
4. Indigenous peoples—Ecology. 5. Traditional ecological knowledge.
6. Geographical perception. I. Johnson, Leslie M. II. Hunn, Eugene S.
    GF90.L256 2009
    304.2'3—dc22

                                                              2009025446

British Library Cataloguing in Publication Data

A catalogue record for this book is available from the British Library

Printed in the United States on acid-free paper

ISBN: 978-0-85745-631-1 (paperback) ISBN: 978-0-85745-632-8 (ebook)

Contents

Part 3: Linkages and Meanings of Landscapes and Cultural Landscapes

Part 4: Conclusions

❧ Figures

∽ Tables

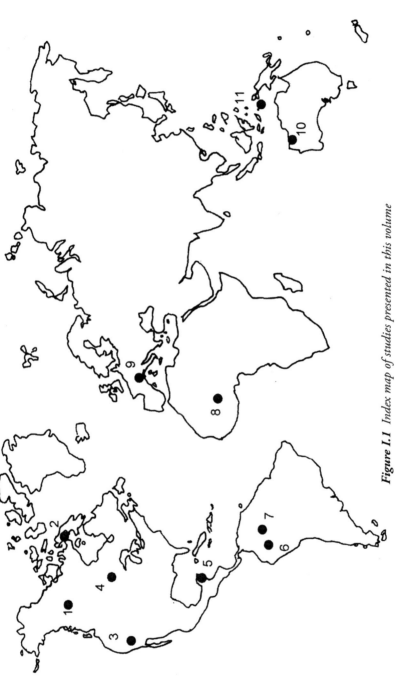

Figure I.1 Index map of studies presented in this volume

1 Kaska Dena (Chapter 10)
2 Ingloolik Inuit (Chapter 9)
3 Southern Paiute (Chapter 12)

4 Iskatewizaagegan Anishinaabe (Chapter 11)
5 Yucatec Maya, Chunhuhub (Chapter 13)
6 Maijuna (Chapter 7)

7 Baniwa (Chapter 5)
8 Fulani (Chapter 4)
9 Savoie (Chapter 8)

10 Yindjibarndi (Chapter 3)
11 Nuaulu (Chapter 6)

Landscape Ethnoecology
Concepts of Biotic and Physical Space

Leslie Main Johnson and Eugene S. Hunn

Overview

The fundamental concern of this volume is landscape. Our focus is on the perception of the land, the parsing of its patterns, and the classification of its constituent parts in local ethnoecological systems, and on the significance of these understandings in the ethnoecology of local groups.

We emphasize landscape as perceived and imagined by the people who live in it, the land seen, used and occupied by the members of a local community. It is a cultural landscape. This notion of landscape has some resonance with "territory," or "country" as used in the Australian literature, but with somewhat greater emphasis on ecological relationships and understandings. Notions from landscape ecology (a spatial patterning of diverse ecosystems or environmental types), cultural geography (climate, landforms, waters, vegetation, and human response and interaction with landscape), and the anthropology of landscape (cultural perceptions, understandings, and meanings of landscape) are all compatible with this understanding of "landscape."

Our particular focus foregrounds cultural understanding of the significant elements of landscape and their ecological entailments. We are interested in the classification, perception, and interaction of local peoples with their homelands and environments and the beings that share their landscape. Our interest is *ethnoecological*, focused on people's knowledge of and interactions with landscape. After some debate, we decided that the term *landscape ethnoecology* was the most appropriate designation for our area of interest. Our work differs from much of the research on place by virtue of its ecological focus, and it is also distinct from ethnogeographic accounts primarily concerned with the recognition and naming of *specific* places (toponymy). It also differs from most ethnobiological research in that its landscape orientation contrasts with a focus on plant or animal species, their naming, use, and so on. Our treatment diverges from other ethnoecological work (e.g., Nazarea 1999) in its particular focus on how people understand their homelands or landscapes, rather than examining land management, conservation, or ecological/subsistence practice, though such issues inform many of the contributions included here. Our approach resembles the ethnoecology of Victor Toledo, who includes landscape knowledge, human practices, and human cosmological beliefs as part of his systems approach to ethnoecology (Toledo 1992, 2002). Although we are not strongly concerned with the spiritual and cosmological realm as a primary focus, for some of our contributors such concerns represent a significant aspect of the ecological understandings of the groups with which they work.

In landscape ecology, *ecotopes* are the smallest units of landscape (Tansley 1939; Troll 1971). We here engage the array of culturally recognized landscape elements, "place kinds" or *folk ecotopes* recognized as significant in the landscape ethnoecology by members of specific local communities or cultural groups. The range of phenomena of interest includes biotic, abiotic, and cultural or anthropogenic types.

We are interested in elements that are distributed repetitively across the landscape, conceptual elements that constitute the biotic and physical space in which peoples live, rather than the rich array of particular places in their homelands, designated by common nouns as opposed to proper nouns of landscape vocabulary. *Folk* or *cultural ecotopes* are not equivalent to cultural understanding of habitat, though kinds of place have entailments, physical and biotic characteristics that may be correlated with habitat-types as conceived in the landscape classification literature (see Daubenmire 1952; Daubenmire and Daubenmire 1968; Franklin and Dyrness 1973; Pfister et al. 1977; Pfister and Arno 1980). The actual correspondence between folk ecotopes and the habitat categories of Western science is an empirical question.

While the classification and naming of plants and animals shows striking cross-cultural similarities (cf. Berlin 1992), ways of recognizing *folk ecotopes* may be more variable between cultures. The reasons for this are diverse. First of all, landform elements are less discrete than populations of living organisms. As Mark and Turk indicate, "Unlike higher plants and animals, which to some large degree are grouped into species by nature, landforms more properly belong to continua" (Mark and Turk 2003: 3). Their contribution to this volume discusses this in greater detail.

Scale is a significant issue in ecology in general and in landscape ecology in particular. The spatial scale relevant for landscape ethnoecology is that which can be observed by people on the ground as they travel through their environments in the course of their normal activities. One of the areas of ambiguity related to scale could be called the substrate problem. Substrates are defined in terms of

[handwritten margin notes: "substrate problem"]

substances, but when does substance become a spatial element? "Sandbar" is an example: a repeating spatial configuration of mineral grains of sand texture. How about "sand" itself? "Sand" is not intrinsically spatial, so only becomes a landscape element when it occurs in particular spatial configurations. Then there is "quicksand," a term that is related to sand as substrate yet implies a specific place that is dangerous and cannot be crossed. An additional example is "moss." On the one hand "moss" may be considered a kind of plant. However, for the Montaignais, an Algonquian-speaking group of sub-Arctic Eastern Canada, "moss" is construed as a kind of earth (Clément 1990)—a substrate—while for the Dene peoples, "moss" may indicate types of forest or muskeg stands that have thick layers of feather moss or sphagnum—an *ecotope* (Johnson field notes). Soils and edaphic types and fine-scale ground cover may have polysemous senses denoting landscape elements within specific landscape ethnoecological systems.

Although landscape elements and their characteristic features are culturally heterogeneous, vary in scope, and lack the discreteness of biological species, they nonetheless reflect aspects of landscape that have biological and, we would argue, adaptive significance. *[margin note: "adaptive significance"]* Folk ecotopes highlight *features of the landscape useful for people making a living off the land*. Landscape is not a *tabula rasa* on which culture elaborates; rather, the relationship between land and classification or understanding of land is a feedback loop that takes in both the potential of the land and *[margin note: "feedback loop"]* human ways of making a living, including human technologies, cosmologies, and knowledge systems.

We will close this discussion by touching briefly on anthropogenic landscapes and ecotopes. From the cases included in this work and elsewhere in the literature (e.g., Alcorn 1981a, 1981b; Deur and Turner 2005) it appears that it is not useful to create a categorical binary contrast between "natural" and "anthropogenic" landscapes because in fact this varies amongst cultures and can best be construed as a *continuum*. The degree of modification of portions of the landscape, and the erection of concepts of "natural" and "anthropogenic," range widely among cultures, as does their distinctiveness or pervasiveness in local contexts. The built environment represents an endpoint in the continuum of "natural" to "anthropogenic" environments, and indeed, many cultures contrast villages or dwellings with environments more dominated by natural vegetation and ecological processes (Johnson 2000).[1]

Significance of Landscape Ethnoecology

Landscape ethnoecology bears on many intellectual and practical concerns. As with ethnobiology and other fields of ethnoscience, examining understandings and classifications of landscape elements cross-culturally illuminates aspects of human cognition and helps to place the "received" perceptions and classifications of Western sciences in perspective, allowing us to question their naturalness or in-

evitability. Perhaps as significant, it underscores the sophistication of local knowledge of landscape and highlights the connections between traditional economies
and management of these lands with ways of thinking about them. The domain
of traditional ecological knowledge of landscape is a rich repository of knowledge
relevant to sustainable development and management of lands and waters. Our
landscape ethnoecological perspective also underscores the holistic construction
of meaning in understandings of landscape, developing a conceptual and moral
dimension of "being in the world" or dwelling in particular environments. Many
traditional or local peoples have a holistic conception of their homelands and the
physical and biological entities that share them, integrating people and specific
knowledge of ecotopes and plants and animals with a more cosmological and
moral understanding of interconnectedness. Some scholars argue that a holistic
ethnoecological conception is inherently more amenable to creating sustainable
environmental relationships than more oppositional or fragmented conceptions.
A landscape ethnoecology perspective thus may enable critical reflection on the
preconceptions and biases that resource managers or development officials bring
with them in their engagement with local communities and their homelands, and
allow those of us in "Western" societies to reflect on our own larger and more
complex global cultural landscapes.

The Concept of Landscape

What is "landscape"? The term, seemingly straightforward, has been used in a
number of contrasting senses in different disciplines. We have presented the approach we have chosen above. As Eric Hirsch (1995: 2) points out in *The Anthropology of Landscape,* one tradition of conceptualizing landscape in the literature
of space and place draws on the artistic conventions of Renaissance and post-
Renaissance European art: landscape as the *viewscape,* or prospect, framed in a
rectangular "window," highly naturalized to those of us reared in European and
Euro-American cultures. In contrast, "landscape" as used by ecologists and geographers emphasizes the array of ecosystems, that is, physiographic and biotic components of an area and their systematic three-dimensional spatial relationships.

In a classic article on landscape ecology, Richard Forman (1982: 35) provides
us with a sense of landscape as understood in ecology. According to Forman: "A
landscape is a kilometers-wide area where a cluster of interacting stands or ecosystems is repeated in similar form." Kevin McGarigal asserts that "from a wildlife
perspective, we might define landscape as an area of land containing a mosaic of
habitat patches, often within which a particular 'focal' or 'target' habitat patch is
embedded" (NRS 222 course website, University of Rhode Island). The scale of
landscape in this conception is linked to the scale of the organisms and habitat
patches under consideration; thus, landscape for frogs or mice may be much
smaller than for bison, caribou, or wolves.

The *cultural landscape* perspective used by archaeologists and cultural resource managers foregrounds relationships of past peoples with environments, especially constructed environments such as terraces, mounds, and infield-outfield systems. Barbara Bender, an archaeologist who has been much concerned with landscape, writes:

> Landscapes are created by people—through their experience and engagement with the world around them. They may be close-grained, worked-up, lived-in places, or they may be distant and half fantasized. In contemporary western societies they involve only the surface of the land; in other parts of the world, or in pre-modern Europe, what lies above the surface, or below, may be as or more important ... The landscape is never inert, people engage with it, rework it, appropriate and contest it. It is part of the way in which identities are created and disputed. (1993: 1)

Cultural resource managers emphasize anthropogenic ecological processes and features such as grove and pasture systems, hedgerows, and culturally significant sites and routes of travel (cf. Andrews and Zoe 1997). Parks Canada, in its website on Aboriginal Cultural Landscapes, writes:

> Indigenous peoples in many parts of the world regard landscape in ways common to their own experience, and different from the Western perspective of land and landscape. The relationship between people and place is conceived fundamentally in spiritual terms, rather than primarily in material terms.
>
> Many Aboriginal peoples consider all the earth to be sacred and regard themselves as an integral part of this holistic and living landscape. They belong to the land and are at one in it with animals, plants, and ancestors whose spirits inhabit it.

Landscape as understood by geographers encompasses spatial relationships among physiographic, biotic, and human elements, including genetic and processual understanding of geomorphology, which is somewhat akin to the perceptions of landscape in landscape ecology. The theory of landscape in geography was especially developed by Carl Sauer (1925, 1963), whose perspective integrates the physiographic and the cultural. According to Sauer ([1925]1963): "Landscape [is] a landshape, in which the process of shaping is by no means ... simply physical. It may be defined, therefore as an area made up of a distinct association of forms, both physical and cultural" (1963: 321). He further states: "The objects which exist together in the landscape exist in interrelation. We assert that they constitute a reality as a whole that is not expressed by a consideration of the constituent parts separately" (1963: 321). In Sauer's perspective, landscape is a result of human management of nature—planned use and unplanned consequences. Human managers must respond to climate, landforms, soils, waters, and vegetation. Sauer saw human experience of the environment—including cognitive, aesthetic, and emotional dimensions—as fundamental. One of his students, Yi-fu

Tuan, developed landscape theory further, elaborating the concept of *topophilia*, human love of place (1977, 1979, 1990).

The now extensive literature on space and place (e.g., Casey 1996; Feld and Basso 1996; Low and Lawrence-Zúñiga 2003) is akin to our focus on landscape and touches on it, but tends to be focused on the meanings and content (often cultural) of specific places and their (cultural) construction, rather than seeking to understand landscape as ecological and interactive.

Approaching Landscape Ethnoecology

This volume represents an early stage in the systematic understanding of human landscapes from an ethnoecological point of view. The various essays included here capture some of this ferment, as scholars seek a vocabulary and theoretical basis for discussing this important nexus of human understanding of lands and homelands. The contributors use a diversity of approaches and focus a variety of disciplinary lenses on people and landscapes, drawing on the perspectives of biological ecology, forestry and land management, cultural anthropology, ethnobiology, political ecology, cultural geography, geographic informatics, and conservation biology. Some of the work is primarily descriptive, more "emic," oriented toward interpreting local conceptions of landscapes and their meanings. Other studies examine in detail correlations of local ethnoecological knowledge of habitat with landscape patterns detected by remote sensing or other methods of ecological sampling. Several studies touch on the significance of local systems of understanding for ecological sustainability. We do not provide a detailed treatment of past human-landscape relations—the province of the archaeologist—but rather address cognitive and practical knowledge that can only be accessed through working with contemporary cultures.

Geographic and Cultural Scope

In this collection we have sought to include a wide range of geographic and cultural settings in which to examine landscape ethnoecology. The chapters here feature broad geographic coverage, including North America, South America, sub-Saharan Africa, Southeast Asia, Australia, and temperate Europe, and a range of traditional economies, including hunters and fishers, hunter-gatherers, pastoralists, swidden horticulturists, and small-scale agriculturists. Northern North American chapters cover sub-Arctic and Boreal peoples and regions (Kaska Dena, Yukon, and Shoal Lake Anishinaabe, Western Ontario) and the Arctic (Igloolik Inuit, Nunavut). Also covered are cultures from a range of arid lands in the North American West (Sahaptin, Columbia Basin, and Paiute, Great Basin), the Sahel (Fulani pastoralists, Burkina Faso), and finally northwestern Australia (Yindjibarndi). Humid tropical regions in both the Americas and Southeast Asia are represented in chapters on the Yucatec Maya of Quintana Roo, the eastern Amazon (Baniwa and Maijuna), and forested environments of insular Southeast Asia (Nuaulu, Eastern Indonesia). Meilleur's study addresses local ethnoecological knowledge in the Alps of southern France (Les Allues, Savoie), bringing to our collection the traditional land knowledge of agrarian Europe.

The landscape perceptions of urban dwellers are only briefly discussed in this *urban* volume in theoretical chapters by Hunn and Meilleur and Mark, Turk, and Stea. Little relevant work has been done in these complex and largely built environments, in part because their linkages are fundamentally nonlocal and their orientation is global. In these environments, as Hunn and Meilleur point out, much of the spatial patterning involves cultural and built environments and is not directly oriented toward the biophysical grid, that is, loosely speaking, the "natural" environment. Our emphasis is squarely on "traditional" societies, a term we use *"traditional"* advisedly. By traditional we do not mean unchanging or "set in stone," as some would have it. Rather, by traditional we refer to social and economic systems relatively independent of global markets, composed of peoples whose livelihoods still depend to a substantial degree on subsistence harvests and who are thus more directly engaged with their natural surroundings than is true of city dwellers.

Some work on attitudes to "nature" in connection with social aspects of urban ecological restoration has been carried out, but not in a systematic way (cf. Dalton 2004 and tangentially Higgs 2003). Although one could argue that urbanized landscapes are the anthropogenic environment endpoint, to deal adequately with the immense complexities and distinctive character of contemporary urban environments is a significant undertaking, and at this preliminary stage relevant work is yet to be done.

Terms and Approaches

A number of terms have been used to describe the ethnoecological elements of biotic and physical space that we call ecotopes. The various contributors to this volume have not been entirely consistent in the terminology they employ. As editors we have done our best to impose some terminological order on the issues we address. Three largely synonymous terms are *habitat, kind of place*, and *biotope*. *habitat kind of place biotope* Habitat forms the framework for Abraão and her coauthors in their chapter on the Baniwa of the Brazilian Amazon. Meilleur, emphasizing the conjunction of plants and features of the physical environment, uses biotopes for the ethnoecological subdivisions of the French Alpine valley of his detailed ethnographic study, a term derivative of the biogeographical and phytosociological literature. Johnson begged the issue, avoiding an a priori decision on what the nature of significant local landscape concepts might be and instead describing an assortment of biophysical terms on a range of scales as "kinds of place." Some of these can be easily conceptualized as habitats or biotopes, and some appear at first blush more

ethnophysiographic or hydrological. Some have proved more elusive to categorize, having to do with hunting or spiritually potent places. Johnson's chief concern was to determine what kinds of places appeared to be ecologically significant
in the landscape understandings of the consultants she worked with.

Mark, Turk, and Stea focused on ethnophysiography, or understandings of
landforms. They carefully examine several specific topographic classes in English
and Yindjibarndi, the language of an aboriginal community of Western Australia,
to illuminate differences. Although they have not explicitly conceptualized their
treatment as ethnoecological, it is clear from their discussions that there are significant ecological and even cosmological entailments to the terms they report.
Conceptually their work occupies one end of a spectrum of landscape element
definitions. The fine-grained focus on Nuaulu forest types by Roy Ellen occupies
the other, being solely concerned with describing types of vegetation. Krohmer's
Fulani work combines biotic and abiotic features of the Sahelian landscape to
describe an explicitly *geoecological* system covering all of the significant habitats
recognized by a traditional West African pastoral people.

Iain Davidson-Hunt and Fikret Berkes describe the landscape of the Anishinaabe of Shoal Lake, Ontario, Canada, as a *cultural landscape,* including topographic and hydrological features, vegetation types, and significant human places
such as camps and fishing sites, as their consultants and collaborators from the
community felt that the separation of human patterns from strictly biophysical
place kinds was not true to their own concepts of their homeland. Johnson's
Kaska chapter also includes terms for cultural kinds of place (camp, trail, lookout) though these are not figured. "Kinds of place," or "place kind generics," as
Athapaskan linguist James Kari called them (Kari and Fall 1987), may be closely
related to specific places, since in certain languages place names characteristically
incorporate such "place kind generics" into place names, as in English "Long
Swamp," "Black Forest," or "Fishhook Bend." Fowler, working with the Paiute,
begins with place names and abstracts from these a set of recognized *kinds* of
place, that is, ecotopes, giving a sense of landscape aesthetics and the significance
of biotic resources in the process.

Organization of the Book

The first section of the book presents two complementary approaches to theoriz-
ing landscape ethnoecology. Hunn and Meilleur take a focused look at landscape
classification, considering perceptual bases for distinguishing classes of landscape
features or *folk ecotopes,* speculating on the purposes such classifications serve
and discussing the relationship of such classifications to ethnobiological classification and to the domain of place names. Mark, Turk, and Stea, coming from
geography and informatics, consider the domain of cultural understanding and
classification of landscape features from the perspective of what they have called

"ethnophysiography," reviewing ethnoscientific and geographic roots and elaborating on distinctions in geographic ontology between different cultural systems of understanding of landscape, and its linkage to other features of culture and cosmology. The series of chapters following this presents detailed case studies of landscape classification. The last section of the book presents a more diverse set of essays that elucidate ethnoecological understanding of landscape from several disciplinary and geographic perspectives.

Notes

1. Natural environments in the pure sense no longer exist anywhere on earth; however, there are environments with minimal human influence that can stand for the "natural" end of the continuum.

References

Alcorn, Janis. 1981a. "Huastec Noncrop Resource Management: Implications for Prehistoric Rain Forest Management." *Human Ecology* 9: 395–417.

Alcorn, Janis. 1981b. "Factors Influencing Botanical Resource Perception among the Huastec: Suggestions for Future Ethnobotanical Inquiry." *Journal of Ethnobiology* 1: 221–230.

Andrews, Thomas D., and John B. Zoe. 1997. "The Idaà Trail: Archaeology and the Dogrib Cultural Landscape, Northwest Territories, Canada." In *At a Crossroads: Archaeology and First Peoples in Canada,* ed. George P. Nicholas and Thomas D. Andrews. Burnaby: Archaeology Press, Archaeology Dept., Simon Fraser University.

Bender, Barbara, ed. 1993. *Landscape, Politics and Perspectives.* Oxford: Berg.

Berlin, Brent. 1992. *Ethnobiological Classification: Principles of Categorization of Plants and Animals in Traditional Societies.* Princeton, NJ: Princeton University Press.

Casey, Edward. 1996. "How to Get from Space to Place in a Fairly Short Stretch of Time: Phenomenological Prolegomena." In *Senses of Place,* ed. Stephen Feld and Keith Basso. Santa Fe, NM: School of American Research.

Clément, Daniel. 1990. *L'Ethnobotanique Montagnaise de Mingan.* Collection Nordicana, No. 53, Cerntre d'études nordiques. Quebec: Université Laval.

Dalton, Zoe. 2004. *Restoration as an Ethnobiological Pursuit: An Integrated Restoration Program for Toronto's Black Oak Savannahs.* Final MA Project, Athabasca University.

Daubenmire, R. F. 1952. "Forest Vegetation of Northern Idaho and Adjacent Washington, and Its Bearing on Concepts of Vegetation Classification." *Ecological Monographs* 22: 301–350.

Daubenmire, R. F., and J. B. Daubenmire. 1968. *Forest Vegetation of Eastern Washington and Northern Idaho.* Washington Agricultural Experiment Station Technical Bulletin 60.

Deur, Douglas, and Nancy J. Turner, eds. 2005. *Keeping it Living: Traditions of Plant Use and Cultivation on the Northwest Coast of North America.* Seattle: University of Washington Press and Vancouver: UBC Press.

Ellen, Roy F. 1993. *The Cultural Relations of Classification: An Analysis of Nuaulu Animal Categories from Central Seram.* Cambridge: Cambridge University Press.

Feld, Stephen, and Keith Basso, eds. 1996. *Senses of Place.* Santa Fe, NM: School of American Research.

Forman, Richard T. T. 1982. "Interaction among Landscape Elements: A Core of Landscape Ecology." In *Perspectives in Landscape Ecology: Contributions to Research, Planning, and Management of Our Environment*, ed. P. Tjallingii and A. A. de Veer. Wageningen: Centre for Agricultural Publishing and Documentation.

Franklin, J. F., and C. T. Dyrness. 1973. *Natural Vegetation of Oregon and Washington*. Corvallis: Oregon State University Press.

Higgs, Eric S. 2003. *Nature by Design: People, Natural Process and Ecological Restoration*. Cambridge: MIT Press.

Hirsch, Eric. 1995. "Introduction." In *The Anthropology of Landscape: Perspectives of Space and Place*, ed. Eric Hirsch and Michael O'Hanlon. Oxford: Clarendon Press.

Hirsch, Eric, and Michael O'Hanlon, eds. 1995. *The Anthropology of Landscape: Perspectives of Space and Place*. Oxford: Clarendon Press.

Johnson, Leslie Main. 2000. "'A place that's good,' Gitksan landscape perception and ethnoecology." *Human Ecology* 28(2): 301-325.

Kari, James, and James Fall. 1987. *Shem Pete's Alaska: The Territory of the Upper Cook Inlet Dena'ina*. Fairbanks: Alaska Native Language Center.

Low, Setha M., and Denize Lawrence-Zúñiga. 2003. *The Anthropology of Space and Place: Locating Culture*. Oxford: Blackwell Publishing.

Mark, David, and Andrew G. Turk. 2003. *Ethnophysiography*. Pre-conference paper for Workshop on Spatial and Geographic Ontologies, 23 September 2003 (prior to COSIT03). Manuscript in possession of the authors.

McGarigal, Kevin, from NRS 223 website: excerpts from background material to FRAGSTATS http://www.edc.uri.edu/nrs/classes/nrs223/readings/fragstatread.htm

Nazaréa, Virginia. 1999. *Ethnoecology, Situated Knowledge/Located Lives*. Tucson: University of Arizona Press.

Parks Canada Website on Aboriginal Cultural Landscapes. http://www.pc.gc.ca/docs/r/pca-acl/sec4/index_e.asp Last updated 26 May 2004, accessed 13 November 2006.

Pfister, R. D., and S. F. Arno. 1980. "Classifying Forest Habitat Types Based on Potential Climax Vegetation." *Forest Science* 26: 52–70.

Pfister, R. D., B. L. Kovalchik, S. F. Arno, and R. C. Presby. 1977. *Forest Habitat Types of Montana*. USDA Forest Service General Technical Report INT-34, Intermountain Forest and Range Experiment Station.

Sauer, Carl. 1925. "The Morphology of Landscape." *University of California Publications in Geology* 2(2): 19–54. Berkeley: University of California Press. Republished in *Land and Life: A Selection from the Writings of Carl Ortwin Sauer*, ed. John Leighly. Berkeley: University of California Press.

———. 1963. *Land and Life: A Selection from the Writings of Carl Ortwin Sauer*. Ed. John Leighly. Berkeley: University of California Press.

Tansley, Arthur. 1939. *The British Isles and Their Vegetation*. Cambridge.

Toledo, V. M. 1992. "What Is Ethnoecology? Origins, Scope and Implications of a Rising Discipline." *Ethnoecológica* 1(1): 5–21.

———. 2002. "Ethnoecology: A Conceptual Framework for the Study of Indigenous Knowledge of Nature." In *Ethnobiology and Biocultural Diversity*, ed. J. R. Stepp, F. S. Wyndham, and R. K. Zarger. International Society of Ethnobiology.

Troll, Carl. 1971. "Landscape Ecology (Geoecology) and Biogeocenology: A Terminological Study." Translated by E. M. Yates. *Geoforum* 8: 43–46.

Tuan, Yi-Fu. 1977. *Space and Place: The Perspective of Experience.* Minneapolis: University of Minnesota Press.

———. 1979. *Landscapes of Fear.* New York: Pantheon Books.

———. 1990. *Topophilia: A Study of Environmental Perceptions, Attitudes, and Values.* Albuquerque: University of New Mexico Press.

PART 1

Theoretical Perspectives

Chapter 1

Toward a Theory of Landscape Ethnoecological Classification

Eugene S. Hunn and Brien A. Meilleur

We propose that landscape ethnoecological classification represents a semantic domain worthy of systematic comparative analysis. A landscape ethnoecological classification is a set of named categories such as "marsh," "cliff face," "old-growth forest," "hedgerow," "mangrove swamp," "oak copse," and "lawn," each of which refers to a perceptually and functionally distinct landscape feature. We propose a comparative analysis of such terminological sets modeled on that which has proved to be productive with ethnobiological (Berlin 1992), ethnoanatomical (Brown 1976), toponymic (Hunn 1996), color (Kay and Berlin 1997), and kinship classifications (Atkins 1974). As with these better-known domains, their successful analysis requires a clear formulation of the formal relationships among the elements classified and an appreciation of the nature of the experiential realms ordered by the classification. We offer the following sketch as an initial step toward that end. Our analysis involves "heroic" simplifications that we hope will prove justified by future results.

Toward a Theory of Landscape Ethnoecological Classification

First, we define a landscape ethnoecological classification as a partition of a "subsistence space" into *patches,* such that every point of that space will fall either within a patch or on the boundary between adjacent patches. Such boundaries may be sharply drawn or diffuse. These patches are tokens of types we prefer to call *ecotopes,* that is, "the smallest ecologically-distinct landscape features in a landscape mapping and classification system." "[E]cotopes are identified using flexible criteria ... a combination of both biotic and abiotic factors, including vegetation, soils, hydrology, and other factors ... In 1945 Carl Troll first applied the term to landscape ecology" (Ellis 2009). This term is roughly synonymous with "kind of land," "biotope," or "habitat," but we prefer ecotope because it does not imply a focus on land forms (versus features of rivers, lakes, or the sea) nor on biological or, more often, botanical markers as definitive. Nor does the term ecotope have the ecological implications of the term "habitat," that is, a home for some particular species of plant or animal, including *Homo sapiens.*

The boundaries between ecotopes, that is, *ecotones,* may be of particular significance to local subsistence practice. As noted, such boundary regions may be more or less distinct. When patches grade one into another broadly, boundaries may not be recognized as such. However, when patches are sharply defined, boundaries may be named as distinct folk ecotopes, e.g., "shoreline" or "forest edge."

Ecotopic patches should map onto closed regions of the earth's surface. Here we are pursuing a structural analogy with ethnobiological classification and nomenclature. In the ethnobiological case names for plants and animals in theory denote categories of similar organisms such that the *biodiversity space* of known living things maps onto a basic, or generic, set of folk biological taxa. That is, each and every individual tree we call a "pine" (i.e., a token) is a member of the pine tree category (the type), and, theoretically, every individual living organism will be classifiable as belonging to one or another named folk biological taxon. This conceptual mapping between names and concepts in the ethnobiological case is complex and "imperfect," as has been widely noted (e.g., Hunn 1982), but by no means random. Might we expect something comparable in the landscape ethnoecological domain? For the sake of argument we would like to offer a series of assertions to that effect.

First it is essential to recognize, as Mark, Turk, and Stea and Meilleur (see also 1986) here note, that there is a fundamental difference between biosystematic classifications and ecological or ecotopic classifications. The former reflect "natural discontinuities" generated by evolutionary processes of speciation (Hunn 1976). The latter reflect more or less continuous patterns of variation along a range of partially independent dimensions, such as soil chemistry and plant associations. As noted in the volume introduction, ecotopic classifications in general do not isolate physiographic, biotic, and cultural significata in defining culturally significant landscape elements. Nevertheless, we believe it is useful to assess potential similarities between ethnobiological and ethnoecotopic classification and nomenclature across a range of languages.

We expect that many folk ecotopes will be characterized by distinctive associations of organisms, particularly of plants (which by virtue of their rootedness will be more predictably associated with particular landscape patches than will animals). We will argue that folk ecotopes should also entail culturally salient ecological distinctions. We are alert to the possibility of a hierarchy of ecotopes (e.g., "forest" < "rain forest" < "temperate rain forest" < "cedar grove"; cf. Abraão et al. on Baniwa forest categories, this volume) but believe that—as in the case of ethnobiological classifications—there should be a *basic-level ecotopic partition* of each local landscape, the elements of which are particularly salient to the people who employ that classification. As with ethnobiological classifications, we might expect that elements at this basic level will be consistently and concisely named in the local language. Elements at a higher or lower level of generality may be named by modifying basic-level ecotope names or by complex descriptive labels (Meilleur, this volume, 1986).

Variation in the number of named ecotopes may reflect differences in the level of analysis. Ellen suggests in his contribution to this volume that some languages may explicitly recognize many lower-level categories that in other languages remain implicit or covert. There is a parallel in ethnobiological classification, in that languages vary in the degree to which they employ "binomials" to recognize folk specific and varietal taxa (Hunn and French 1984).

We expect also that the distinction between general-purpose and special-purpose categories, elaborated in theoretical discussions of ethnobiological classification (e.g., Hunn 1982), will also be relevant to landscape ethnoecological classification. For example, "sacred place" would appear to be a special-purpose rather than a general-purpose ethnoecological category, since "sacred places" may coincide with a variety of ecotopes (see Johnson's account here of Kaska landscape classification).

Furthermore, given that we are dealing here with a spatial partition, the formal properties will be "partonomic" rather than taxonomic (Brown 1976), that is, relations of contiguity are more fundamental than relations of similarity (Meilleur, this volume, 1986). We also expect that the ecotopes will be structured around prototypical regions (Berlin 1992: 24).

We propose the following hypotheses for test:

- Ecotopes are "natural categories" in that particular species of plants and/or animals will be predictably associated with certain ecotopic patches. However, they are often intergradient rather than discontinuous, which problematizes their systematic recognition.
- People cannot do with just any landscape ethnoecological classification, but will adopt and maintain systems of distinctions that maximize the spatial predictability of local biotic and other resources.

Where Do Place Names Fit?

So far we have mentioned two intersecting classifications, that of plants and animals (the ethnobiological), about which we know quite a lot (Berlin 1992), and that of ecotopes, about which we know little. But there is a third semantic realm that must be integrated into this plan of investigation, to wit, the toponymic, the system of geographic place names that is recognized in every society. We believe that landscape ethnoecological classifications function to integrate efficiently the information captured within the ethnobiological and ethnogeographic domains. For example, Fowler's analysis here of Southern Paiute landscape concepts emphasizes the intimate relationship between these domains. A key issue here is the nature of the system of systems by which the three classifications are integrated.

Biological taxa and ecotopes are *spatially distributed* types, that is, each species and each ecotope occur repeatedly across space. Place names, by contrast, denote—as proper names—unique spots on the landscape. Named places do not exhaustively partition space (except for the special case of nation states and their administrative subdivisions), but rather are scattered across it, often with much "empty" space in between (Hunn 1996). We have found that in many languages place names are binomial expressions in which ecotope names serve as the "head" element, e.g., "Long Swamp," "Fork's Prairie," "Walden Pond." However, some languages do not employ such transparent constructions, e.g., Sahaptin of the Columbia Plateau of northwestern North America (Hunn 1996), or a mix of the two approaches is used.

Landscape ethnoecological classifications, like ethnobiological and toponymic systems, are recognized in every human language. Why should this be? First, biological species categories are recognized presumably because they are motivated by compelling perceptual discontinuities and because the organisms distinguished by name differ one from another in useful ways (Hunn 1982). In fact, there is strong evidence that humans are innately programmed to recognize nomenclaturally within their subsistence space on the order of 500 each of basic plant and animal categories (Levi-Strauss 1966; Berlin 1992). Places are named presumably because such focal points of the landscape preserve in memory critically important information needed to locate and acquire resources, including, of course, plants and animals. These are elements of what Mithen identifies as the "natural history intelligence," one of three "multiple intelligences" critical for the evolution of modern *Homo sapiens* (2006: 62). Place names also index social relations and emotional ties at the foundation of personal identity (Basso 1996) and may represent spiritual anchors and legal claims to the land (Thornton 1995). There is evidence that people will name in the neighborhood of 500 places also within their subsistence space (Hunn 1996).

Hypothetically, if people knew which of 500 named plants and 500 animals occurred at each of 500 named places, there would seem to be little need to recognize and classify ecotopes, since species could be located simply by canvassing one's toponymic inventory. However, we believe that naming ecotopes saves mental energy and enhances the efficiency of subsistence activities by facilitating the integration of these two massive data bases, the ethnobiological and the toponymic. To appreciate this point, consider the following thought experiment.

If we recognize 500 plants and 500 animals, that equals 1,000 kinds of organisms. If, in addition, we recognize 500 named places, we will have 500,000 (1000 × 500) bits of information about the environment to keep track of. On the other hand, if we were to define a few dozen ecotopes such that the organisms and places were evenly distributed among them (each plant and animal and each place uniquely associated with one and only one ecotope), the task of locating a particular organism at a particular place would be substantially simplified.

Of course, these assumptions represent an ideal case that one would never encounter in the real world. In particular, more realistically, ecotopes with biotic content will differ from one another in their organismic inventories to some less-than-perfect degree. Nor will organisms be distributed uniformly within ecotopes. Nor will places necessarily fall within a single ecotope. However, if there is a substantial statistical correlation between species and ecotopes, the efficiency of recognizing even less-than-perfectly distinct ecotopes should justify their recognition.

Potential ecotopic distinctions may be ranked in terms of the predictive power they offer (Meilleur, this volume, 1986). We have suggested that landscape ethnoecological classification is characteristically "natural." We believe this to be the case because the spatial distribution of organisms (especially of plants) is strongly determined by a few fundamental topographic, climatologic, and edaphic dimensions, most notably those that determine the availability of (1) solar energy, (2) water, and (3) mineral nutrients. Among these factors, clearly, are latitude, elevation, aspect, and placement with respect to global and regional atmospheric currents. Also salient are geological factors that determine soil chemistry. Finally, patterns of human disturbance will play a key role (Meilleur, this volume, 1986; Krohmer, this volume).

We expect that ecotopic distinctions will be recognized explicitly in the local vernacular vocabulary in order of their predictive power as defined above. As is typically the case with ethnobiological names, terms naming basic-level categories and those that are more inclusive will be given simple or complex but "unproductive" names (cf. Conklin 1962). More specific ecotopic categories will be named by productive compound expressions. Examples include plant associations named for indicator species, such as the Kaska *gǫ́dze tah* 'among pines' and *gat tah* 'among spruce' (Johnson, this volume); Anishinaabe *okwokizowaag agimakoog* 'black ash grove' and *okwokizowaag geezhigoog* 'cedar grove' (Davidson-Hunt and Berkes, this volume); Sahaptin *patat-naq'it* 'ponderosa pine [forest] edge' (Hunn 1996).

Other basic ecotopes may be differentiated by productive morphosyntactic processes such as reduplication to generate two or more related terms naming variants of a single theme, as in the case of Sahaptin *tnan-naq'it* 'cliff base' versus *tnán* 'cliff' and the Burkina Faso differentiation of hills of various sizes: e.g., *waamnde* is a mountain or large hill while *baamngel* is a smaller hill; *tilde* is an elevation lower than *waamnde,* while the diminutive variant *tilel* is a very small hill (Krohmer, this volume). Though these pairs of terms are not related as type and subtype, the diminutive variants are clearly defined in terms of the unmarked terms of the set, a pattern described by Hunn and French as "horizontal" versus "vertical" subordination in the Sahaptin ethnobiological nomenclature (1984). Distinctions may remain implicit or covert, defined by the intersection of named basic terms drawn from intersecting dimensions. Ellen's account of the Nuaulu system is exemplary: "[T]here is evidence for subtle and extensive understanding of variation, and of the ecological properties of different vegetative associations. However, only a small number of categories are shared and consistently organized, and there is a low degree of formal lexicalization" (Ellen, this volume).

Finally, academic biogeographic classifications based upon climatological, hydrological, geological, phytosociological, and successional patterns will tend to correspond with landscape ethnoecological classifications. When ecological classifications—whether academic or folk—*do not* so correspond, conceptual emphases of particular cultural significance—symbolic or historical—likely motivate such "departures" from a "natural" ecological partition. In our hypotheses we have emphasized "practical" or "empirical" considerations, but we are open to the possibility that landscape elements defined by reference to aesthetic, emotional, or spiritual qualities may play an important role.

Discussion and Examples

The Geo-hydro-edapho-logical Substrate Problem

We seek to define a semantic domain that might appear to defy recognition on the basis of shared substantive attributes, as it includes elements as diverse as "piney wood," "eddy," "salt lick," "cliff," "hedgerow," and "road." That is, the named elements that partition space are predicated not only on botanical and zoological associations, but may also highlight geological, hydrological, and edaphic features (which in turn constrain biotic species associations), and reflect both the "natural" and the "built" environment.

A key question is whether multiple spatial partitions will be recognized as corresponding to the several substrates we have noted—that is, will we find that cultures distinguish biotopic, geotopic, hydrotopic, and anthropotopic partitions as separate and distinct from an ecotopic partition? Mark, Turk, and Stea here define "ethnophysiography" in terms of the geotopic dimension. Aporta describes a system of ecotopes characterized by snow and ice. Ellen contrasts a number of studies focusing on forest ecotopes. However, we are inclined to the view that it should be possible to recognize a single multidimensional landscape ethnoecological partition without regard for the predominant type of feature that may serve to define elements of the partition. Furthermore, most elements of a given ecotopic partition will likely exhibit some biotic aspect. For example, the geotope "talus" in the Columbia Plateau homeland of the Sahaptin-speaking Indians is named **pshwá-pshwa**, literally 'many rocks', and recognized as the unique habitat of a rare plant, *Lomatium minus*, **nak'únk** in John Day River Sahaptin. This plant is sought for its edible roots. Sahaptin speakers also recognize **shám** 'lithosol patch' as habitat for several staple root food plants.

Hydrotopic features may likewise serve to target important biotic resources. For Sahaptin again, **xaat'áy** 'shallow stream bed of pale, flat rock' indicates a site sought for spear fishing, as migrating salmon are clearly exposed in such situations. In Micronesia, Palauan fishermen recognize a convergence zone of sea currents downstream from islands that they call **hapitsetse.** This zone is characterized by exceptionally rough water at certain seasons that can be hazardous to fishermen in canoes, but it also helps define the daily movements of certain prize fish such as tuna (Johannes 1981: 101–109). Aporta's fascinating analysis of the relevance of sea ice patterns for Inuit subsistence practice provides another example. Krohmer's meticulous account of Fulani classification of their desert environment illustrates the multidimensionality of these classifications.

Anthropogenic Ecotopes

Anthropogenic features also mark resource locations and predictable resource associations. For example, in the French Alps of Savoie, Alluetais peasants call a cultivated field rockpile **le mordjé.** Waste rock is cleared from plowed fields and piled on the field borders. The rock is useless, but after some years a distinctive biota develops on these rockpiles, characterized by several species of brambles (*Rubus* spp.) and elderberry (*Sambucus nigra*), both of which provide edible berries in season, and a species of ash (*Fraxinus excelsior*) with multiple uses (Meilleur 1986: 84, 98). Huastec Maya cultivate dozens of herbal medicines in an area they call **wal eleeb** 'dooryard edge', the margin of the cleared space they maintain around their houses.

It may be the case that every ecotope is to some extent "anthropogenic." Prairies in the Pacific Northwest of North America depend upon systematic burning (Boyd 1999). The Huastec forest **te'-lom** is often planted to coffee or carefully tended to encourage useful forest species (Alcorn 1984). This is reminiscent of the Yucatec Mayan **pet kot,** forest patches now known to be thoroughly anthropogenic (Gómez-Pompa, Flores, and Sosa 1987). Posey has argued likewise that certain forest patches in southern Amazonian savannahs were established by local Indians quite intentionally (1984).

We suspect that communities will vary along a continuum of intensity of local resource management between hunter-gatherers and agriculturalists to modern urbanites. We expect that agriculturalists will be more likely to recognize anthropogenic ecotopes than hunter-gatherers, since farmers interact more intensively with their local landscape. Our several previous examples would seem to bear this out. What might be called a *monocrop ecotope* should be characteristic of agricultural systems with a commercial emphasis. Names for such ecotopes are often productive binomial compounds, such as "apple orchard," "banana plantation," or "wheat field." The basic-level categories are "orchard," "plantation," and "field," appropriately modified by specifying the species to which the plot is devoted. Alluetais terms of this sort include *le tsnavyé* 'hemp plot' and *la vnyé* 'vineyard'. Huastec terms are *pakab-lom* 'sugar-cane field', *weey-lom* 'henequen plantation', and *lanaax-lom* 'orange grove'. This usage is not limited to such anthropogenic plots however, as *te'-lom* is a forest, literally 'tree-land' (Alcorn 1984), and in Alluetais, *lèz arkosè* is both the plural of 'green alder' (*Alnus viridis*) and a site at which green alder grows in dense thickets on wet mountain slopes (Meilleur 1986). Even hunter-gatherers may name plots dominated by single plant species, as in the Sahaptin *tawshá-tawsha* 'many sagebrush' and *wiwnú-wiwnu* 'many huckleberries', each formed by reduplicating the name of the predominant species of the association, though the Sahaptin ecotope 'huckleberry patch' may have been shaped by the systematic use of fire to enhance the productivity of this valued biotic resource.

Are There Urban Ecotopes?

Our analysis obviously foregrounds rural subsistence-oriented communities. As with the study of ethnobiological classification systems, modern urban societies lurk in the background as exceptions that prove the rule. How might an urban landscape be organized conceptually by the urban "native"? Spradley's classic ethnography of Seattle's homeless (1999) suggests one answer. He found that the semantic domain most highly elaborated for these "urban nomads" was that of "flops," that is, places to sleep.

Clearly our emphasis on biotic and other subsistence resources does not readily generalize to the urban core. Critical resources for urbanites are elements of the social or built environments; less often are they of the non-urban (wild or rural) environment. Still, we may recognize such urban ecotopes as "park," "parkstrip," "yard," "lawn," "garden," "P-patch" (a small urban community garden), and "zoo." These are likely to be less salient than such environmental features as "playground," "mall," "intersection," "cloverleaf," and "high-rise." We might enquire in this regard whether these latter "urbotopes" are conceptually equivalent to the ecotopes of more "natural" human environments, a question parallel to the comparison of folk biological taxa and categories of artifacts, such as "furniture," "vehicle," and "building" (cf. Brown et al. 1976; Wierzbicka 1984).

Ecotopes as Multipurpose Categories

Ecotopes are multipurpose concepts, good not only for finding key plant and animal resources, but also for marking social and spiritual spaces and their boundaries. To cite just a few examples: Mixtepec Zapotec (Oaxaca, Mexico) speakers respect wetlands and distinguish several, such as *guièl* 'lake', *gòdz* 'marshy spot', and *xlià︎n* 'spring'. Such water sources may be "enchanted" and are thus approached with a mixture of awe and fascination. The enchanting spirits may take offense and depart, leaving these spots dry, as happened a few years ago to one of San Juan Mixtepec's two main water sources, an event attributed to the rash of forest fires that plagued the region that year. In this vein, Sahaptin speakers consider landslides, or *txápnash,* a spiritual resource, for here one is likely to encounter *shúkat,* literally 'knowledge', a spirit ally that may take animal and human form. Yet these features are also ecotopes with distinctively valuable flora and fauna. San Juan Mixtepec lakes and marshes support a sharply distinct flora, while Sahaptin slide areas are known to harbor certain species of game.

Conclusions

We do not claim to have discovered the phenomenon of ecotopic classification. Consider the comments of three of our predecessors, an anthropologist, a botanist, and a geographer. In a 1946 article, the anthropologist David Thomson described an Australian Aboriginal classification system, that of the Wik Monkan. According to Thomson, the Wik Monkan classify "the country into 'types' based on its geographical and botanical associations, as critically as any ecologist" (1946: 165). In fact, "so detailed and accurate is their knowledge of these areas that they note the gradual changes in marginal areas as one association merges into another and they often use distinctive names not listed here, for each transitional area … [they are also] … able to relate without hesitation the changes in fauna and in food supply in each association in relation to the seasonal changes" (1946: 166). Wik Monkan use a domain-specific prefix, roughly meaning 'place', to help distinguish the lexical set of 'types of country' from other domains of natural phenomena.

Likewise, the botanist Harley Harris Bartlett noted in 1936 that the Maya recognized natural groupings of plants. Wherever he traveled, he found that the native people distinguished and named plant communities as well as noting the dominant species within each. Such categories were used by his consultants to predict the whereabouts of resources while out gathering and hunting. Though unfamiliar with the local vegetation, Bartlett found the habitat categories of the local people immediately comprehensible. In fact, he was able to employ them as would a native classifier to predict where certain plant species would be found.

In 1973, the human geographer Bernard Nietschmann, in his ethnographic monograph *Between Land and Water,* described patterns of Nicaraguan Miskito Indian environmental classification and use, documenting over twenty folk "biotic communities": "The Miskito recognize many biotopes, mostly in terms of structural composition. They perceive the relation of specific … species to certain biotopes and direct their [resource]-getting activities accordingly" (1973: 168).

Table 1.1 Summary of landscape ethnoecological categories of ten cultures

Group	Ecotopes	Other categories	Source
Alluetais, France	20	+34 other landscape features	Meilleur 1986
Huastec Mayan	30	ecotopes +19 milpa stages	Alcorn 1984
Koyukon, Alaska	33	including all ecotopes	Nelson 1983
Sahaptin, NW US	44	including all ecotopes	Hunn 1990
Mixtepec Zapotec	31	ecotopes + 3 milpa stages	Hunn 2008
Wola, New Guinea	13	vegetation associations only	Sillitoe 1998
Yanyuwa, Australia	47	including all "land units"	Baker 1999:46
Kaska, Yukon Terr.	60	including all "kinds of place"	* Johnson
Fulani, Burkina Faso	102	all landscape units	*Krohmer
Baniwa, Brazil	88	forest types < 4 broad categories	*Abraão et al.

*essays included in this volume

The examples we have reviewed to date (see Table 1.1 and several chapters below) support our initial claims as follows: ecotopic classifications are widely reported and include roughly twenty-five basic-level ecotopic categories. The total number of topographic feature terms, however, may range to 100 or more, depending on how the terminological set is delimited. These ecotopic classifications partition the subsistence space of the community by creating a mosaic of ecotopic patches (which may be linked by transitional zones). Local people use these ecotopic categories to guide them in locating resources and otherwise journeying through their home territories. Ecotopic concepts are rarely very general or very specific in application; most appear to be basic-level categories, though instances occur of a shallow hierarchy of named ecotopes. Named places are located with respect to basic-level ecotopic categories.

The systematic comparative study of landscape ethnoecological semantic domains is just beginning. This domain is of special interest because it involves the conceptual coordination at a higher level of abstraction of two basic domains of environmental knowledge, that of species and that of places. Together these three domains constitute a system of systems with clear adaptive value.

References

Alcorn, Janis. 1984. *Huastec Mayan Ethnobotany.* Austin: University of Texas Press.

Atkins, John R. 1974. "On the Fundamental Consanguineal Numbers and Their Structural Basis." *American Ethnologist* 1: 1–31.

Baker, Richard. 1999. *Land Is Life, From Bush to Town: The Story of the Yanyuwa People.* St. Leonards, NSW, Australia: Allen and Unwin.

Bartlett, H. H. 1936. "A Point of View and a Method of Procedure for Rapid Field Work in Tropical Phytogeography: Botany of the Maya Area." *Miscellaneous Papers* 1–13. Carnegie Institution.

Basso, Keith H. 1996. *Wisdom Sits in Places: Landscape and Language among the Western Apache.* Albuquerque: University of New Mexico Press.

Berlin, Brent. 1992. *Ethnobiological Classification: Principles of Categorization of Plants and Animals in Traditional Societies.* Princeton, NJ: Princeton University Press.

Boyd, Robert (ed.). 1999. *Indians, Fire and the Land in the Pacific Northwest.* Corvallis: Oregon State University Press.

Brown, Cecil H. 1976. "General Principles of Human Anatomical Partonomy and Speculations on the Growth of Partonomic Nomenclature." *American Ethnologist* 3: 400–424.

Brown, Cecil H., John Kilar, Barbara J. Torrey, Tipawan Truong-Quang, and Phillip Volkman. 1976. "Some General Principles of Biological and Non-biological Classification." *American Ethnologist* 3: 73–85.

Conklin, Harold C. 1962. "Lexicographical Treatment of Folk Taxonomies." In *Problems in Lexicography.* Publication 21, Indiana University Research Center in Anthropology, Folklore, and Linguistics, ed. F. W. Householder and S. Saporta. Bloomington, IN.

Ellis, Erle C. 2009. "What Are Ecotopes?" Laboratory for Anthropogenic Landscape Ecology. http://www.ecotope.org/about/ecotopes

Gómez-Pompa, A., S. Flores, and V. Sosa. 1987. "The Pet Kot: A Man-made Tropical Forest of the Maya." *Interciencia* 12(1): 10–15.

Gove, Philip Babcock, editor-in-chief. 1986. *Webster's Third New International Dictionary of the English Language.* Springfield, MA: Merriam-Webster.

Hunn, Eugene S. 1976. "Toward a Perceptual Model of Folk Biological Classification". *American Ethnologist,* 3(3): 508-524.

———.1982. "The Utilitarian Factor in Folk Biological Classification." *American Anthropologist* 84: 830–847.

———. 1990. *Nch'i-Wána "The Big River": Mid-Columbia Indians and Their Land.* Seattle: University of Washington Press.

———. 1994. "Place-names, Population Density, and the Magic Number 500." *Current Anthropology* 35: 81–85.

———. 1996. "Columbia Plateau Indian Place Names: What Can They Teach Us?" *Journal of Linguistic Anthropology* 6: 1–26.

———. 2008. *A Zapotec Natural History.* Tucson: University of Arizona Press.

———. n.d. "Plants and Animals as Semantic Points of Reference in Sahaptin Ethnotoponymy." Manuscript in author's possession, Department of Anthropology, University of Washington, Seattle.

Hunn, Eugene S., and David H. French. 1984. "Alternatives to Taxonomic Hierarchy: The Sahaptin Case." *Journal of Ethnobiology* 4: 73–92.

Johannes, R. E. 1981. *Words of the Lagoon.* Berkeley: University of California Press.

Johnson, Leslie Main. n.d. "Trail of Story: Gitksan Understanding of Land and Place in Northwestern British Columbia." Paper presented to the Session on "Ethnoecology and Kinds of Place: An Examination of Understanding of Landscape" at the American Anthropological Association Annual Meeting, Philadelphia, Pennsylvania, 4 December 1998.

Kay, Paul, and Brent Berlin. 1997. "Science ≠ Imperialism: There Are Nontrivial Constraints on Color Naming." *Behavioral and Brain Sciences* 20: 196–201.

Lévi-Strauss, Claude. 1966. *The Savage Mind.* Chicago: University of Chicago Press.

Martin, Gary J. 1993. "Ecological Classification among the Chinantec and Mixe of Oaxaca, Mexico." *Etnoecológica* 1(2): 16–33.

Meilleur, Brien A. 1984. "Une recherche ethnoecologique en Vanoise." *Trav. Scien. Du Parc Nat. de la Vanoise* 14: 123–133.

———. 1985. "Gens de montagne, plantes et saisons." *Le Monde Alpin et Rhodanien* 1: 1–79.

———. 1986. "Alluetais Ethnoecology and Traditional Economy." PhD diss., Department of Anthropology, University of Washington, Seattle.

Mithen, Steven. 2006. "Ethnobiology and the Evolution of the Human Mind." In *Ethnobiology and the Science of Humankind,* ed. Roy Ellen. *J. Roy. Anthrop. Inst.* (N.S.), S45–S61.

Nelson, Richard K. 1983. *Make Prayers to the Raven: A Koyukon View of the Northern Forest.* Chicago: University of Chicago Press.

Nietschmann, Bernard. 1973. *Between Land and Water.* New York: Seminar Press.

Posey, Darrell. 1984. "A Preliminary Report on Diversified Management of Tropical Rainforest by the Kayapó Indians of the Brazilian Amazon." *Advances in Economic Botany* 1: 112–126.

Sillitoe, Paul. 1998. "An Ethnobotanical Account of the Vegetation Communities of the Wola Region, Southern Highlands Province, Papua New Guinea." *Journal of Ethnobiology* 18: 99–126.

Spradley, James P. 1999. *You Owe Yourself a Drunk.* Prospect Heights, IL: Waveland Press.

Thomson, David. 1946. "Names and Naming in the Wik Monkan Tribe." *Journal of the Royal Anthropological Institute* 76: 157–168.

Thornton, Thomas. 1995. *Place and Being among the Tlingit.* PhD diss., Department of Anthropology, University of Washington, Seattle.

Wierzbicka, Anna W. 1984. "Apples Are Not a Kind of Fruit." *American Ethnologist* 11: 313–326.

Williams, Nancy M. 1986 [1982]. "A Boundary Is to Cross: Observations on Yolngu Boundaries and Permission." In *Resource Managers: North American and Australian Hunter-Gatherers,* ed. Nancy M. Williams and Eugene S. Hunn. Canberra: Australian Institute of Aboriginal Studies.

Chapter 2

Ethnophysiography of Arid Lands
Categories for Landscape Features

David M. Mark, Andrew G. Turk, and David Stea

Introduction

How do people understand their environment? How do they remember it? How do they communicate this knowledge to others? These questions all address abilities that are essential to human existence. The environment may be thought of as a continuum, populated by objects of various sizes, and shaped and maintained by various processes, some small, some large, some close, and some distant. Ethnoecologists have studied many aspects of local knowledge of the environment, but much of this work has concentrated on elements of the proximity, such as plants and animals, while the larger and more distant components of the environment have received less attention. The proximity of course is extremely important, but the landscape, composed of places to stand, to live, and to find resources, also is absolutely fundamental, and we believe that the landscape needs its own ethnoscience.

Ethnophysiography is a recently defined field of study that seeks to understand cultural differences in conceptualizations of landscape, focusing in particular on physical components such as landforms, water features, and vegetation assemblages, via, for instance, comparisons between the meanings of terms that people from different cultures use to refer to the landscape and its components (Mark and Turk 2003). Ethnophysiography is motivated by a number of questions. When people look at a natural landscape, do they see it as filled up with features (objects) such as hills, lakes, and woodlands? Or do they simply see it as a continuous landscape? Perhaps they take an intermediate conceptualization, seeing scattered features over a continuous landscape. Perhaps they mainly see materials and the shapes that they take. Are the boundaries of individual landscape features contingent on feature class definitions? What things (entities, regions, objects, features, places) in the landscape are named so that they can be talked about? Of those things, which get *common* (generic) names (that is, things that are considered to belong to *kinds* and denoted by nouns or noun phrases) (always, sometimes, never), and which get *proper* (individual) names (always, sometimes, never). Do the proper names contain generic nouns that denote what kind of thing the named entity is? Are some landforms labeled with action phrases (verbs or verb phrases) representing what humans or animals do in the presence of these landforms? To put this more specifically, do the identification, delimitation, and classification of landscape features vary across cultures, landscape, languages, or individuals, and if so, what is the nature of that variation?

Landforms, water bodies, and other natural landscape features are excellent subject matter for investigating the relationships among categories, language, and natural variation of the environment in a more general sense. On the one hand, the range of earth materials and the processes that shape them are more or less the same all over the world, so areas with similar geologic structures, similar climates, and similar geologic histories may be expected to have very similar suites of landforms—a common baseline in physical reality. Yet, being inorganic, landforms and water bodies are not strongly organized into *kinds* by geomorphic processes. This gives more opportunity for different cultures and languages to organize near-identical landscapes into very different feature classes, categories, and instances. In these ways, the constraints on ethnophysiography differ markedly from established fields of ethnoscience such as ethnobiology, in which nature is organized by evolutionary history into species, yet the species are different in different parts of the world (Berlin 1992; Medin and Atran 1999). Thus the work has potential to advance our understanding of categorization in general.

In the remainder of this chapter, we first outline the nature of the geographic domain and follow this with a review of some previous work on anthropology of landscape and on landscape categories. We then describe the methods we are using in our case studies and present some results from our studies in northwestern Australia. Lastly, we draw some conclusions, discuss practical implications of the work, and present plans for further research.

Theoretical Background

The Geographic or Landscape Domain

Does a *geographic* or *landscape* domain *exist* as a separate domain of reality, in the way that the subject matter of, say botany or chemistry exists? That is, are larger patterns and forms of landforms or vegetation mutually more similar to each other than they are to their smaller geologic or biological complements? This question has intrigued and at times obsessed geographers, since it goes to the heart of the validity of geography as an academic discipline. Perhaps the phenomena that we think of as 'geographic' have no fundamental coherence, and have just been collected together in an ad hoc fashion for arbitrary or historic reasons. Or perhaps the geographic domain consists of merely the larger parts of various other domains.

A third possibility is that geographic things form a coherent *cognitive* subdomain. Some support for the existence of a coherent geographic domain within common-sense knowledge has been provided by Smith and Mark (2001). As part of their research on geographic categories, they asked large numbers of unspecialized undergraduate students at the State University of New York in Buffalo to list examples of geographic features or objects. Results suggest that, to these subjects, the word "geographic" does denote a coherent, familiar domain. The examples they gave most often were large, natural features and objects affixed to, or part of, the Earth's surface. In order, the ten most frequent examples given were mountain, river, lake, ocean, valley, hill, plain, plateau, desert, and volcano.

Some theoretical support for the coherence of such landscape features as a natural domain of common-sense knowledge is provided in Granö's book *Pure Geography* (Granö 1997). Originally published in German in 1929 and in Finnish in 1930, and not available in English until the 1990s, Granö's book presents the central thesis that the human perceptual environment should be the core subject matter of geography. Granö further divides the perceptual environment into two distinct zones: the *proximity* and the *landscape,* and he claims that these zones are apprehended in different ways: "The perceived environment can thus be divided quantitatively into two major parts on the basis of distances in the field of vision, that is, the proximity, which we perceive with all our senses, and farther away the landscape, which extends to the horizon and which we perceive by sight alone" (Granö 1997: 19).

The lists of entity types that students in recent studies volunteered as typical of "geographic" things are exactly the most prominent feature types of the landscape identified by Granö. These included students from Guatemala as well as from the United States and several European countries. An intriguing implication is that the essence of these geographic entities is that they are in the distance: a mountain *qua* mountain is away from us, in the visual landscape, and when we stand on it, our proximate environment is filled not with 'mountain', but with rocks, trees, snow etc. The more distant elements of the human perceptual environment are the main subject matter of ethnophysiography.

Ontology of the Geographic Domain

As originally defined and employed in philosophy, the term "ontology" deals with the nature of reality, with what exists, at a high level of abstraction. Ethnophysiography depends on an ontology of this sort to provide a framework within which meanings of everyday terms for landscape elements can be defined: high-level concepts such as land and water, boundary, concave and convex, vertical and horizontal, large and small, flowing and still, deep and shallow, permanent and intermittent, etc., would allow one to characterize the nature of what can be measured, delimited, and classified in the real world or in images of it. It should be possible to develop a formal notation, based on such high-level concepts, that would allow linguists and landscape specialists to record the definitions of terms in a largely language-independent way. Such an ontology could provide an etic grid for recording observations, analogous to a Munsell color chart, or to the Lin-nean classification in ethnobiology, although not as clearly structured.

Ontologies in the information systems sense, on the other hand, are formaliza-tions of conceptualizations, suitable for implementation in information systems (Gruber 1993; Guarino and Giaretta 1995). The conceptualization involved usu-ally is an expert's view of the important aspects of the domain that the informa-tion system will represent, but it could be a philosopher's "ontology" as described above, or a collection of beliefs about some domain, in this case the landscape. In information systems ontologies, conceptualizations embedded in the infor-mation system should be faithful to the conceptualizations held by the people whose knowledge is to be represented, i.e., they should reflect *emic* conceptual-izations. Information systems ontologies are dominated by terms, organized into taxonomic hierarchies, that form the basis of data dictionaries. Ethnophysiogra-phy provides methods that can be used to document the conceptualizations of landscape to be represented in geographic information systems that are able to provide semantic interoperability between traditional and scientific knowledge systems. Terms discovered, documented, and defined in an ethnographic study of landscape can be formalized to produce data dictionaries and other imple-mentable representations.

The physiographic domain introduces a critical issue that is rarely encountered in other domains of knowledge: individual instances of geographic features often have inherently vague or graded boundaries, and the delimitation of individual entities may depend on their category (Smith and Mark 1998). As Smith and Mark (1998, 2001, 2003) have pointed out, typical geographic entities such as mountains or rivers are not bona fide objects in the usual sense, but are more properly considered to be parts of the Earth's surface. The categorization and de-limitation processes are not independent, so a different set of categories projected onto the same landscape might produce a different partition into geographic ob-jects. Whether geographic features are delimited from their surroundings by crisp boundaries or by transition zones is an important element of their definitions.

A number of issues arise regarding the definitions of geographic entities and their categories. Most landforms, such as mountains and valleys, are identified mainly by shape (form) and typically have graded or transitional boundaries.

However, some landforms are bounded by shorelines (e.g. islands), and still others by a mixture of shorelines and topographic forms (e.g., capes). All landscapes features have *affordances* (Gibson 1979) that provide opportunities for human beings, but water in the landscape is privileged by its being essential for human life and by especially distinctive affordances. It seems highly likely that all cultures and languages pay attention to the ways in which water can exist in the landscape, and languages are likely to have terms that distinguish various kinds of water bodies and their parts. An ontology of water in the landscape would include dichotomies or qualitative dimensions such as flowing or not; size; seasonality of surface water; seasonality of flow; origin; presence or absence of surface outlets; etc. Assemblages of vegetation, such as forests, woods, grasslands, etc., also fall within the physiographic domain in most contexts. Some languages also have specific terms for flat areas depending on whether the surface is covered by rock, sand, grass, or other substances. Whereas the domains of topographic forms and of watercourses may seem distinct from each other to an English speaker from a humid region, in arid lands these categories may merge into a broad spectrum of low-lying places that may collect water during rainy periods.

A critical distinction for ethnophysiography as a field is that natural inorganic domains are not organized by nature into kinds to the same degree that biological entities are so organized. Thus geomorphology does not provide the baseline or grid of categories against which ethnophysiographic categories can be evaluated, in the way that scientific taxonomies of plants and animals provide baselines for ethnobotany and ethnozoology. Unlike higher plants and animals, which to some large degree are grouped into species by nature, landforms more properly belong to continua. We do not deal here with types of plants in the landscape, which of course is biological, but vegetation assemblages (e.g., woodlands, forests) also exhibit gradational boundaries or form continua (cf. Curtis 1959; Bray and Curtis 1957). Water is certainly ontologically distinct from land, but the sizes and shapes of lakes or islands do not naturally fall into discrete categories with absent intermediate cases, in the same way that kinds of trees or birds fall into such groups. This provides both an opportunity for languages, cultures, and individuals to vary much more in their categorization of geomorphologically very similar landscape elements, and a methodological challenge for ethnophysiography, due to the lack of an independently defined baseline.

Previous Anthropological and Linguistic Work on Landscape

Lived-In Landscape

It is useful to review the ways in which people interact with their physical environments in order better to understand people's conceptions of landscape—

the way lived experience makes space into place (Hirsch and O'Hanlon 1995; Tuan 1974). James, Hockey, and Dawson (1997: 6) cite Bourdieu (1977), Caws (1984), Stoller (1989), Ingold (1991), and Richards (1993) concerning the way an individual's conceptions are grounded in everyday practices: "The view of such authors, though variously expressed, is that a purely cognitivist vision of human agency underplays the individual's direct engagement with a social and material world and fails to account for the ways in which that engagement might actively contribute to or shape representational knowledge itself."

This association may concern practical landuse related attributes (affordances) of landscape. For instance, in the Indigenous Australian language Yindjibarndi, the term **wangguri** refers to a box canyon type of landscape feature, which is a useful place into which to drive kangaroos so that they are trapped and easier to spear. Layton (1997: 132) discusses the terms for landscape (country) elements in the language of the Alawa people (of Northern Australia): "Country can also be referred to according to the dominant species in the woodland. **Mandiwaja** is scrubby wattle country, such as grows on sandstone hills or rock outcrops (the 'object' which the Alawa call **Mandiwaja** ...) ... **Mandiwaja** country is good for hunting emu."

The association between people and landscape may be of a deeper kind, which could perhaps be termed 'spiritual'. For many indigenous peoples such associations include reference to creation concepts, from the time that the Yindjibarndi refer to as "when the world was soft." In discussing a chapter by Layton in their book, James et al. (1997: 7) mention that he "bids us address the distinction between representations that carry a direct reference to an external, locally situated material reality—a hole in a rock, a track, a river valley—and representations that are self-referential, which carry meaning only to the extent that they make sense within the framework of a culturally specific knowledge-base. This might be a sacred site which marks the passage of a totemic creature." Halfway around the world, Basso (1996) describes the close relationship between landscape, wisdom, and morality among Western Apaches.

In order to understand a language (and do justice to translation) it is important to appreciate as fully as possible the true nature of the concepts to which words and phrases refer. Layton (1997: 140–141) suggests that "Alawa discourse represents the landscape as the embodiment of animate agencies, whereas we represent it as the product of blind forces . . . If the criterion for complete translation is that we render even the causal theories of the other cultures familiar, then complete translation of Alawa representations is impossible."

Categories for Landscape Elements

We have not as yet found any published study prior to our own studies whose primary focus was common-sense categories for landscape elements and their definitions, though Kofod's 2003 paper comes close (see below). What we find instead are several essays on *toponyms* (geographic names) or publications on ethnoecology that provide short sections about landscape-scale phenomena. We also find a few works on toponyms that discuss generic terms for landscape features, and geographic categories. Undoubtedly there are more such publications that touch on geographic categories; however, the ones we have encountered so far are presented here to convey the flavor of previous work, intellectual antecedents, and the variety of approaches used.

Wilbur Zelinsky's (1955) regional cultural geography essay on the distribution of generic English-language terms in the place names of the northeastern United States clearly is a contribution to understanding of common-sense geographic categories. More recently, Jett (1997) analyzed Navajo-language place names in Canyon de Chelly, Arizona, and provides considerable detail regarding the generic landscape terms that appear in such names. In 1957, in a completely different context, Voegelin and Voegelin used a new approach in surveying the Hopi language, working systematically through a series of semantic domains; the very first domain that they presented was the "Domain of Topography," listing many ethnophysiographic terms under subdomains of convexities, longitudinal depressions, and oval openings in the earth. Shortly after that monograph appeared, in yet another unrelated domain of scholarship, urban designer and planner Kevin Lynch (1960) illustrated the importance of language in spatial conceptualization with examples drawn from Sapir's 1912 article on language and environment: "They [the Southern Paiute] have single terms in their vocabulary for such precise topographical features as a 'spot of level ground in mountains surrounded by ridges' or 'canyon wall receiving sunlight' or 'rolling country intersected by several small hill-ridges.' Such accurate reference to topography is necessary for definite locations in a semi-arid region" (Lynch 1960: 132).

Yi-fu Tuan (1974) introduced the concept of *topophilia,* the love of place, the emotional bond between people and their home landscapes. Tuan illustrated his thesis with many examples, including one case study regarding the different relations of several cultures to the landscapes of northern New Mexico. A decade later, Pinxten, Van Dooren, and Harvey (1983) conducted a detailed study of spatial knowledge in Navajo culture, and spatial reference in the Navajo language. Their book is almost entirely about spatial relations and abstract concepts, but it also contains an interesting and detailed analysis of the meanings of nine landscape terms in the Navajo language (Pinxten et al., 1983: 87–91).

Since 1990, several anthropologists and ethnographers have conducted work in ethnoecology that included aspects of the environment at geographic or landscape scales. Hunn's 1990 book, *Nch'i-Wána, "the Big River": Mid-Columbia Indians and Their Land,* contains a detailed section on landscape terminology (Hunn 1990: 89–97), including an oblique drawing of a mountainside and valley with various features sketched and terms shown. Hunn suggests that the rich set of landscape terms in the Sahaptin vocabulary may result from "a long period of stable residence on this stretch of river" (Hunn 1990: 97), a principle later asserted by Kofod in an Australian context (see below).

In another study, Beaucage, working with an organization called Taller de Tradición Oral del CEPEC, examines ethnoecological concepts in relation to agriculture and land use in the Lower Sierra Norte de Puebla, in Mexico (Taller de Tradición Oral and Beaucage 1996; Beaucage and Taller de Tradición Oral 1997). Although mainly about ethnobotany, Beaucage begins his discussion of the Nahua view of the environment by listing and defining what he calls ethno-topographic terms, which use a human body metaphor for their landscape, with the mountains as the head. Baker's (1999) work on Yanyuwa traditional knowledge in northern Australia contains detailed treatments of terminology for kinds of landscape elements in their tropical coastal environment. Johnson (2000) surveys the ethnoecology of Gitksan people of northern British Columbia and includes several references to topographic feature terms and generic places, including their connections to important plants and to land use.

Perhaps the most advanced ethnographic treatment of landscape terms and categories is provided by Kofod (2003), who writes about what she calls the "topographical nominals" in the Gija (Kija) language in northwestern Australia. Kofod presents detailed lists of particular terms for landscape elements. In the context of the Gija native title land claim, she makes the case that having a detailed lexicon of terms that fit the landscapes in a particular region is evidence that the people "belong" in that landscape.

Research in spatial cognition and geographic information science, meanwhile, has largely ignored geographic entity types and categories. An exception is an essay by Mark (1993) that attempts to provide a theoretical basis for geographic categories, giving an example of misaligned landscape categories in two rather closely related languages (English and French). Smith and Mark (2001) extend this work in the context of geospatial ontology, but emphasize cross-cultural similarities in categories rather than differences. The likelihood of important cross-linguistic differences in geographic categories was the impetus for Mark and Turk's (2003) initial efforts in "ethnophysiography," and for the research project reported in this chapter.

Researchers at the Max Planck Institute for Psycholinguistics (MPI) in Nijmegen have also been active in research related to landscape and language. MPI's Language and Cognition Group have recently concluded a set of case studies of landscape terms (and some place names) in nine languages in a wide variety of geographic locations (although all in tropical regions) (Burenhult 2008). In the introduction to this collection of studies, Burenhult and Levinson (2008: 1) discuss the theoretical basis of the work and its relationship to ethnophysiography. They review the results of the case studies and state that "[t]he data point to considerable variation within and across languages in how systems of landscape terms and place names are ontologised. This has important implications for practical applications from international law to modern navigation systems."

Ethnophysiography Case Studies

Methodology for Case Studies

For our ethnophysiography case studies, the authors are using a methodological approach that was first applied successfully in a case study with the Yindjibarndi language and people (Mark and Turk 2003; Mark, Turk and Stea 2007). After obtaining permission to conduct the research from both university ethics committees and tribal or local authorities, the first stage of the methodology involves the compilation of a draft list of landscape terms in the target language from available bilingual dictionaries relating English and the target language. Each English-language landscape term in word lists for English collected using subjects in Buffalo, New York (Smith and Mark 2001), is looked up in the English-to-X part of the bilingual dictionary or dictionaries available, and entered into a table. The table is sorted by the target-language terms, and duplicate target-language terms (if any) are consolidated. Then the approximate or provisional English equivalents for each word are determined from the X-to-English part of the dictionary. Lastly in this dictionary-based phase, a provisional superordinate semantic category for each term is determined from the English word equivalents of the term and entered in another column in the table. Broad semantic categories include convex, concave, flat, or vertical landforms, flowing or static water bodies, vegetation assemblages, etc. It is important to note that, at this phase, semantic interpretations of the terms in the target language are based *only* on generalizations over the semantics of the English equivalents or definitions as recorded by the compilers of the dictionaries. Experience with Yindjibarndi has shown that, at least in some cases, the linguists who originally compiled these bilingual dictionaries were implicitly assuming that the indigenous concepts were the same as those that underlie English. Particular note is taken of apparent one-to-many or many-to-many relations among terms of the two languages.

The next stage in the refinement process involves interviews and discussions with bilingual consultants or informants. In the work with Yindjibarndi, Mark and Turk used a white board while discussing meanings of terms, refining the meanings, discussing differences between terms with similar meanings, and eliciting additional terms missing form the list. They also made extensive use of photographs of landform examples in Yindjibarndi country and took additional photos as they learned relevant dimensions of Yindjibarndi landscape terminology.

We are using a cross-linguistic, comparative approach to extend the scope and applicability of the research by examining environmental classification of a similar semi-arid or desert environment in an unrelated language. The simi-

larity of the landforms and appearance of Yindjibarndi country to landscapes
of northern New Mexico and Arizona suggested an initial comparison of terms
and concepts used in Yindjibarndi with terms and concepts used in the Navajo
(Diné) language.

As the work has developed, we have found it more and more important to
interact with our informants directly in the field. During recent fieldwork in
New Mexico and Arizona, we have spent more time with local people in the field.
During field trips to familiar areas of their territory, the participants were asked
to describe the landscape (in their language and in English) and to discuss the
terms for features. This material was audio recorded, and GPS coordinates and
photographs of significant features were taken. This integrated material was then
examined for landscape terms, compared to the term lists from the dictionaries, and used to provide a more informed understanding of landscape concepts
of that language group and of how "generic" landscape terms relate to proper
names used for landscape features. This approach has proven very productive,
providing information about, and examples of, definitions and landscape category differences.

The work to date has indicated the importance of looking beyond simple
nouns for generic landscape terms. Conceptualizations of landscape may also
be expressed using verbs (e.g., water falling down) or prepositions (water next
to a bank). This pattern is especially common in Athapaskan languages, which
include the Navajo (Diné) language. Compound words (flat-topped mountain)
may also be used. Hence, it is important to record passages of language, rather
than just words, and for this to be as "natural" as possible. In sum, the objective
of the methodology is to utilize ethnographic approaches, such as those long used
in ethnobiology, to elicit both the distributions of words associated with landscape forms that occur in the cultural geography of language groups and their
linguistic and cultural contexts.

Insights from the Yindjibarndi Case Study, NW Australia

Our first ethnophysiographic case study was carried out with the Yindjibarndi
people, who are an Indigenous (Aboriginal) group living in the state of Western
Australia, near the northwestern corner of Australia. Until the nineteenth century, the Yindjibarndi people lived mostly along the middle part of the valley of
what Europeans named the Fortescue River, and on adjacent uplands (Tindale
1974). However, they have been progressively displaced from their "country"
over the last 150 years as part of the colonial process, having been obliged to
make way for, initially, pastoral activities, and in more recent times, the mining
industry (Rijavec et al. 1995). Many Yindjibarndi now live in the small town
of Roebourne (Yirramagardu), near the coast, in the Pilbara region of Western
Australia, together with Ngaluma and Banjima people. Yindjibarndi belongs to

the Coastal Ngayarda language group, within the South-West group of the Pama-Nyungan languages (SIL 2001). The 2006 Australian Census data indicate that there were about 320 speakers of the Yindjibarndi language, of whom about 230 live in and near Roeburn. In Mark and Turk (2003) and Mark, Turk and Stea (2007), the authors presented preliminary results of an ethnophysiographic study of landscape concepts employed by Yindjibarndi-speaking people.

The Yindjibarndi live in a landscape with no permanent rivers. It appears that the same physical reality is parsed into objects, and the objects categorized and named with common nouns, according to schematizations that are substantially different from the ontology that implicitly underlies English language terms such as "river" or "creek." Consider the feature shown in Figure 2.1 (from Mark and Turk 2003). The English-language proper name of the feature is "Dawson Creek." In English-language dictionaries, rivers and creeks are usually defined as being streams of water. In English, the above feature is conceptualized as the bed of a river, even when no water is present. English terminology for watercourses implicitly gives primacy to the water that forms them, even if that water flow is intermittent or ephemeral.

This feature would be referred to by the term **wundu** in Yindjibarndi. The Yindjibarndi-English dictionaries (Wordick 1982; Anderson 1986; Anderson and Thieberger n.d.) list the English equivalent of **wundu** as "river(bed), gorge".

Figure 2.1 *A dry watercourse in country north of Jindawarrina*

Our field research indicates that the **wundu** is the watercourse, the place where water sometimes flows. Thus the term **wundu** is not equivalent to the terms *river* or *creek*. A **wundu** is an almost-always-dry place where water occasionally flows, while the words 'river' and 'creek' refer to features normally composed of flowing water and which necessarily have beds and banks. **Wundu** is more nearly equivalent to the terms *wash* or *arroyo* as used in the American southwest. When water does flow in a **wundu** such as Dawson Creek, usually during heavy rains produced by a tropical cyclone, the flood (of water) is referred to as a **mang-kurdu**, and if there is a small patch of water lying in the **wundu** after light rain it is referred to as **bawa** (the generic term for water). In both cases, there is a name for the water when it is there, distinct from the channel in which it lies or flows. The situation is schematized quite differently in Yindjibarndi and English, perhaps based on prototypical environmental conditions in the home regions of these languages.

The situation for **wundu** calls into question an ontology of natural landscapes that first divides hydrography (water-related features) from topography (land forms), and then classifies the water features as flowing or standing and the topographic features as convex or concave, vertical or horizontal, etc. In arid landscapes with no permanent streams, most concave features contain water only rarely, and the hydro/topo distinction itself dries up, so to speak. Thus it is not surprising to learn that the Hopi language, also from a region with no permanently flowing streams, also has a word, **pööva,** that is described in English as meaning "riverbed, wash, gully, arroyo; path for runoff water" (Hill 1998). Perhaps **pööva** means exactly the same as **wundu,** perhaps not—further ethnographic fieldwork in both Arizona and Australia will be needed to document the schematizations that under-lie the definitions of terms for landscape features in these environments.

Significant differences in concepts for convex landscape features were also found in the Yindjibarndi case study. In the English-Yindjibarndi section of Anderson's (1986) Yindjibarndi-English dictionary, Anderson lists **marnda** for "mountain" and **bargu** for "hill". In the Yindjibarndi-English section of An-derson, the complete list for **marnda** is "rock, mountain, metal, hard material, money", and for **bargu** is just "hill". However, Von Brandenstein (1992) lists the meanings of **marnda** as 'hill, mountain, metal, horseshoe, ore, rock, stone', while **bargu** is not listed at all. When Mark and Turk conducted interviews with fluent Yindjibarndi speakers in Roebourne, they found that **marnda** was the general term for most hills and for mountains, as well as mountain ranges and ridges—almost any convex topographic features of moderate to large size was a **marnda** (see Figure 2.2). Use of **bargu,** on the other hand, was restricted to small rough hills or mounds. Comments by Yindjibarndi speakers on landscape pho-tographs confirmed that the division between **marnda** and **bargu** is at a much smaller size threshold than the mountain/hill division for English.

Figure 2.2 *This hill near Roebourne was termed a* **marnda** *by our Yindjibarndi informants.*

Much more detailed analysis of Yindjibarndi terms for landscape features is provided in Mark and Turk (2003) and Mark, Turk, and Stea (2007). The initial phase of this case study (2002 to 2007) provides strong support for the basic hypothesis of ethnophysiography—that people from different places and cultures use different conceptual categories for geographic features. A preliminary pictorial dictionary of Yindjibarndi landscape terms has been prepared (Turk and Mark 2008), and copies have been provided to community members for feedback. Further fieldwork will lead to a revised landscape dictionary, which will be useful in the community for teaching the Yindjibarndi language to children.

Issues, Future Work, and Practical Applications

Some Outstanding Issues

Several questions have arisen in the course of our research to date. The first is that of "exhaustiveness"—i.e., how can we be confident that we have identified all possible terms for a given landscape form or all landscape forms for a given term, much less all landscape terms in the lexicon? A term for a landform may be a single word, an agglutinated word, or a descriptive phrase. Second, how much variability is there in matching terms and land forms, or in other words, do the speakers of language X in subregion Y use different terms than speakers of the same language in subregion Z? Third, how has the usage of terms changed over time? Might certain distinctive labels for more subtle landscape variations have been replaced by a single term, especially as people became more disconnected from the land and from traditional activities? Fourth, what is the relative importance of landforms (whether viewed in plan or elevation) to the culture and lifeways of a particular group of people? How are landform terms embedded in or related to "creation stories," traditional tales, and people's ethnoscience? Fifth, are there temporal or seasonal variations, such that the term used for a landform seen at night might be different from that used for the same landform seen in the day, or a snow-covered mountain perhaps differently referred to than the same mountain devoid of snow? Sixth, are different landform words used by people who differ in age, gender or gender roles, social class, or relative adherence to tradition? Finally, what is the impact of the gender, age, or culture group of the researcher/investigator upon the kind or amount of information received?

Clearly, some of these issues inhere in the landforms and others in the human participants (resident groups or researchers). In identifying and compiling a vocabulary (and eventually a dictionary) of landforms, some of these issues must be taken into account. We are investigating a range of elicitation techniques so as to adopt approaches that address these issues.

Additional Methods

The validity of ethnophysiographic findings may be compromised during the collection, recording, and presentation phases of the research. Carrying out the interview in the "real world" among the landscape features being discussed (or at least using a rich collection of photos) presumably improves the validity of the information obtained. Walking the country in open-ended field interviews with indigenous participants is perhaps too time-consuming to be practical, though it would carry the significant advantage of the identification and immediate labeling of significant landscape features. Conducting field interviews from moving vehicles, on the other hand, may add an unnatural twist to the interview context. A time-efficient compromise would be to give respondents a limited list of landscape terms and ask them to point out exemplars during the automobile tours.

We have employed several variants of visual methods in landscape research. Our respondents have been asked to label individual photographs of landforms, or to establish landform categories by sorting large numbers of photographs into groups. Whereas these methods are economical of both money and time, photographs cannot fully simulate field observation. Yet another approach would be to have members of the target group take photographs, show them to their elders, and request oral descriptions—which, in some cultural groups, may result in extensive stories about the significance of a given landform, relating it to actual history or to traditional stories. The advantage of this emic approach to visual ethnography is that it is more culturally sensitive. An advantage of the use of photographs supplied by the researchers would be that the same photos could be shown to the members of different groups, raising the possibility of cross-cultural comparison.

The main purpose of an emic photographic substudy is to examine the narrative context in which terms indicative or descriptive of landscape forms occur in normal discourse in different languages (e.g., Navajo or Yindjibarndi) so as to reveal the interrelation of landscape, language, and culture. This approach to research provides important contextual information for landscape labels and can enhance understanding of the significance of landscape labels obtained by other means.

Practical Implications of the Research

This research is expected to provide teaching materials to be used in language preservation and promotion. Other practical implications of the research arise in the contexts of cartographic and geographic data standards, automated reasoning systems, and geographic software interoperability. A naïve view of geographic categories implicitly asserts that categorizations are universal across all cultures, languages, and landscapes. Under this naïve view, entity types such as mountains, rivers, and lakes exist in the world, whether they are in the mind-independent world itself, or at least in universal human concepts. Many of us learned vocabularies in other languages just that way—as a look-up table of word equivalences, plus a few memorized exceptions. But such a view of translation requires a universal set of categories with corresponding language-independent mental concepts, available to be labeled simply by language. For the realm of landscape, this is an area that must be investigated empirically.

Such a view may be reasonable for biological entities such as cats or oaks, where there may be so-called *natural kinds* in the world (Keil 1989), and likewise for artifacts such as chairs or cars, where the kind may exist before the entity is constructed specifically to be of that kind. But for inorganic natural domains such as physiography, the questions of universality of categories are more open (Mark and Turk 2003; Smith and Mark 2003). Malt (1995) presents an excellent review of the different ways in which categories are studied and understood in anthropology and in psychology, and discusses the relative strengths of non-uniformities in nature versus cognitive processes in determining categories for a domain. Geographic categories might be adjusted to fit the variations of a particular landscape, or differences in categories may reflect priorities in the way a community interacts with the landscape. They also may differ because of contingencies of linguistic and cultural history, or even simply due to chance.

The results of our research so far suggest that at best, a term substitution approach to categories in information standards can work only at a superordinate level of abstraction where categories may be more nearly universal. But even this is called into question when such a seemingly fundamental distinction as topography versus hydrography (land features vs. water features) is less clear in the desert, where surface water may be ephemeral. This research project will continue to document the extent of this non-alignment of basic landscape categories by studying languages that are members of completely different language families but spoken by people who live in similar landscapes. Results will allow us to make recommendations for developing multilingual, multicultural geographic information standards among markedly different groups.

Ethnophysiography research can also have practical implications for the interoperability of geographic information systems (GIS). For example, the Australian Federal Court has recognized the native title rights of the Ngarluma and Yindjibarndi peoples of Western Australia (NNTT 2003). This decision is expected to lead to an enhanced joint management agreement covering land use and environmental issues for some areas covered by the claim, including Millstream National Park, which includes very significant sacred sites. Joint management of cultural and natural resources would be facilitated by interoperable companion GISs (or complementary layers within a single GIS), one using English terminology and national mapping feature codes (AUSLIG 2003) and the other using Yindjibarndi terms and concepts for landscape features (Mark and Turk 2003).

Acknowledgments

Mark and Stea's research on this project has been supported in part by the U. S. National Science Foundation (NSF) through collaborative research grants BCS-0423023 and BCS-0423075, with Ashley McNabb as research assistant at Texas State University. Turk's work on the project has been supported in part by a Murdoch University REGS grant and by the above NSF grant. Initial work on the project was supported by NSF grant BCS-9975557 to David Mark and Barry Smith. Carmelita Topaha has been an extremely valuable consultant to the Navajo portion of the project and has recruited other Navajo participants. We thank the Navajo Nation's Historic Preservation Department for issuing Cultural Resources Investigation Permit No. C0513-E to allow portions of the research to be conducted on the Navajo Reservation. Many members of the Roeburn community, the cultural organization Juluwarlu, and the Pilbara Aboriginal Language Centre (Wangka Maya) provided valuable assistance regarding the Yindjibarndi language.

References

Anderson, B. 1986. *Yindjibarndi Dictionary.* Photocopy.

Anderson, B., and N. Thieberger. n.d. *Yindjibarndi Dictionary.* Document 0297 of the Aboriginal Studies Electronic Data Archive (ASEDA) Australian Institute of Aboriginal and Torres Strait Islander Studies, Canberra, ACT 2601, Australia.

AUSLIG. 2002. *Feature Codes used by the Gazetteer of Australia.* http://www.ga.gov.au/map/names/featurecodes.jsp (accessed 23 June 2004).

Australia Bureau of Statistics. 2006. "2006 Census Quickstats: 6718 Postal Area." http://www.censusdata.abs.gov.au/ABSNavigation/prenav/ViewData.

Baker, R. 1999. *Land Is life: From Bush to Town, the Story of the Yanyuwa People.* Sydney: Allen and Unwin.

Basso, K. H. 1996. *Wisdom Sits in Places: Landscape and Language Among the Western Apache.* Albuquerque: University of New Mexico Press.

Beaucage, P., and Taller de Tradición Oral del CEPEC. 1997. "Integrating Innovation: Traditional Nahua Coffee-Orchard (Sierra Norte de Puebla, Mexico)." *Journal of Ethnobiology* 17(1) 45–67.

Berlin, B. 1992. *Ethnobiological Classification: Principles of Categorization of Plants and Animals in Traditional Societies.* Princeton, NJ: Princeton University Press.

Bourdieu, P. 1977. *Outline of a Theory of Practice.* Cambridge: Cambridge University Press.

Bray, J. R., and J. T. Curtis. 1957. "An Ordination of the Upland Forest Communities of Southern Wisconsin." *Ecological Monographs* 27(4): 325–349.

Burenhult, N., ed. 2008. "Language and Landscape: Geographical Ontology in Cross-linguistic Perspective." *Language Sciences* (special issue) 30: 135–150.

Burenhult, N., and S. C. Levinson. 2008. "Language and Landscape: A Cross-linguistic Perspective." In *Language and Landscape: Geographical Ontology in Cross-linguistic Perspective,* ed. N. Burenhult. *Language Sciences* (special issue) 30.

Caws, P. 1984. "Operational, Representational, and Explanatory Models." *American Anthropologist* 76 (1): 1–10.

Curtis, J. T. 1959. *The Vegetation of Wisconsin: An Ordination of Plant Communities.* Madison: University of Wisconsin Press.

Gibson, J. J. 1979. *The Ecological Approach to Visual Perception.* Boston, MA: Houghton-Mifflin.

Granö, J. G. 1997 *Pure Geography.* Baltimore, MD: Johns Hopkins University Press. [Book originally published in German in 1929 and in Finnish in 1930.]

Gruber, T. R. 1993. "A Translation Approach to Portable Ontology Specifications." *Knowledge Acquisition* 5: 199–220.

Guarino, N., and P. Giaretta. 1995. "Ontologies and Knowledge Bases: Towards a Terminological Clarification." In *Towards Very Large Knowledge Bases,* ed. N. J. I. Mars. IOS Press.

Hill, K. C., editor-in-chief. 1998. *Hopi dictionary = Hopiikwa lavaytutuveni: A Hopi-English Dictionary of the Third Mesa Dialect with an English-Hopi Finder List and a Sketch of Hopi Grammar.* Tucson: University of Arizona Press.

Hirsch, E., and M. O'Hanlon, eds. 1995. *The Anthropology of Landscape: Perspectives on Place and Space.* Oxford: Clarendon Press.

Hunn, E. S., with J. Selam and family. 1990. *Nch'i-Wána, "the Big River": Mid-Columbia Indians and Their Land.* Seattle: University of Washington Press.

Ingold, T. 1991. "Against the Motion (1)." In *Human Worlds are Culturally Constructed: Group for Debates in Anthropological Theory.* Manchester: Dept. of Anthropology, University of Manchester.

Iverson, P., with M. Roessel. 2002. *Diné: A History of the Navajos.* Albuquerque: University of New Mexico Press.

James, A., J. Hockey, and A. Dawson, eds. 1997. *After Writing Culture: Epistemology and Praxis in Contemporary Anthropology—ASA Monograph 34.* London: Routledge.

Jett, S. C. 1997. "Place-Naming, Environment, and Perception Among the Canyon de Chelly Navajo of Arizona." *Professional Geographer* 49(4): 481–493.

Johnson, L. M. 2000. "'A Place That's Good': Gitksan Landscape Perception and Ethnoecology." *Human Ecology* 28(2): 301–325.

Keil, F. C. 1989. *Concepts, Kinds, and Cognitive Development.* Cambridge, MA: The MIT Press.

Kofod, F. 2003. "My Relations, My Country: Language, Identity and Land in the East Kimberley of Western Australia." In *Maintaining the Links: Language, Identity, and the Land* (Proceedings of the Seventh Conference of the Foundation for Endangered Languages, Broome, Western Australia, 22–24 September 2003), ed. J. Blyth and R. McKenna Brown.

Layton, R. 1997. "Representing and Translating People's Place in the Landscape of Northern Australia." In *After Writing Culture: Epistemology and Praxis in Contemporary Anthropology—ASA Monograph 34,* ed. A. James, J. Hockey, and A. Dawson. London: Routledge.

Lynch, K. 1960, *The Image of the City.* Cambridge, MA: MIT Press.

Malt, B. C. 1995. "Category Coherence in Cross-Cultural-Perspective." *Cognitive Psychology* 29(2): 85–148.

Mark, D. M. 1993. "Toward a Theoretical Framework for Geographic Entity Types." In *Spatial Information Theory: A Theoretical Basis for GIS,* ed. A. U. Frank and I. Campari. Berlin: Springer-Verlag, Lecture Notes in Computer Sciences 716.

Mark, D. M., and A. G. Turk. 2003. "Landscape Categories in Yindjibarndi: Ontology, Environment, and Language." In *Spatial Information Theory: Foundations of Geographic Information Science,* ed. W. Kuhn, M. Worboys, and S. Timpf. Berlin: Springer-Verlag, Lecture Notes in Computer Science 2825.

Mark, D. M., A. G. Turk, and D. Stea. 2007. "Progress on Yindjibarndi Ethnophysiography." In *Spatial Information Theory,* ed. S. Winter, M. Duckham, L. Kulik, and A. Kuipers. Berlin: Springer-Verlag, Lecture Notes in Computer Science 4736.

Medin, D. L., and S. Atran, eds. 1999. *Folkbiology.* Cambridge, MA: MIT Press.

NNTT. 2003. *Australian National Native Title Tribunal:-Rreport on Recent Decisions.* http://www.nntt.gov.au/media/1057275393_2456.html (accessed 23 June 2004).

Pinxten, R., I. Van Dooren, and F. Harvey. 1983. *Anthropology of Space: Explorations into the Natural Philosophy and Semantics of the Navajo.* Philadelphia: University of Pennsylvania Press.

Richards, P. 1993. "Natural Symbols and Natural History: Chimpanzees, Elephants and Experiments in Mende Thought." In: *Environmentalism: The View from Anthropology—ASA Monographs 32,* ed. K. Milton. London: Routledge.

Rijavec, F., N. Harrison, and R. Soloman. 1995. *Exile and the Kingdom.* Documentary film. Roebourne, Western Australia: Ieramugadu Group Inc. and Film Australia.

Sapir, E. 1912. "Language and Environment." *American Anthropologist* 14(2): 226–242.

SIL. 2001. "YINDJIBARNDI: A Language of Australia." *Ethnologue: Languages of the World*, 14th ed. http://www.ethnologue.com/show_language.asp?code=YIJ (accessed December 2001).

SIL. 2003. "NAVAJO: A Language of the USA." *Ethnologue: Languages of the World*, 14th ed. Summer Institute of Linguistics. http://www.ethnologue.com/show_language.asp?code=NAV (accessed 12 August 2003).

Smith, B., and D. M. Mark. 1998. "Ontology and Geographic Kinds." In *Proceedings: 8th International Symposium on Spatial Data Handling* (SDH'98), ed. T. K. Poiker and N. Chrisman. Vancouver: International Geographical Union.

Smith, B., and D. M. Mark. 2001. "Geographic Categories: An Ontological Investigation." *International Journal of Geographical Information Science* 15(7): 591–612.

Smith, B., and D. M. Mark. 2003. "Do Mountains Exist? Towards an Ontology of Landforms." *Environment and Planning B: Planning and Design* 30(3): 411–427.

Stoller, P. 1989. *The Taste of Ethnographic Things–The Senses in Anthropology.* Philadelphia: University of Pennsylvania Press.

Taller de Tradición Oral del CEPEC and Beaucage, P. 1996. "La bonne montagne et l'eau malfaisante, Toponymie et practique environmentales chez les Nahuas de basse montagne (Sierra Norte de Puebla, Mexique)." *Anthropologie et Sociétés* 20(3): 33–54.

Tindale, N. B. 1974. *Aboriginal Tribes of Australia: Their Terrain, Environmental Controls, Distribution, Limits, and Proper Names.* Canberra: Australian National University Press.

Tuan, Y.-F. 1974. *Topophilia: A Study Of Environmental Perception, Attitudes, and Values.* Englewood Cliffs, NJ: Prentice-Hall.

Turk, A. G. and D. M. Mark. 2008. *Illustrated Dictionary of Yinjibarndi Landscape Terms.* Informal Publication, Murdoch University (Australia).

Voegelin, C. F., and F. M. Voegelin. 1957. *Hopi Domains: A Lexical Approach to the Problem of Selection.* Memoir 14 of the International Journal of American Linguistics. Baltimore, MD: Waverly Press, Inc.

Von Brandenstein, C. G. 1992. *Wordlist from Narratives from the North-west of Western Australia in the Ngarluma and Jindjiparndi Languages.* Canberra, ASEDA Document #0428.

Wordick, F. J. F. 1982. *The Yindjibarndi Language.* Pacific linguistics, Series C., no. 71. Canberra: Dept. of Linguistics, Research School of Pacific Studies, Australian University.

Zelinsky, Wilbur. 1955. "Some Problems in the Distribution of Generic Terms in the Place-Names of the Northeastern United States." *Annals of the Association of American Geographers* 45: 319–349.

PART 2

Landscape Classification: Ecotopes, Biotopes, Landscape Elements, and Forest Types

Landscape Perception, Classification, and Use among Sahelian Fulani in Burkina Faso

Julia Krohmer

Introduction

Pastoralists in a Changing World

The Fulani are one of the largest groups of cattle pastoralists in Africa. In historical migrations they spread from Senegambia over the entire West African savannah landscapes into Sudan and the Central African Republic. More recently most have settled as agropastoralists (Frantz 1993; Azarya 1999), a crucial change compared to their original nomadic way of life. In view of this fundamental change, the question arises whether and how the broad realm of traditional knowledge, not least the knowledge of the landscape in which they live, is affected. This knowledge represented the foundation of their mobile way of life and the basis for acting in the highly variable environmental conditions of their homeland (Schareika 2003). Is it not needed in the newly changed way of life and thus forgotten? Do sedentary Fulani perceive their environment in a different way from those who are still nomads?

Traditional Environmental Knowledge in Science and Practice

Traditional ecological knowledge has long been neglected by academic science (e.g., Colding and Folke 1997; Bassett and Crummey 2003). Only recently has a more favorable perception of traditional knowledge been widely promoted, parallel to a comparable trend in development cooperation, where, after a long period of ignoring and devaluing local understanding, there is also a growing sense that traditional knowledge merits more attention (e.g., Richards 1985; Warren,

Slikkerveer, and Brokensha 1995; Scott 1998; Bassett and Crummey 2003; Schareika 2003). Recently, development strategies are increasingly incorporating this knowledge as well as the perspectives and needs of the so-called target groups, and which are based on local participation in codesigning the projects (e.g., Thébaud, Grell, and Miehe 1995; Scoones 1996; Sturm 1999a, 1999b; Borrini-Feyerabend et al. 2000; Robbins 2003). Indigenous activities and strategies are comprehensible to outsiders only when they are acquainted with the underlying knowledge systems. So far, this connection is rarely made in development research and application (Johnson 1980; Niamir 1990; Schareika 2003).

It is becoming increasingly clear that indigenous activity and thus indigenous knowledge are indispensable to long-term sustainable management of resources and the preservation of ecological diversity (e.g., Haverkort and Millar 1994; Reichhardt et al.1994; Tiffen, Mortimore, and Gichuki 1994; Toledo, Ortiz, and Medellín-Morales 1994; Kyiogwom, Umaru, and Bello 1998; Etkin 2002; Ross and Pickering 2002; Lizarralde 2004). Many investigations, however, restrict their analysis to particular sections of the natural environment, like vegetation (e.g., Reiff 1998; Moritz and Tarla 1999) or soils (e.g., Dialla 1993; Kolbe 1994; Krogh and Paarup-Laursen 1997); only a few refer to the complete traditional environmental landscape classification systems. However, some studies published in recent years (e.g., Johnson 2000; Shepard et al. 2001; Scarpa and Arenas 2004) show that these systems are anything but inferior to scientific systems with regard to completeness and accuracy. Often, they are even more differentiated than the current scientific systems (see Martin 1993; Toledo et al. 1994; Boni and Gaynor 1996; Kyiogwom et al. 1998; Shepard, Yu, and Nelson 2004).

The Fulani are one of the better-known nomadic groups with respect to "classical" ethnological topics (ethnicity, social organization forms, intra- and interethnic relations, traditions, history, etc.) in the postcolonial era (Azarya 1999). This is reflected in a more-than-abundant literature. But ethnoecological investigations that reflect the Fulani's detailed knowledge of landscape units and ecosystems, and especially pasture, are still generally lacking. Exceptions include Moritz and Tarla (1999; see also Moritz 1994), who present a geoecological pasture classification of Fulani in Cameroon, and Schareika (2003), who describes phenological vegetation phases in particular, and also reviews the most important environmental units of the Wodaabe Fulani of southwest Niger.

Objectives

In view of this research gap, the investigations for my PhD research (Krohmer 2004), which included the results presented here, aimed at documenting the different aspects of the Fulani's traditional environmental knowledge as completely as possible. This chapter presents two main aspects of that work:

- a detailed description of the traditional environmental classification system of traditional Fulani pastoralists in the Sahelian Zone, with special emphasis on the value for pasture of the delineated landscape units (*ecotopes* in the sense of this volume);
- a comparison of the names of environmental units used by the autochthonous Fulani in the Sahel with those used by immigrant Fulani and by autochthonous Fulani in the North Sudanian Zone to reveal how landscape elements encountered in a new environment are conceptualized.

Study Area and the Fulani Who Live There

The study area is situated in northern Burkina Faso in the province of Oudalan, near the border of Mali. Annual precipitation is between 400 and 500 mm and falls in a two- to five-month summer rainy season, typical for the Sahel. The dominant vegetation is grass and thornshrub with scattered savannah woodlands. Cattle herding is the most important economic activity of the region. Agriculture is limited by low precipitation. With fewer than ten inhabitants per square kilometer (Claude, Grouzis, and Milleville 1991: 51), the region is one of the most sparsely populated regions of Burkina Faso (average forty-two inhabitants per square kilometer). Two different Fulani groups (Jelgoobe and Gaobe) represent approximately 25 percent of the population. About three-quarters of the local population still lives a predominantly or partly nomadic life (Claude et al. 1991: 53).

The actual study area is defined by the operating range (that is, the experiential realm of the community) of the Jelgoobe-Fulani of Férériwo in "normal" years. The Jelgoobe of Férériwo have lived for several generations in the region and at this place, and have hardly changed their lifestyle in this time. They live chiefly on cattle breeding: the milk of their Zebus is an essential part of the daily nutrition throughout the year, and those young bulls not used for breeding are sold to cover the cash needs of the family. They also possess goats and sheep, which are sold if necessary; their meat, unlike that of cattle, is eaten. As is true for all Fulani pastoralists, the herd represents a reserve for bad times: the larger it is, the better the chances that in case of a drought, enough of the herd will survive to be able to quickly reconstitute a new herd. Cattle breeding is also the most important marker of identity of the Fulani herders. Their second economic occupation is the cultivation of millet for their own needs. In the ideal case, when they harvest sufficient pearl millet (*Pennisetum glaucum*) to nourish the family the whole year, they have to sell only a few bulls, and the herd grows. Otherwise additional millet must be bought.

The Jelgoobe live in loose multigenerational family groups in spacious village-like camps called **wuros** (temporary or permanent community settlements consisting of several households), with approximately 100 to 150 inhabitants. Within this settlement the individual families—father, mother and children—each live in one of the traditional, hemispherical huts. The **wuro** composition changes frequently, since the individual families often temporarily join other camps belonging to the same extensive clan. Several times per year the entire **wuro** is moved a few kilometers, according to the pasture situation in the bush and the short-term need to stay close to the fields respectively. Approximately the same settlement sites are frequented every year. A transhumant migration to southern regions is ventured only in catastrophic years, when absolutely no forage and water remains in the Sahel. For more detailed information on Sahelian Fulani traditions and way of life see Barral (1977), Riesman (1974), Barral and Benoit (1976), Bartelsmeier (2001), and Krohmer (2002).

Methods of Eliciting Fulani Traditional Environmental Classification

The local environmental classification (landscape classification) of the Fulani groups studied was recorded as completely as possible. It was planned originally to record only the nomenclature of the vegetation units, but it was soon apparent that the classification system of the Fulani functions differently from ours: thus a vegetation section that is quite characteristic in our eyes and always present at particular locations may not, however, be used by the Fulani for naming or delimitation of the unit; they may instead designate the unit after the soil, in which case the information about the vegetation found there is only implicitly included. Therefore, the complete landscape classification system was recorded, particularly since this permits a much fuller understanding of the Fulani's environmental knowledge system. There is no standard approach for this (Scarpa and Arenas 2004).

We proceeded as follows: for each newly encountered unit, we first asked "What do you call this place?" to get an idea of the main criteria. Then we asked for a description of the unit. When a particularly characteristic vegetation community or other noteworthy attributes were not mentioned by the Fulani, we asked in addition whether these were of importance and if they were distinguished by a particular name. In that way, the bulk of the classification system was compiled with some of the most experienced herders. All the gathered information was analyzed and examined repeatedly in single and group interviews, while accompanying the herders on their daily herding trips, and during directed field trips to different terrain types. As many different informants as possible were consulted, belonging to different age groups and with a range of expertise concerning the natural environment. Rare units are naturally documented to a lesser extent; but this concerned only a few units. If different descriptions occurred for one unit, the "majority opinion" was ultimately adopted, but the minority opinion is mentioned in the description that follows. Terms were recorded in Fulfulde, using standard linguistic notation for the language.[1]

Traditional Classification of their Natural Environment by the Fulani

For Fulani in all regions of West Africa, an environmental dichotomy is recognized between settlements and their immediate surroundings, designated **wuro** (house, settlement), and the unoccupied range, the bush, designated **ladde** (see also Bierschenk 1997; Boesen 1997; Schareika 2003; de Bruijn and van Dijk 1995; Bartelsmeier 2001; Riesman 1974). While most other ethnic groups perceive the bush more as a place of threat or danger for humans, the Fulani, at least those who still are dedicated cattle breeders, conceive of the bush as their true homeland, where they feel comfortable with their herds. **wuro** is the area within which the woman provides for the well-being of the family, while **ladde,** the more virgin the better, is the realm of the men with their animals. The Fulani know and classify this bush and the entire environment in a very precise way. The following categories are relevant for the classification:

- Relief
- Soil: color and texture, as well as its content of water, stones, and nutrients
- Water: quantity and duration of its presence
- Vegetation: density, height, and dominant species
- Human influence
- Animal influence

Often several criteria—weighted to different degrees—are taken into account in the classification of one unit. Therefore the units form more an interlaced system with frequently overlapping categories than a hierarchical classification, in contrast to the classic taxonomic hierarchic model for ethnobiological taxa. We also may call it a geoecological system. Composite designations, which reflect the main criterion and one or more differentiating subcriteria, are very frequent. The main criterion forms the first part of the name, and the subcriteria follow (e.g., **cukkuri came** is a grove (main criterion) dominated by *Pterocarpus lucens* (subcriterion), forming a binomial name). The criteria, according to which the delimitation of the units was made, were determined partly from direct information from the Fulani, partly from translation of the names, and partly from translation of the descriptions of the units. The order in which the groups of units are presented here does not reflect a Fulani perspective. When possible, units within each group are treated in descending order (e.g. from larger to smaller topographic elevations).

The evaluation of pasture value concerns not only quantity and fodder value of the existing vegetation, but includes everything the Fulani consider of importance for their cattle. This may include factors such as available water, dry sites where the cattle can stand during the rainy season, protection from flies, shade, and so on. For the Fulani it is important not just to nourish their cattle well, but also to control their pastoral activities to promote the general well-being of their animals. Therefore the values indicated in the tables (from 3, very important, to 0, unimportant) reflect various criteria. This means that one cannot presume that a unit with a pasture value rating of 1 will be insignificant. Rather, the fodder and water supply during large periods of the year may be guaranteed only by ingeniously combining the available—even small—advantages of all units, to attain the best possible supply at each moment of the year.

The Landscape Classification System of the Sahelian Jelgoobe

The Sahelian landscape is dominated by two broad environmental types: large dune ridges running parallel to each other in an east-west direction, and the vast, almost flat peneplain called *glacis* in French, which is interrupted here and there by intermittent watercourses. This dichotomy is also reflected in the Fulani classification which differentiates between **seeno,** the dune ranges, and *ferro,* the *glacis.* I group the landscape units described below into several broad categories: units characterized by specific topographic features, units influenced by water, units characterized by soil properties, units characterized by vegetation, and finally zoogenic units.

Topographic or Landform Units

This category comprises all units which are characterized by their special surface form, thus any kind of topographic elevations or depressions. Vegetation does not play a significant role in characterizing these units, but sometimes the soil material is important. Note that the depressions possessing a watercourse, even intermittent, belong to the water-influenced units, which constitute the next section.

Elevated features

waamnde, baamngel

waamnde (plural *baamle*) designates a mountain or a large hill, regardless of the composition (laterite and/or rocks of the pre-Cambrian basement), and whether it consists of bedrock or debris. Slope steepness is also irrelevant. This designation is not applied to the dunes, however high or steep they may be, because these always fall under the category **seeno** (see below).

baamngel (pl. *mbamkoy*) designates a small hill with the same characteristics as *waamnde* but smaller dimensions. It often lies alone. The material is almost irrelevant, but cannot be sand.

tilde, tilel

tilde designates an elevation thatis lower and smaller than *waamnde.* The composition is not significant; in contrast to *waamnde* or *baamngel,* it may be

Table 3.1 Topographic Units

Unit	Brief description by the Fulani	Pastoral value during the year*			
		a	b	c	d
waamnde	large topographic elevation, mountain, standing above the surrounding landscape	?	?	?	?
baamngel	diminutive of *waamnde*; accordingly smaller mountain	1	0	0	0
tilde	hill, consisting of sand or, more rarely, rock; designates also summit of dunes	3**	1	1	3
tilel	diminutive of *tilde*; small hill	2	1	1	2
sallere	lateritic hill of various height and size, may be eroded to different degrees; also: steepened summit area of *caddi*	1	0	0	0
caddi	low elevated area consisting of lateritic crust outcrops, covered with coarse laterite chunks	1	0	0	1
callel	almost flat elevated area consisting of lateritic crust outcrops, smaller laterite chunks than on *caddi*	1	0	0	0
hukaawo	oblong, small elevated area on the peneplain	?	?	?	?
luggere	smaller depression strongly varying in size, may occur on all soil materials	2	2	0	0
luggol	small, oblong depression, mostly on dunes	2	2	0	0
naDDere	very shallow depression, mostly on dunes	1	1	0	0

*Legend: a = rainy season; b = late rainy season; c = cool dry season, d = hot dry season,
3 = very important, 2 = important, 1 = less important, 0 = unimportant;
** all four values apply to sandy hills or dune summits; for other cases, values are the same as for *baamngel*

a sand hill. The designation is often used for the steep top of the huge dunes. The term seems to be used more frequently for sandy elevations rather than for stony ones, although informants do not express that explicitly. *tilel* (diminutive of *tilde*) designates a very small hill that may stand isolated in an open space, for example the *tilel hakkunde kollaaDe korkaa'ye* 'small hill in the midst of adjoining *kollaaDe* (see pg. 72) covered with rough lateritic gravel'.

sallere, callel, caddi, hukaawo

Units designated as *sallere* are always lateritic[2] hills, which are variable in height and weathering. Some informants use this name also for the sometimes steeper central area of *caddi*.

The meaning of *caddi* cannot be exactly defined; some informants describe it as flat or only very slightly elevated, in any case lower than *sallere,* while others use the term to describe a clear elevation. All agree however on the material, a more or less eroded lateritic crust, and on the fact that it is covered by rough laterite chunks whose diameter is larger than fist size. There is often a typical vegetation assigned to the unit, characterized by *Schizachyrium exile* and *Andropogon fastigiatus,* which are typical in the Sahel for laterite stone sites and which the Fulani collectively refer to as **woDeriiho.** The bigger the laterite chunks are, the more fine soil matter, transported by wind and water, can accumulate between them, which is favorable for the establishment of plants. A large **caddi** is called **caddi mawndi.**

callel (pl. **calloy**) corresponds to the diminutive of **sallere.** The term designates a very low hill composed of a lateritic crust, on which the laterite blocks are smaller than on **caddi.** The associated vegetation is the same as for **sallere.** Finally **hukaawo** designates a low (lower than 2 m), oblong elevation in the peneplain. According to the informants, the soil type is of no importance. Often it is aeolian sand.

Pastoral value of elevated features

Elevated features are not significant for pasture due to their usually rocky condition and the resulting low vegetation coverage as well as the often poor accessibility for cattle. Only in the hot dry season do cattle systematically visit lower elevated features like **caddi** in order to eat the excrement of small rodents, which are frequently found there, as a salt substitute. Cattle do not like to stand in water, so during and after heavy rains herds rest on these small elevations, waiting for the water to run off.

Depressions

luggere

This term designates smaller depressions of different depth and extent. They may be situated on dunes (**luggere seeno**) or in the peneplain (**luggere ferro**). If the clay content in the soil is high enough, they may hold water for a short time, but this is of no importance for the classification (see the next section for units for which water *does* represent a classification criterion). Since the soil is always slightly more humid in depressions than in their surrounding area, small groves may grow there. If this is the case, the unit is called e.g. **luggere DooDe (DooDe** = the tree species *Combretum glutinosum*) or **luggere jelooDe** (*jelooDe* = *Guiera senegalensis*), but only as long as it is a matter of a few individuals. If the groves are bigger and reach the size and density of a **guyfal** or **cukkuri** (see pp. 67–69), the unit is named **luggere guyfal** or **luggere cukkuri.**

A distinctive subfeature of **luggere** is **luggol,** also a small depression, usually situated on the dunes (**luggol seeno**), but elongated, which is expressed by the ending -**ol.**

naDDere

This term designates also a depression, but of clearly smaller extent and depth than **luggere**. This unit occurs both on the dunes (**naDDere seeno**) and—more rarely—in the peneplain (**naDDere ferro**). However it never reaches a size that permits the growth of woody species. Finally the variant **naDDere seeDa** "a little bit **naDDere**" is a very small, almost imperceptible depression.

PASTORAL VALUE OF DEPRESSIONS

All depressions are relevant, if only because they hold water at least for a short period after a rain and thus constitute within the Sahelian landscape a dense grid of watering places, permitting cattle to satisfy their thirst at will. In the Jelgoobe's opinion this is very favorable for the cattle's well-being and thus their general condition. Furthermore, it permits herders to access and use pastures that are otherwise not within reach due to lack of water. Another benefit is that depressions are pastured only after the soil has dried; the grasses and herbs growing here then are still fresh and green because of the water storage capacity of the depressions, while they have already dried up in the surrounding areas.

Water-Influenced Units

The Jelgoobe differentiate a multiplicity of water-related units, both with flowing water and with standing water. All water resources are precisely categorized. It is noticeable that almost all running water features end with the syllable *-ol* which expresses their elongated form.

FLOWING WATER

weendu, pogowol, gooruwol, palol

weendu (pl. **beeli**) designates a river of regional importance. The plural **beeli** is also a place name designating the only river of the region. A **weendu** never runs completely dry. At most it is reduced during the dry season to small pools (**welde**) in the riverbed. Soil characteristics and accompanying vegetation are irrelevant for naming water features.

pogowol (also: **poguwol**) is part of a **weendu;** some informants assert that it is a lengthy wide stretch of a river. Others, however, state that **pogowol** designates a temporary watercourse, long but not wide, and therefore unable to hold very much water. Thus, with a heavy rain, the adjacent flats may be inundated. The same is true of **gooruwol**, a watercourse with dimensions between **weendu** and **palol** (see below). Though vegetation and soils do not help define **poguwol** or **gooruwol**, **palol** designates a medium-sized temporary watercourse and the associated vegetation, such as a gallery forest, a vegetated inundation surface, etc. A small **palol** is called a **palel. caltol palol** is a tributary branch of such a temporary stream.

Table 3.2 Water-Influenced Units

Unit	Brief description by the Fulani	Pastoral value during the year*			
		a	b	c	d
weendu	large river, never running dry completely	1	1	3	3
gooruwol	quite deep, temporary watercourse, a little deeper than **pogowol**	1	2	2	2
pogowol	narrow temporary watercourse, cannot take very much water; sometimes also: oblong, broadened section of a river	1	1	0	0
palol	temporary watercourse of middle size, including its accompanying vegetation	3	3	1	1
palel	small **palol**	1	1	1	1
caltol palol	a branch of a **palol**	?	?	?	?
gurunfuntuwol	deep erosion gully, in dunes or peneplain, regardless of its width	1	1	0	0
dogginirgol	shallow erosion gully, mostly on sandy ground	1	1	0	0
ciwtorgol	erosion gully with some deep sections that hold the water even after the rain	1	0	0	0
njaareendiwol	wide, shallow **gurunfuntuwol**, always with sandy ground	1	1	0	0
feto	natural pond of any size	0	1	3	3
welde	particularly deep section of a stagnant water	0	0	0	3
coofol	a branch of a stagnant water	?	?	?	?
gasel	man-made stagnant water	0	0	3	3
hokuluuru	small pool, small **feto,** contains water up to one month after rainfall	2	2	1	0
deeku'yal	small, shallow tarn, contains water only a few days after rainfall	1	1	0	0
cutorgol	very small waterhole, contains water only immediately after rainfall	1	0	0	0
godowaare	puddle, exists only immediately after rainfall	1	0	0	0
gedeeru	small spring, running dry in the dry season	n.a.	n.a.	n.a.	n.a.
mamasiiru	spring, wet even during the dry season	n.a.	n.a.	n.a.	n.a.
yayre	*mare* (e.g. Mare d'Oursi)	0	0	3	3

*Legend: a = rainy season; b = late rainy season; c = cool dry season, d = hot dry season, 3 = very important, 2 = important, 1 = less important, 0 = unimportant

gurunfuntuwol, dogginirgol, ciwtorgol, njaareendiwol

There are several small features that contain water only immediately after rain, and only for a period of hours. The development of such landscape features by widening, deepening, or headward erosion is precisely observed and designated by the Jelgoobe; the term *gurunfuntuwol* designates a deeply incised gully of indefinite width, which may develop both on dunes and in the *glacis*. Soil conditions do not play a role in their development and designation. Usually strong headward erosion is observed. For *dogginirgol,* this is not the case; this term designates a shallow, scarcely incised rill. Some informants describe this unit simply as a shallow *gurunfuntuwol,* with the soil playing no role, others stress that *dogginirgol* always develops on sandy soil. Immediately after a rain event the unit *ciwtorgol* can be found: it corresponds to a *dogginirgol* in whose deepest sections some water still remains. Finally, we have to mention *njaareendiwol* (from *njaareendi,* 'sand'), which designates a special form of *gurunfuntuwol*; it is larger and deeper than most, and always over a deep sandy bank, consisting of the sand deposited when the current weakens.

STAGNANT WATER

feto, welde, coofol, gasel

The term *feto* (pl. *pete*) designates a water hole of natural origin. Its dimension ranges from a pond with a diameter of some meters to a lake several hundred meters in diameter. *pete* are rather deep, so that they dry out only in the late dry season. They often occur at the base of dunes. Particularly deep sections of a *feto,* and of other stagnant pools, are called *welde. coofol* designates a tributary branch of a *feto.* This term is used only for larger water features.

feto may be differentiated even further, depending on its surrounding woody vegetation: *feto barkeehi (f. kojole, f. kooli,* etc.) is predominantly surrounded by *Piliostigma reticulatum* (or *Anogeissus leiocarpus* or *Mitragyna inermis* respectively). A watering hole that is entirely man-made is called *gasel* (see below).

hokuluuru, deeku'yal, godowaare, cutorgol

hokuluuru is a small stagnant pool; it may also be designated as a small *feto.* It is always situated in the peneplain, often on relatively stony soil, which however contains a lot of clay in its upper horizons and can therefore hold water up to one month after a rain. The unit *deeku'yal* is almost the same but smaller and therefore holds rainwater only for some days. Even smaller are *godowaare,* a large puddle existing only immediately after rain events, and *cutorgol,* a very small water hole ('a place where the water reposes').

SPRINGS

In the Sahel, springs are rare. There are two different types: *gedeeru,* a small depression containing a water source that fails in the dry season. The second type of spring is a ***mamasiiru,*** a water hole fed from groundwater and running dry only very late in the dry season.

yayre

This term is not uniformly applied. Young Jelgoobe use it for a seasonally flooded depression with almost no trees or bushes. The margins are dominated by Poaceae and Cyperaceae. For the elders, this term designates large, temporary lakes, none of which occur in the study area around Férériwo. The nearest is located at the village of Oursi, about thirty kilometers distant. Younger Jelgoobe usage corresponds to that of the Fulani in other regions. This suggests that younger herders may be influenced by their travels in other areas or by migrant herders from other regions.

PASTORAL VALUE OF WATER-RELATED FEATURES

For cattle, such features are important mainly as watering places. Depending upon size and underground soil, the water holding capacity may strongly diverge in quantity and time. Flat watercourses like ***dogginirgol*** or ***gurunfuntuwol*** contain water only for a short time after a rain and are insignificant for cattle herding. Larger rivers are much more important, e.g. ***weendu*** or ***palol***, as well as larger pools such as ***feto***, as they may provide water far into the dry season.

However, water features play not only an important role in view of their suitability as watering places, but also because they store water in the soil after drying out. This permits the herbaceous plant cover to remain green and fresh at least during the first part of the dry season, when it is a very important fodder reserve. Especially the large *mares* (French: large, flat, temporary seas, usually at the basement of large dunes; Fulfulde: ***yayre***) are therefore of enormous importance for local cattle owners.

Given that the Sahelian landscape is largely defined by the absence of water, it is surprising to see the variety of water features that are recognized by the Fulani. This emphasis on water features is due to the scarcity of water. The life of a Sahelian cattle herder is shaped by the search for water.

Units That Are Characterized by Soil Properties

Features characterized by soil properties are defined more by soil type than soil texture. Soil types usually contain information about the ecological situation that is only rarely implicated by the soil material (texture). The Jelgoobe soil classification corresponds closely to that used in soil geography (FAO UNESCO 1997 Soil Survey Staff 1975). Poorly developed, stony raw soils (like ***korkaa'ye, caddi, cakuwaari, sanngo***) correspond to Leptosols. Clayey ***bolaawo,*** which hardens like stone when dry, corresponds to the Vertisols. Fulani classify these in more detail, according to the stone content. ***seeno*** and related terms correspond to the

Table 3.3 Units Characterized by Soil Properties

Unit	Brief description by the Fulani	Pastoral value during the year*			
		a	b	c	d
njaareendi	soil type sand	–	–	–	–
seeno	dune	3**	0***	0***	3
seeno ladde	'bush dune' (not cultivated, only pastured)	3**	0***	0***	3
seeno (ley) gese	cultivated dune	0	0	0	2
seeno heso	'young dune', not settled, not cultivated	3**	0	0	3
seeno hiingo	'old dune' (in the sense of "used"), settled and/or cultivated	0	0	0	2
seeno hoyguruure	dune that was settled very long ago, recognizable by potsherds, foundation vestiges and pisoliths	?	?	?	?
seentere (ceentere)	small hill consisting of sand, situated in the peneplain, developed from an eroding sand overlay	2	2	1	1
sentatiire	small accumulation of sand in the peneplain	2	2	1	1
ceentel	oblong, very small dune; also: sand overlay in the *glacis*	2	1	1	1
ceenel	small, low dune	2	1	1	1
bolaawo	vast plain with dark clayey soil, keeps the water a long time	0	1	2	2
loopal	pure clay	0	1	2	2
sinngaawo	lateritic crust without any vegetation	0	0	0	0
hukaawo	lateritic site with laterite boulders the size of a fist (or bigger)	1	1	0	0
korkaa'ye	lateritic soil, predominantly small to medium-size laterite pebbles	1	1	0	0
cakuwaari	lateritic soil densely covered by predominantly fine laterite gravel (pisoliths)	1	1	0	0
sanngo	lateritic soil, predominantly very fine laterite gravel	1	1	0	0
tepaare	white rocks	0	0	0	0
guraawo	surface covered with bright gravel	?	?	?	?
hoyguruure	ancient iron melting place, covered with iron cinder (*ndon'yikiire*) and laterite chunks	0	0	0	0
moonde	salt- and potassium-rich soil of salt lick areas	3	2	1	0
hanhade	a surface of a few square meters with compressed, salty-bitter tasting soil	0	0	0	0
puunDi	very small (<1 m²), strongly compressed, dark surface with salty-tasting soil	0	0	0	0

*Legend: a = rainy season; b = late rainy season; c = cool dry season, d = hot dry season,
3 = very important, 2 = important, 1 = less important, 0 = unimportant
** at the beginning of the rainy season; *** is saved for the late dry season

Arenosols (sandy soils). Only the Fluvisols, with material deposited by water, fail to correspond to the Jelgoobe classification system. Such soils are not separately designated by the Fulani as a type of soil.

Vegetation is less important for demarcating soil units. This is the case because the Jelgoobe know exactly which plants grow in each soil unit, so they do not have to explicitly mention this.

DUNES AND SANDY SOILS

The Jelgoobe differentiate between *njaareendi,* sand, per se, and the several features with names derivative of the term *seeno* (dune, pl. *ceene*). Due to their relief, dune ridges might also be classified as geomorphological units; but here they are treated with the soils, since the pedological qualities are most important for the Jelgoobe, especially with regard to pasture quality. The fact that they are elevated is less important.

seeno

This term signifies dune. It thus designates the usually west-east running dune ridges that are typical for the Sahel of Burkina Faso. They consist of Tertiary sand, and pedogenesis has, depending upon age and land use, variably progressed. The soil can also be hard surfaced or possess a small amount of loam or silt. From the mere term *seeno* it is not evident whether an old or a young dune ridge is meant, nor which plants species can be found there, nor if the dune is predominantly pastured or cultivated. However, dunes are further differentiated: "virgin" dunes are uncultivated and far from any settlement. These are called *seeno ladde* (*ladde* bushland). A further classification criterion for the Jelgoobe is the fact that the species that grow here are different from those on cultivated dunes, where, according to them, the soil cannot hold moisture for long, and therefore grasses and herbs dry up early. Similar characteristics (different species, higher soil moisture) are assigned to *seeno heso,* a young dune which is neither settled nor cultivated but may be pastured. *seeno hiingo* (old dune) is a dune that has long been cultivated and/or settled. The term *seeno gese* (*gesa* field, pl. *gese*) is used as well. The distinction *heso/hiingo* has nothing to do with the geological age of the dune (recent dunes 20,000 years, old dunes 40,000 years) but refers only to the duration and mode of its use by people.

A formerly settled dune section, occupied in historical or even archaeological times and long abandoned, is named *seeno hoyguruure* (or *koyguruuje,* pl.) by the Jelgoobe. Such sites can be recognized by settlement relics such as artifacts, iron cinder, stone foundations, pisoliths, and so on.

Dunes or dune sections may differ from others also by their vegetation. *seeno cukke* (pl. of *cukkuri* see below), e.g. is a dune with several dense groves, usually situated in small depressions, since their slightly higher soil humidity allows the establishment of woody species. It is the same for *seeno guyfe* (pl. of *guyfal*), with groves that are smaller and less dense. *seeno bedereeje* is a dune characterized by a regular occurrence of the salient woody species *Euphorbia balsamifera* (Fulfulde: *bedereehi*, pl. *bedereeje*). In one case not far away from Férériwo, this designation has become established as a distinct toponym. In no case was a dune section classified or designated according to its herbaceous vegetation, as might be expected when viewing certain sections dominated by just one or two species.

seentere, sentatiire, ceentel, ceenel

The units *seentere, sentatiire,* and *ceentel* as well as *ceenel* are derived from the word *seeno*. All are diminutives of this term. In the broadest sense they may be regarded as synonyms, since they all designate an accumulation of sand in the *glacis* (peneplain); however, the depth and spatial extent of the sand may vary considerably. Informants stated that *seeno* is much larger than *seentere* and *seentere* clearly larger than *ceentel*; however, this did not always prove true on the ground. In detail the units were described as follows: a sand layer from half a meter up to several meters is called *seentere ferro* as well as *ceentel ley ferro.* Uneven wind and water erosion processes may reduce it to a small-scale mosaic consisting of small relatively thin patches of sand interspersed with *glacis* soils. This mosaic is also called *seentere ferro* or *ceentel ley ferro. seentere* (also *ceentere*) alone designates a slight sandy elevation, which in the *glacis* exceeds the surface level of its surroundings; therefore its height may vary. This unit may also be named *sentatiire. sentatiire golome, s. kelle,* and *s. came* designate small accumulations of sand or sandy surfaces, with woody vegetation that is dominated by *Stereospermum kunthianum, Grewia bicolour* or *Pterocarpus lucens,* respectively. *ceentel* may be a long-extended small dune in the midst of the *glacis.* At the same time this term may simply designate the sand layer. In addition, *ceentel daneyel* (small white dune) is a low, not very large dune, with vegetation that dries up and turns whitish after the end of the rains. In Férériwo it is also a toponym.

It is quite difficult to gain any sense of real proportions from these descriptions. One could presume that *ceentel* is always smaller than *seeno,* because the first is a grammatical diminutive of the latter, but in fact this depends very strongly on the subjective evaluation of the informants: what one of them calls *ceentel,* another calls *seentere,* and so on.

PASTORAL VALUE OF SANDY SOILS

The sandy soils are highly valued by Fulani as pasturelands because of the numerous highly prized fodder plants growing there, despite the recent reduction in cover of the much esteemed fodder grass *Andropogon gayanus* (*rannyere*). It has been replaced by predominantly annual species such as *Schoenefeldia gracilis* (*raneriiho*), *Cenchrus biflorus* (*kebbe*), *Tribulus terrestris* (*tuppe*), etc. The herbaceous Fabaceae *Alysicarpus ovalifolius* (*sinkaare*) or *Zornia glochidiata* (*dengeere*) are also highly valued fodder species. The dunes are particularly important during two periods of the year. The first is at the beginning of the rainy season, when the often loamy soils of the *glacis* retain so much water that they are saturated and swampy, and the cattle avoid grazing there. The sandy dune soils with their higher permeability absorb the water better and permit the seeds of various herbaceous species (e.g. *Tribulus terrestris*, various grasses) to sprout quickly even during the rainy season. For this reason, the dune ridges are at this season the first usable pasture, which is of vital importance to the cattle since they are emaciated from the long dry season and need to find fresh fodder quickly. The second period is the late dry season, when the herds are led predominantly to the dunes for grazing because the vegetation of the *ferro* has been almost entirely cropped down and what is left possesses almost no nutritive value. Moreover, according to the Jelgoobe, many of the dune species, even when dry, have high nutritional value (**semmbe,** 'strength'); therefore the dunes are systematically conserved as pasture reserves for this season. The smaller sandy soil units as well as the *glacis* surfaces with sand layers are also esteemed as pasturelands, due to the useful fodder species that can be found there; however, unlike the dune ridges their use is not limited to certain seasons.

Clay soils

bolaawo

Clay soils are relatively frequent in the Sahel, mostly on the *glacis* and in depressions. In the surroundings of Férériwo, however, they are found only approximately 20 km away and therefore are not among the regularly visited pasturelands. In this case, the unit is called **bolaawo**. The term designates a vast plain with dark, clayey soil, which may consist of pure clay but usually contains an admixture of silt and sand. In the rainy season and even a certain period after, it is not inundated but soaked with water, due to the clay content. Also a more or less elevated soil skeleton is mentioned as typical. Some informants refer to **bolaawo** as **kollangal,** which actually is a consolidated, stony surface with poor vegetation or unvegetated. This is however plausible if one considers the appearance of the **bolaaje** (plural of **bolaawo**) during the dry season; when all grasses and herbs are dried or grazed, the meager trees—only very few species can survive the strongly changing water availability of clayey soils—tend to be overlooked because they lack leaves, and the whole surface is quite bare.

loopal

In contrast to **bolaawo, loopal** is an almost pure clay soil. It is always situated within periodically inundated depressions (see **palol**), which is why those sites are often simply summarized under **palol,** without any specific designation for the soil. **loopal** is an extremely hydromorphic soil. For most woody species such soil conditions are problematic; therefore, only some particularly adapted trees or bushes are to be found here, and herbaceous species constitute the dominant life form. Like the **bolaaje, loope** (pl. of **loopal**) are an important dry season pasture reserve.

There are also other sites with loamy to clayey soils that are not classified as **loopal,** since other criteria are the center of attention. This applies to the unit **jaayal** (see pg. 70), which is classified and designated according to its herbaceous vegetation, as well as to the unit **palol,** which belongs to the units influenced by water (see water-influenced units).

Pastoral value of clayey soils

All **bolaaje** are highly valued pasture units, since many valuable pasture grasses grow there. According to the Fulani they are particularly nutritious due to the soil properties. The most important pasture time is the dry season. In the rainy season and a certain time after, the cattle avoid these soils because they do not like to stand in the mud. Therefore these soils are only pastured when dry and thus represent an important pasture reserve.

Stony and lateritic soils

As elsewhere in West Africa, lateritic soils or soils dominated by little-decomposed lateritic crust are widespread in the Sahel. This is, however, limited to the *glacis.* The Jelgoobe differentiate the laterite units very exactly, especially the raw skeletal soils with limited soil formation. The most important criterion is the size of the skeleton fragments.

sinngaawo

This term designates the almost completely vegetation-free lateritic crust that is not yet eroded. Among the Jelgoobe it is well known that other ethnic groups formerly used this crust to extract iron. However, this is no longer the case, at least in the Sahel.

The units **sallere, caddi,** and **callel** (see pp. 55–56) are closely related to **sinngaawo.** The Jelgoobe emphasize the topography of these elements; they are treated above in the section on topographic units.

hukaawo, korkaa'ye, cakuwaari

These three terms designate raw stony soils (lithosols and regosols), with soil skeleton elements of variable size. **hukaawo** designates soils that are dominated by rough lateritic boulders, fist-size or larger. This unit can be also slightly elevated, but if so it is assigned to the topographic relief units. **korkaa'ye** soils contain a high proportion of small to medium size laterite pebbles. **cakuwaari** contains much fine lateritic gravel (pisoliths), which accumulate gradually at the surface due to erosion, covering it in a dense layer. Such locations may be slightly raised over the level of the surrounding sand and may or may not expose a lateritic crust, but these criteria are not emphasized as characteristic; neither is the almost entirely absent vegetation.

sanngo

With **sanngo,** the laterite gravel is even finer. The vegetation cover is very limited due to the exposed bedrock and dense gravel surface. The usual vegetation consists only of few species adapted to these difficult soil conditions. **sanngo** can be subdivided into **sanngo woDeewo** (with red laterite gravel) and **sanngo raneewo** (with white quartz gravel). **sanngo** without further specification generally designates the red **sanngo.**

Pasture value of stony lateritic soils

Stony lateritic soils have limited pasture value and only are used in the rainy season, when they permit cattle to stand on a dry place in an otherwise inundated landscape. In addition, at this time some interesting fodder species may sprout in the fine soil material accumulated between the laterite blocks.

Saline soils

moonde, hanhade, puunDi

Salt licks, which were regularly visited traditionally, are called **moonde** and consist of dark sand rich in salt and minerals (especially potassium). However, in the study area no such place could be found. The nearest is approximately 30 km away near Oursi. According to informants, this site is no longer visited. Cattle are given salt bought at the market instead.

hanhade (possibly from **haaDa** to be bitter; also **haanhadde**) designates an area only a few square meters extent, probably saline, in the midst of a **kollangal** (consolidated open space). According to informants, this soil resembles **moonde.** This apparent similarity however does not indicate anything about its suitability as a source of minerals; rather this soil is harmful for cattle, who eat it when suffering an acute lack of salt. This may cause digestive problems and even death. It is the same with **puunDi,** typically a very small site with strongly consolidated, dark soil with high salinity. The genesis of the latter two units could not be clarified.

Soil condition

The Jelgoobe use several different terms to describe the condition of a soil. **leydi keyri** (**leydi** soil) designates an intact, undegraded soil. **leydi tampi** or **leydi mburndi** is tired soil. **leydi mbanndi** or **leydi waati** refers to an irreversibly dead soil. When evaluating soil quality, the Jelgoobe consider primarily the condition, composition and extent of vegetation cover, and the consolidation of the soil surface.

Units Characterized by Vegetation

Woody formations are differentiated in a very detailed way. Herbaceous vegetation is often summarized under other units. For example, if it is located on a site where soil or relief are decisive for the classification, the vegetation remains unnamed. However, Jelgoobe herders know the associated vegetation implicitly. This applies for example to different grass communities on the dunes, or to the herbaceous layer in woodland units.

WOODLAND UNITS

cukkuri

cukkuri (pl. *cukke*) is a sharply limited, usually dense grove of variable size. Often it is to be found at sites with slightly higher soil moisture. It may consist of several woody species, in which case no further designation is used, or it may be dominated by one or two species, as for example **cukkuri came, cukkuri jelooDe,** or **cukkuri gungume** (a **cukkuri** dominated by *Pterocarpus lucens, Guiera senegalensis,* or *Combretum micranthum,* respectively). However, this species-related designation can be used even if the respective species does not dominate in terms of cover percentage; its dominance may be purely visual, as when some particularly tall individuals of the species exceed the height of the grove canopy. **cukke** may be found isolated or clustered on the dune ridges (**cukke dow seeno**) and in the *glacis* (**cukke ley ferro**).

Table 3.4 Units Characterized by Vegetation

Unit	Brief description by the Fulani	Pastoral value during the year*			
		a	b	c	d
cukkuri	sharply delimited, generally dense grove of various size	0	2	1	2
yaha-warawol	oblong *cukkuri*	0	2	1	2
guyfal	small, not very dense, more or less circular grove, growing on dunes	0	2	1	2
toggere	small group of trees or shrubs with arbitrary species composition	1	1	1	1
duunde *(e.g.,* **kojole***)*	monospecific group of several big individuals (e.g., *Anogeissus leiocarpus*) around a small pond	0	1	1	1
juulaafuuje kojole	group of *Anogeissus leiocarpus*	0	1	1	1
leDDe dow weendu	woody vegetation that accompanies a larger river (*weenDu*); also gallery forest; often too dense for pasture	0	1	1	1

(continued)

Table 3.4 Units Characterized by Vegetation *(continued)*

Unit	Brief description by the Fulani	a	b	c	d
jaayal	area always situated on the *glacis* and almost free of woody species; it varies in size from some m^2 to several acres	1	3	2	1
jaayal loope	*jaayal* with especially clayey soil	0	3	2	1
jaayal pagguri	*jaayal* dominated by *Panicum laetum*	1	3	2	1
jaayal ndiiriiri	*jaayal* dominated by *Echinochloa colona*	1	3	2	1
jaayal raneriiho	*jaayal* dominated by *Schoenefeldia gracilis*	1	3	2	1
jaayal lu'e na'i	*jaayal* dominated by *Dactyloctenium aegyptium*	1	3	2	1
jaayal kollangal	degraded *jaayal,* whose clayey soil has been partly eroded, with strongly consolidated soil and almost without vegetation	1	1	0	0
alhaali jaayal, *jaayal seeDa*	degraded *jaayal* whose herbaceous layer is getting thin; the soil is partly washed away	1	2	1	0
jayri	open space between, e.g., two *cukkuri,* grassy or not; may also be a glade inside a *cukkuri* or *guyfal*	2	3	1	2
jaayal pamaral	small *jaayal*	1	3	2	1
kollangal	very flat area with consolidated soil, without vegetation	0	0	0	0
kollangal ferro	open space (without vegetation) on the peneplain	0	0	0	0
kollangal seeno	open space on a dune	0	0	0	0
kollangal bolaawo	open space on the clayey plain	0	0	0	0
kollangal kaaje	open space covered with gravel	0	0	0	0
kollangal korkaa'ye	open space covered with pisoliths	0	0	0	0
kollangal danewal	the "real" *kollangal,* which has a whitish surface, not loamy or clayey, without any vegetation	0	0	0	0
kollangal huDo	grassy open space between shrubberies or groves, on the point of degrading	2	1	0	0
kollaaDe (k. ferro, *bolaawo, …)*	mosaic of vegetation and small open spaces	2	1	0	0
kollaaDe seeno, *kartaale*	mosaic of vegetation and small open spaces, on dunes	2	1	0	0
karal	very vast open space with consolidated soil, without sand accumulation or stone chunks	0	0	0	0

Legend: a = rainy season; b = late rainy season; c = cool dry season, d = hot dry season, 3 = very important, 2 = important, 1 = less important, 0 = unimportant

The latter (*cukke ley ferro*) corresponds in Férériwo to the woody formation *brousse tigrée* ('tiger bush'), which is typical for the Sahel. *cukkuri palol* is a very dense grove along a temporary watercourse, where cattle cannot graze due to its impenetrability. A stratigraphic element is *cukkuri dowdow*. Despite a very dense canopy, here stems are not so close together; thus it is accessible for cattle. *cukkuri seeDa* is a strongly degraded grove. Only a person familiar with the place will recognize its former importance. *cukkuri guyfal* represents the smallest *cukkuri* variant. It is a dense, circular grove, smaller than the other types of *cukkuri*. These two characteristics are transitional to the next group of woody units.

yaha-warawol

is an oblong *cukkuri,* often accompanying a small depression (*palel*). Species composition plays no role. Literally, this is a grove that "comes and goes," possibly referring to the fact that it grows on both sides of the depression.

guyfal

This term (pl. *guyfe*) designates a small, not very dense, more or less circular grove on dunes (*guyfal dow seeno*) or on the *glacis* (*guyfal ley ferro*). It may also be called *cukkuri pamarel* (small *cukkuri*). The maximum size is defined by the fact that one "can walk around it." It is so sparsely vegetated that cattle can pasture within it. Like *cukkuri* it may be dominated by a single species (e.g. *guyfal jelooDe* dominated by *Guiera senegalensis*). A *guyfal DooDe* may be dominated numerically by *Guiera senegalensis,* but nevertheless be named for some particularly prominent specimen of *Combretum glutinosum* (*DooDe*). *guyfal cukkuri* is a larger circular *guyfal,* a transitional unit to the larger *cukkuri. guyfe yaha-warawol* refers to several small groves in line as if on a chain, whether on dunes or in the peneplain.

toggere, duunde, leDDe dow weendu

Other woodland units are less often encountered. *toggere* is a small grove of trees or bushes with arbitrary species composition. Two other units are defined by *Anogeissus leiocarpus* (Combretaceae): *duunde kojole,* a group of large *A. leiocarpus* close together beside a pond, and *juulaafuuje kojole,* the same without the pond.

An azonal unit is *leDDe dow weendu,* the woody vegetation bordering a large river.

Pastoral value of woodland units

Woodland units are important for herders and their cattle for two reasons. On the one hand, many woody species are important for fodder, especially in the dry season; the leaves, fresh or dried, provide substantial nutrients. Jelgoobe stress that during the dry season, woody species are important not only for their cattle but also for themselves, for the leaf sauce that accompanies their millet gruel and contains all necessary vitamins. In this context, *Pterocarpus lucens, Cadaba farinosa, Maerua crassifolia,* and *Boscia angustifolia* are the most important species.

A second value of woodlands for cattle is the shady microclimate they offer, which shelters herbaceous ground cover that often comprises important fodder species (such as *Brachiaria* species or *Panicum laetum*), so that they may remain fresh until the beginning of the dry season. During this season, wooded groves in the *glacis* are systematically visited for pasture. Only those thickets that are too dense or thorny for cattle to enter are of no value for grazing.

UNITS CHARACTERIZED BY THE ABSENCE OF WOODY VEGETATION

jaayal (from Fulfulde wide, open; pl. *jaaye*) designates a surface almost free of woody species, always situated on the *glacis,* which may extend from a few to several hundred square meters. It is predominantly flat, sometimes with an almost imperceptible depression inundated only briefly after rains. Due to its loamy or clayey soil, water is retained longer than in sandy or consolidated soils nearby. The high clay content is evident also in the fissures that open in the dry season. If it is particularly high in clay, the Fulani speak of a *jaayal loope,* a clayey *jaayal.* Due to these soil conditions and absence of competition from woody species, *jaayal* has a dense grass cover and thus is a pasture unit highly esteemed at all seasons owing to the predominance there of excellent fodder grasses, which are highly nutritious both fresh and dry.

This unit is further differentiated by the Jelgoobe according to species dominance: *jaayal pagguri* (*jaayal* with *Panicum laetum* dominant, but also with different *Brachiaria* species present; *pagguri* is a collective name for these species); *jaayal ndiiriiri* (*jaayal* with *Echinochloa colona*); *jaayal raneriiho* (*jaayal* with *Schoenefeldia gracilis*); *jaayal lu'e na'i* (*jaayal* with *Dactyloctenium aegyptium*) etc. A site covered exclusively by *Panicum laetum* may be called **BalBalndi.** The species present permit the Jelgoobe to deduce the slightly varying soil characteristics within the *jaayal.* A *jaayal raneriiho* is clearly drier than a *jaayal pagguri.* Jelgoobe may also deduce how long a particular area will be useful for pasture.

As for other units (e.g. *cukkuri*), the quality of the habitat is also significant for the classification of *jaayal: jaayal kollangal,* for example, is a strongly degraded *jaayal,* the original clay-rich soil partially washed away and the remainder consolidated and largely without vegetation. *alhaali jaayal* (in Fulfulde 'looks like *jaayal*' or a kind of *jaayal*) and *jaayal seeDa* ('a little bit *jaayal*') are further possibilities for designating the quality of the area. These last two designate degraded or degrading surfaces with thin grass layers and clear signs of soil erosion. Another term may also express degradation: *jaayal pamaral* (small *jaayal*) may be used both for a *jaayal* reduced in size by progressive edge degradation and for a naturally small *jaayal.*

The term *jayri* (from the Fulfulde term for far, wide) may designate both an open space between e.g. two *cukke*, grass-covered or not, and a clearing within a *cukkuri* or *guyfal. jayri ferro* is an open tree savannah in which copses, grass-lands and open spaces alternate.

PASTORAL VALUE OF UNITS THAT LACK WOODY VEGETATION

Units lacking woody vegetation are all extraordinarily important pasturelands. Their dense herbaceous layer consists almost exclusively of both fresh and dry high-quality pasture species, mainly grasses, e.g. *Panicum laetum, Echinochloa colona,* or *Schoenefeldia gracilis.* The loam- or clay-rich soils, which store rainwater for a long time, are pastured mainly in the early dry season, once the surface has dried. *jayri* has less soil-humidity and constitutes a pasture unit of special value, particularly from the late rainy season until the late dry season, due to the presence of both grasses and woody plants and great species diversity. At times humans directly compete with cattle for resources here, as the tasty grains of the grass *Panicum laetum* may be harvested in quantity over large areas for human consumption.

UNITS CHARACTERIZED BY THE ABSENCE OF ANY VEGETATION

Although the presence of rather dense vegetation is a prerequisite for Jelgoobe cattleherding, vegetation-free ranges are very precisely differentiated.

kollangal

The term *kollangal* designates a very flat, consolidated, vegetation-free surface. The unit may be differentiated in diverse ways, e.g. after the soil type. Thus *kollangal bolaawo* is a consolidated surface in a *bolaawo* (see description above); *kollangal seeno* is an identical surface on a dune, and *kollangal ferro* is in the *glacis. kollangal kaaje* (*kollangal* with gravel) is an open space densely covered with gravel of any rock material. *kollangal korkaa'ye* refers expressly to a surface covered with finer laterite gravel, even if the soil beneath the surface is loamy or sandy. *kollangal danewal* (white *kollangal*) is the "genuine" *kollangal.* It exhibits a whitish surface, is neither loamy nor clayey, and is completely free of vegetation. The term *kollangal huDo* (*kollangal* of the grass, *kollangal* with grass), which at first seems to be contradictory, is used for a grass-covered open space between groves undergoing degradation. *kollangal jaaBe* (= *kollangal* with *Ziziphus mauritiana*) also at first appears to be contradictory: This is a toponym designating a site at which there once grew a stand of *Ziziphus mauritiana*, which has since disappeared, so that today there is a *kollangal.* The name reminds one of the vegetation history of the place. This example demonstrates how environmental change is incorporated in traditional classification, which thereby is constantly updated.

kollaaDe

If the term *kollangal* is used in the plural, **kollaaDe,** this always contains the meaning of a mosaic, i.e., an open space surrounded by vegetation, in contact with other open spaces. Different geographic scales may be involved, from just a few square meters to several hectares. The Jelgoobe in Férériwo use **kollaaDe** without further differentiation for locations on the *glacis*, where the sand cover is eroded and interrupted by numerous small open spaces and the sand on the covered surfaces forms small, grass-covered elevations (**ceentel** or **tilel**). Such areas may also be called **kollaaDe ferro. kollaaDe bolaawo** and **kollaaDe seeno** designate similar mosaics in a clayey plain and on a dune, respectively. **kollaaDe seeno** may also be called **kartaale.**The highly valued forage grass *Schoenefeldia gracilis,* when it dominates such a mosaic (**kollaaDe raneriiho**), is the only species singled out to differentiate **kollaaDe.**

karal

karal is a very large, consolidated open space that preferably exhibits a whitish surface without accumulations of sand or rock. Some informants use this term also for a very large **kollangal.**

PASTORAL VALUE OF UNVEGETATED UNITS

As expected, vegetation-free units possess little pasture value. However, the **kollaaDe** interdigitated among other units form an exception. If these other units support valuable species, they may be esteemed as pasture, particularly during the late rainy season.

Anthropogenic Units

Anthropogenic units are sites of abandoned settlements and/or cultivated areas. Except for agricultural fields, all anthropogenic units in the Sahel are relatively small in size and therefore of no consequence for pasturing. They are treated here summarily.

gese, puyagaare

The fields, **gese** (sg. **ngesa**), on which pearl millet (*Pennisetum glaucum*) is planted every year, are situated exclusively on old, stabilized dunes in the study area. These soils are favorable for millet, particularly with regard to water-retention capacity. In contrast, the soils of the *glacis* are not suitable for cultivation, not even in sandy areas. Jelgoobe state that this sand layer is not deep enough to store sufficient water for field crops.

puyagaare designates a fallow field and was found only once in the study area. Since millet fields on dunes regularly fertilized by cattle usually permit annual cultivation,[3] fallows in the classic sense scarcely occur here (see also Riesman

Table 3.5 Anthropogenic Units

Unit	Brief description by the Fulani		Pastoral value during the year*			
			a	b	c	d
ngesa (pl. *gese*)	Field	cultivation	0	2	3	0
pu'yagaare	Fallow	cultivation	3	2	1	0
gasel	not very large pond, dug by men	water	1	1	3	0
barasi	big, deep artificial pond	water	1	1	1	3
woyndu	deep well	water	0	0	2	3
Bulli	well situated on a site with a not very low ground water table	water	0	0	2	3
bille	recently abandoned but regularly resettled sites	settlement	1	1	0	0
hoyguruure	settlement sites on the dunes abandoned a very long time ago	settlement	0	0	0	0
mbuneeri	dune soil with many artifacts and other remains of settlement	soil	0	0	0	0
ladde yeerumbereere, *ladde Baleere*	bush not influenced by people in any way	absence of human influence	3	3	3	3

Legend: a = rainy season; b = late rainy season; c = cool dry season, d = hot dry season,
3 = very important, 2 = important, 1 = less important, 0 = unimportant

1974). Note also that the term for fallow field is borrowed from a neighboring language, the Mooré.

gasel, barasi, woyndu, Bulli

gasel is a medium-sized pond dug by people. A large and deep artificial pond is a *barasi* (from French *barrage* artificial lake). *barasi* may represent the last open water at the end of the dry season and are thus of great interest for cattle herders. Other artificial water sources are the *woyndu,* deep groundwater wells, and *Bulli,* wells that are dug for groundwater near the surface. People and animals often spend the late dry season in proximity to such *Bulli* after all other surface water has evaporated.

bille, hoyguruure, mbuneeri

In the study area, former human settlements are frequently encountered. These may have been abandoned for variable periods of time. *bille* (sg. *winnde*) have been recently abandoned and are likely to be regularly settled again. *hoygu-*

ruure are ancient settlements on dunes. Frequently, iron cinder residues indicate traditional iron smelting. According to Jelgoobe informants, the inhabitants of these settlements belonged to other ethnic groups that have long since vanished from the region. The consolidated sandy soil of these old settlements, often interspersed with pisoliths and bits of broken glass, is called **mbuneeri.** Ancient dune settlements are **seeno koyguruuje.**

A "negative" anthropogenic unit that exists thanks to the absence of any human impact is **ladde yeerumbereere** (also **ladde Baleere**). This "virgin" bush, apparently unaltered by human activity, is universally described as the best of all pasture. In the Sahel it may be found today only far from open water, at places with groundwater levels too deep for wells dug by hand.

PASTORAL VALUE OF ANTHROPOGENIC UNITS

The only anthropogenic units besides water sources of importance for herders and their animals are the **gese,** dune ridge fields. After the harvest, cattle forage there for several weeks or months on the stubble, a valuable, nutrient-rich fodder. This is an important fodder supplement particularly in the dry season, when pasture in the bush is scarce. Fallow fields, which would also be good pasture, scarcely occur in the study area.

Compared with other climatic regions in West Africa, in the Sahel relatively few landscape elements are attributed to direct human influence. This is remarkable, given that in the Sahel there is scarcely a single place not regularly used by people, particularly by their omnipresent cattle. However, from a Jelgoobe perspective, human influences on the landscape are rarely as important for landscape classification as "natural" criteria.

Zoogenic Units

Zoogenic units occur frequently throughout the study area but are always of limited extent. For the Fulani, they are of little importance; I mention them here for the sake of completeness.

gotol

gotol (pl. *goti*) designates a cattle track. *goti* cross the whole Sahelian landscape.

horndolde, waande, ton'yolde, nyonkolde, roŋaare

horndolde is the underground nest of harvester ants. The consolidated surface above such nests is free of vegetation. Such nests present a danger to cattle if they should break through into these subterranean galleries. In times of famine, they may be dug up to harvest the grass seeds stored by the ants as a cereal substitute.

Table 3.6 Zoogenic Units

Unit	Brief description by the Fulani	Pastoral value during the year*			
		a	b	c	d
jaayal	area always consolidated and almost free of woody species; it varies in size from some m² to several acres	1	3	2	1
gotol	cattle path	0	0	0	0
horndolde	subterranean nest of harvesting ants	0	0	0	0
waande	active termite mound	0	0	0	0
ton'yolde	inactive, uneroded termite mound	0	0	0	0
nyonkolde	inactive, half-eroded termite mound	0	0	0	0
roɲaare	died off, almost totally eroded termite mound	0	0	0	0

Legend: a = rainy season; b = late rainy season; c = cool dry season, d = hot dry season,
3 = very important, 2 = important, 1 = less important, 0 = unimportant

waande are occupied termite mounds, found everywhere across the landscape. They have no use for humans or cattle, nor do they pose a danger. The same is true of *ton'yolde,* an abandoned, but not yet eroded termite mound, and *nyonkolde,* a partly eroded, but still clearly recognizable termite mound. *roɲaare* is an almost completely eroded mound, which is perceived as a great danger for humans and animals: such features are only very slightly elevated above the surrounding surface and are most readily recognized by the absence of vegetation. Jelgoobe think that such abandoned nests are often occupied by dangerous spirits, *jinnaaji,* which may take revenge by causing diseases or other misfortunes if a person steps on the mound, even unintentionally.

Overview of All Sahelian Landscape Units and Discussion in a Larger Context

In all, 100 different landscape units were documented in the study area, as follows: eleven topographic units, twenty-one hydrographic units, twenty-four soil units, twenty-eight units defined by vegetation (of which eleven were characterized by the absence of vegetation), ten units characterized as due to anthropogenic influence (or its absence), and six units attributed to zoogenic influence (see Table 3.7). A comparison of this classification with that of two other autochthonous Fulani groups in West Africa (see Figure 3.1) reveals two facts.

First, the quantitative presence of an environmental feature may be reflected in the precision and the diversity of the classification units. A landscape with a

Table 3.7 Overview of the number of environment classification units in the Sahelian investigation area and two comparison regions.

Units characterized by ...	Sahel	North Sudanian region	South Sudanian region
Relief	11	5	12
Water	21	15	16
Soil	24	22	16
Vegetation (and its absence)	28	13	16
Anthropogenic influence	10	15	16
Zoogenic influence	6	5	4
Total	100	75	80

(See Figure 3.1 for locations of North and South Sudanian regions).

strongly varied topography will accordingly be classified by a wide diversity of carefully differentiated classification units referring to topographical characteristics. This applies to the study area in Northern Benin (see Krohmer 2004), where twelve relief units are distinguished. By contrast, in southeastern Burkina Faso, with a far less accentuated relief, only five topographic units were recognized (Krohmer 2004). In most parts of the Sahelian study area described in this chapter, the relief is not pronounced, but during the long dry season, when vegetation is sparse or even not existing (because it is all dried up or has been eaten by cattle and small ruminants), the topographic elements are perceived very precisely, as is demonstrated by the eleven relief units of the Sahelian Fulani classification system. Soils likewise are classified in great detail, likely for the same reason.

Second, it appears that the scarcity of an environmental element enhances its importance for the inhabitants and motivates close observation, which is then reflected in a detailed classification in response to what one might call "ecological salience." This applies to the water and vegetation units in the Sahel. Thus, the number of vegetation units (including those characterized by a lack of vegetation) is almost twice as large as in the two other regions, although the vegetation in the other regions seems to be (and in fact is) much more diverse. The water sources likewise are carefully distinguished. In Benin, water is readily available all year. Thus fewer water units are recognized there than in the Sahel, where open water is scarcely to be found during half the year. This shows that scarce elements may be classified in more detail and abundant elements in less detail than might be expected.

Finally, it is interesting to consider how human influences on a landscape are perceived: although all three study areas are ancient cultivated landscapes and therefore profoundly influenced by human activity, the classification systems

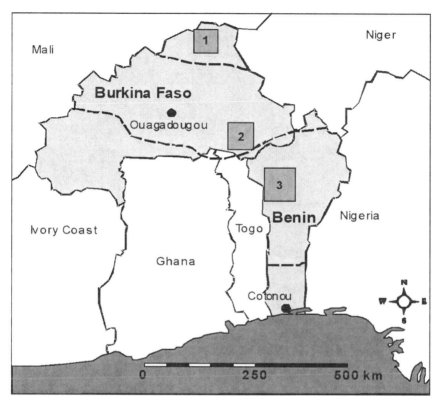

Figure 3.1 *Overview of the three investigation regions in West Africa (Krohmer 2004). (1 = situated in the Sahel, 2 = situated in the North Sudanian zone, 3 = situated in the South Sudanian zone)*

show very different numbers of anthropogenic units. In the Sahel, this number is much smaller than in the other regions, which might suggest that these Fulani perceive their impact on the environment as less important than do other Fulani. However, in the other two regions cultivated areas are omnipresent, and their various fallow stages are all named, whereas in the Sahel fallow fields are virtually nonexistent. This might explain why there are so few human-influenced units in the Sahel; however, it would still be interesting to know why the undeniably strong influence of cattle herding on the landscape is not reflected in the classification system, for example by some designations for variably heavily grazed (and therefore variably degraded) pasturelands. One possible reason for this is that overgrazing and degradation due to rising grazing pressure have appeared only in the last decades—and that the Fulani classification system takes some time to render an account of such a change. An indication that this adaptation process is already beginning are the different terms used to describe soil conditions (***leydi*** *keyri,* healthy soil; *leydi tampi / mburndi,* tired soil; *leydi mbanndi / waati,* dead soil). The elder herders reported that in former times, there was no such thing as dead soil; nowadays, they have to use that term quite often to describe a place. Thus, the classification system seems to be flexible, to adapt itself to a changing landscape.

Concluding Remarks

To quote Eugene Hunn, traditional ecological knowledge (TEK) is "a work of art, a symphony of understanding, and a scientific contribution to human knowledge of nature" (Hunn 2002: 7). Though in this chapter we have only considered the traditional landscape classification, leaving aside the Fulani's rich ethnobotanical knowledge and their deep understanding of ecological processes, the data collected in this study fit quite well in this beautiful description. The Fulani's landscape and environmental knowledge is deep and comprehensive, and shared by all members of the respective community—just as the knowledge of how to perform a symphony is shared by the members of an orchestra, to take up Hunn's metaphor. And like the musicians, with everyone contributing his or her part, every member of a Fulani community contributes knowledge to the collective opus of environmental knowledge.

Moreover, the Fulani classification system is the equal of scientific classification, when it comes to accuracy and completeness. A comparison between Fulani and scientific vegetation classification revealed a high concordance between units of these two categories (Krohmer 2004).

To finish, let us consider again the initial question: do sedentary Fulani perceive their environment in a different way than do those who are still nomads? The surveys conducted with five different Fulani groups in West Africa, occupying a range of situations between nomadic and the sedentary lifestyles, showed that they do not. As long as they continue to rear cattle, they use and perceive the surrounding landscape in a quite similar, detailed way. Differences in classification between the groups result from differences in the regional landscapes rather than from different perceptions. We therefore believe that the key factor in changing their view on landscape is the abandonment of herding rather than the adoption of a sedentary way of life.

Acknowledgments

I would like to gratefully acknowledge all the Fulani in Burkina Faso and Benin who patiently and generously shared their knowledge with me. Special thanks to Gnanando Seydou in Natitingou, Benin, who has been my assistant and interpreter during more than four years in the field and who, with his knowledge and personal contacts, contributed a lot to this work. I finally thank the German Research Foundation DFG (SFB 268) for its financial support and the public authorities of Burkina Faso and Benin for their permission to conduct research in their countries. And I am very grateful to Leslie Main Johnson for inviting me to write a chapter for this volume, as well as for very carefully reviewing (together with Eugene Hunn) and editing the manuscript.

Notes

1. For terms in Fulfulde, standard linguistic symbols are used: **B** designates an implosive b, **D** an implosive d, **ŋ** a velar nasal (pronounced as *ng* in "sing"), and **'y** an implosive y.
2. Laterite is a rock-like surface formation in hot and wet tropical areas that is enriched in iron and aluminium and develops by intensive and long lasting weathering of the underlying parent rock. In the Sahel, the lateritic surfaces originate from more humid periods in the past.
3. After harvest, the whole ***wuro,*** including the cattle, moves to the harvested fields; cattle manure fertilizes the soil sufficiently to permit millet cultivation at the same place the following year.

References

Azarya, V. 1999. "Introduction: Pastoralists under Pressure." In *Pastoralists under Pressure? Fulbe Societies Confronting Change in West Africa,* ed. V. Azarya, M. Breedveld, M. de Bruijn, and H. van Dijk. Leiden, Boston, Cologne: Brill.

Barral, H. 1977. Les populations nomades de l'Oudalan et leur espace pastoral. – Travaux et documents de l'ORSTOM 77. Paris: ORSTOM.

Barral, H., and M. Benoit. 1976. Nature et genre de vie au Sahel: L'année 1973 dans le nord de la Haute-Volta. Ouagadougou: ORSTOM.

Bartelsmeier, A. 2001. Fulbe und Rinder. Mensch-Tier-Beziehung und Lebenswelt mobiler Tierhalter im Sahel Burkina Fasos. – Sozialökonomische Schriften zur Ruralen Entwicklung, Vol. 125. Kiel: Wissenschaftsverlag Vauk.

Basset, T. J., and D. Crummey. 2003. "Contested Images, Contested Realities: Environment and Society in African Savannas." In *African Savannas: Global Narratives and Local Knowledge of Environmental Change,* ed. T. J. Basset and D. Crummey. Oxford: James Currey Ltd; Portsmouth: Heinemann.

Bierschenk, T. 1997. Die Fulbe Nordbenins. Geschichte, soziale Organisation, Wirtschaftsweise. Hamburg: LIT.

Boesen, E. 1997. "Identité et démarcation: les pasteurs peuls et leurs voisins paysans." In *Trajectoires peules au Bénin,* ed. T. Bierschenk and P.-Y. Le Meur. Paris: Karthala.

Boni, Y., and N. Gaynor. 1996. "Classification Traditionnelle des Pâturages et Modalités de Gestion dans l'Atakora du Nord-Est. – Projet Promotion de l'Elevage dans l'Atakora (PPEA)." Natitingou.

Borrini-Feyerabend, G., M. T. Farvar, J. C. Nguinguiri, and V. Ndangang. 2000. *La Gestion Participative des Ressources Naturelles: Organisation, Négociation et Apprentissage par l'Action.* Heidelberg: GTZ und UICN, Kasparek-Verlag.

Claude, J., M. Grouzis, and P. Milleville. 1991. *Un espace sahélien. La Mare d'Oursi, Burkina Faso.* Paris: Éditions de l'ORSTOM.

Colding, J., and C. Folke. 1997. "The Relations Among Threatened Species, Their Protection, and Taboos." *Conservation Ecology* 1(1): 6–20.

de Bruijn, M., and H. van Dijk. 1995. *Arid Ways: Cultural Understandings of Insecurity in Fulbe Society, Central Mali.* Amsterdam: Thela Publishers.

Dialla, B. E. 1993. "The Mossi Indigenous Soil Classification in Burkina Faso." *IK Monitor* 1(3), http://www.nuffic.nl/ciran/ikdm/1-3/articles/dialla.html.

Etkin, N. L. 2002. "Local Knowledge of Biotic Diversity and its Conservation in Rural Hausaland, Northern Nigeria." *Economic Botany* 56(1): 73–88.

FAO-UNESCO. 1997. *Soil Map of the World, Revised Legend.* Wageningen: FAO.

Frantz, C. 1993. "Are the Mbororo'en Boring, and Are the Fulbe Finished?" *Senri Ethnological Studies* 35: 11–34.

Haverkort, B., and D. Millar. 1994. "Constructing Diversity: The Active Role of Rural People in Maintaining and Enhancing Biodiversity." *Etnoecológica* 2(3): 44–60.

Hunn, E. S. 2002. "Traditional Environmental Knowledge: Alienable or Inalienable Intellectual Property." In *Ethnobiology and Biocultural Diversity: Proceedings of the 7th International Congress of Ethnobiology,* ed. J. R. Stepp, F. S. Wyndham, and R. K. Zarger. Athens: University of Georgia Press.

Johnson, D. H. 1980. "Ethnoecology and Planting Practices in a Swidden Agricultural System." In *Indigenous Knowledge Systems and Development,* ed. D. Brokensha, D. M. Warren, and O. Werner. Lanham, MD, and London: University Press of America.

Johnson, L. M. 2000. "'A Place That's Good': Gitksan Landscape Perception and Ethnoecology." *Human Ecology* 28(2): 301–325.

Kolbe, D. 1994. *Exploitation rationnelle des ressources naturelles dans le cadre de la Gestion des Terroirs au Bam / Burkina Faso.* Bayreuth: Universität Bayreuth .

Krogh, L., and B. Paarup-Laursen. 1997. "Indigenous Soil Knowledge Among the Fulani of Northern Burkina Faso: Linking Soil Science and Anthropology in Analysis of Nature Resource Management." *GeoJournal* 43: 189–197.

Krohmer, J. 2002. "Die Fulbe und ihr liebes Vieh." In *Leben in Westafrika,* ed. A. Reikat. Frankfurt am Main: Plexus.

———. 2004. "Umweltwahrnehmung und –klassifikation bei Fulbegruppen in verschiedenen Naturräumen Burkina Fasos und Benins (Westafrika)." PhD diss., Faculty of Biology, Goethe-University Frankfurt. (http://publikationen.ub.uni-frankfurt.de/volltexte/2005/508)

Kyiogwom, U. B., B. F. Umaru, and H. M. Bello. 1998. "The Use of Indigenous Knowledge in Land Classification and Management Among Farmers in the Zamfara Reserve." *Giessener Beiträge zur Entwicklungsforschung* 25: 220–227.

Lizarralde, M. 2004. "Indigenous Knowledge and Conservation of the Rain Forest: Ethnobotany of the Bári of Venezuela. " In *Ethnobotany and Conservation of Biocultural Diversity.* ed. J. S. Carlson and L. Maffi. Advances in Economic Botany 15. New York: New York Botanical Gardens.

Martin, G. J. 1993. "Ecological Classification Among the Chinantec and Mixe of Oaxaca, México." *Etnoecológica* 1(2): 14–31.

Moritz, M. 1994. *Yake, yaere wi'eto no yaere: Pastoralists' Perceptions of and Adaptation to Rangeland Degradation of the Lagoone Floodplain, Far-North Cameroon.* Maroua: UICN.

Moritz, M., and F. Tarla. 1999. "Fulani Pastoralists' Perceptions and Perspectives on Rangeland and Its Degradation in Northern Cameroon." *Mega-Tchad Bulletin* 99(1/2). (http://www .uni-bayreuth.de/afrikanistik/mega-tchad/Bulletin/Bulletin)

Niamir, M. 1990. *Herder's Decision-Making in Natural Resources Management in Arid and Semi-arid Africa.* Rome: FAO.

Reichhardt, K. L., E. Mellink, G. P. Nabhan, and A. Rea. 1994. "Habitat Heterogeneity and Biodiversity Associated with Indigenous Agriculture in the Sonoran Desert." *Etnoecológica* 2(3): 13–30.

Reiff, K. 1998. "Das weidewirtschaftliche Nutzungspotential der Savannen Nordwest-Benins aus floristisch-vegetationskundlicher Sicht." In *Geo- und weideökologische Untersuchungen in der subhumiden Savannenzone NW-Benins. Karlsruher Schriften zur Geographie und Geo- ökologie,* Bd. 1, ed. M. Meurer. Karlsruhe: Institut für Geographie und Geoökologie der Universität Karlsruhe.

Richards, P. 1985. *Indigenous Agricultural Revolution: Ecology and Food Production in West Africa.* London: Unwin Hyman Ltd.

Riesman, P. 1974. *Société et Liberté chez les Peul Djelgôbé de Haute-Volta. Essai d'anthropologie introspective. Cahiers de l'Homme,* Nouvelle Série XIV. Paris and The Hague: Mouton and Co.

Robbins, P. 2003. "Beyond Ground Truth: GIS and the Environmental Knowledge of Herders, Professional Foresters and Other Traditional Communities." *Human Ecology* 31(2): 233–253.

Ross, A., and Pickering, K. 2002. "The Politics of Reintegrating Australian Aboriginal and American Indian Indigenous Knowledge into Resource Management: The Dynamics of Resource Appropriation and Cultural Revival." *Human Ecology* 30(2): 187–214.

Scarpa, G. F., and P. Arenas. 2004. "Vegetation Units of the Argentine Semi-arid Chaco: The Toba-Pilagá Perception." *Phytocoenologia* 34(1): 133–161.

Schareika, N. 2003. *Westlich der Kälberleine. Nomadische Tierhaltung und naturkundliches Wissen bei den Wodaabe Südostnigers. Mainzer Beiträge zur Afrika-Forschung* 9. Hamburg: LIT.

Scoones, I. 1996. *Living with Uncertainty: New Directions in Pastoral Development in Africa.* London: Intermediate Technology Publications.

Scott, J. 1998. *Seeing Like a State: How Certain Schemes to Improve the Human Condition Have Failed.* New Haven, CT: Yale University Press.

Shepard, G. H., Jr., D. W. Yu, M. Lizarralde, and M. Italiano. 2001. "Rain Forest Habitat Classification Among the Matsigenka of the Peruvian Amazon." *Journal of Ethnobiology* 21(1): 1–38.

Shepard, G. H., Jr., D. W. Yu, and B. Nelson. 2004. "Ethnobotanical Ground-Truthing and Forest Diversity in the Western Amazon." In *Ethnobotany and Conservation of Biocultural Diversity,* ed. J. S. Carlson and L. Maffi. Advances in Economic Botany 15. New York: New York Botanical Gardens.

Soil Survey Staff. 1975. *Soil Taxonomy.* Agriculture Handbook 436. Washington, D.C.: US Government Printing Office.

Sturm, H.-J. 1999a. "Weidewirtschaft in West Afrika." *Geographische Rundschau* 51(5): 269–274.

———. 1999b. "Temporal and spatial patterns of exploitation of pasture resources in the sub-humid savanna zone." *Proceedings of the 6th International Rangeland Congress, Townsville.*

Thébaud, B., H. Grell, and S. Miehe. 1995. *Recognising the Effectiveness of Traditional Pastoral Practices: Lessons from a Controlled Grazing Experiment in Northern Senegal.* Paper No. 55. London: International Institute for Environment and Development.

Tiffen, M., M. Mortimore, and F. Gichuki. 1994. *More People, Less Erosion: Environmental Recovery in Kenya.* Chichester: Wiley, for Overseas Development Institute.

Toledo, V. M., B. Ortiz, and S. Medellín-Morales. 1994. "Biodiversity Islands in a Sea of Pasturelands: Indigenous Resource Management in the Humid Tropics of Mexico." *Etnoecológica* 2(3): 30–44.

Warren, D. M., L. J. Slikkerveer, and D. Brokensha. 1995. *The Cultural Dimension of Development: Indigenous Knowledge Systems.* London: Intermediate Technology Publications.

Chapter 4

Baniwa Vegetation Classification in the White-Sand *Campinarana* Habitat of the Northwest Amazon, Brazil

Marcia Barbosa Abraão, Glenn H. Shepard, Jr.,
Bruce W. Nelson, João Cláudio Baniwa,
Geraldo Andrello, Douglas W. Yu

Introduction

Anthropological approaches to landscape have paid special attention to how local environmental knowledge is imbued with individual and collective histories,
symbols, values, and memories (Basso 1996; Hill 1989; Hirsch and O'Hanlon
1995; Santos-Granero 1998; Strang 1997). In studying the biotic elements of
indigenous environmental knowledge, ethnobiologists have traditionally focused
on the classification and use of plants and animals (Berlin, Breedlove and Raven 1974; Berlin 1992; Conklin 1954; Ellen 1979; Hunn 1977). More recently,
systems of indigenous landscape classification have been the subject of increasingly sophisticated interdisciplinary studies focusing especially on topographic,
soil, and vegetation typologies used by local peoples (Fleck and Harder 2000;
Krohmer 2004; Parker et al. 1983; Halme and Bodmer 2007; Shepard and Chicchon 2001; Shepard et al. 2001; Shepard, Yu, and Nelson 2004a; Sillitoe 1996).
In some cases, indigenous communities recognize more landscape diversity at the
local level than do tropical ecologists working in the same regions (see Shepard
et al. 2001).

Such findings are no surprise to ethnobiologists and a few nonconformist taxonomists who have been documenting the sophisticated environmental knowledge of indigenous societies for decades (Berlin 1984; Bulmer 1974; Conklin
1957; Diamond 1966; Posey, Frechione, and Eddins 1984). Indeed, indigenous

and other local landscape concepts are sometimes incorporated into scientific classifications, though this intellectual debt is rarely acknowledged explicitly (Encarnación 1993; Pires and Prance 1985). Only recently have mainstream ecologists and conservationists been called to task to consider seriously the role of indigenous peoples as partners in research and conservation of tropical forests (Chapin 2004; Sheil and Lawrence 2004). For their part, ethnobiologists and environmental anthropologists have also made wider use of quantitative techniques of data collection and analysis to render their results more accessible, comparable, and robust to the broader scientific community (Martin et al. 2002; Phillips 1996; Sillitoe 1996; Toledo and Salick 2006; Zent 1996). The classification of landscape-scale diversity patterns in Amazonia has proven especially challenging (Tuomisto 1998) and offers a special opportunity for interdisciplinary collaboration between social and natural scientists as well as with indigenous communities themselves (Shepard et al. 2001; Shepard et al. 2004a).

In this study, we present the cumulative results of ongoing research into the landscape classification system of the Arawakan-speaking Baniwa people of the Upper Rio Negro, Brazil. After presenting a general overview of Baniwa landscape classification, we provide more detailed analysis of the Baniwa's sophisticated system for characterizing different vegetation types within the white-sand savanna-forest matrix, or *campinarana,* that predominates in the northwest Amazon. Ten Baniwa informants provided a repertoire of 150 indigenous-language plant names (hereafter, "species") during interviews about perceived botanical composition in fourteen predominant white-sand vegetation types. Informant consensus regarding the presence of these species was used as a proxy for the species' actual abundance for each vegetation type. Species assemblages and patterns of gradual floristic change across the fourteen vegetation types were analyzed using cluster analysis, indicator species analysis, and ordination by nonmetric multi-dimensional scaling. These tools from quantitative ecological community analysis confirm the ecological validity of Baniwa landscape knowledge while providing further insights into how this knowledge is structured, and how it relates to specific biotic and abiotic elements of the *campinarana* landscape. The methods applied here would likewise prove useful for rapid, cost-effective assessment of conservation priorities of plant communities in regions where, as among the Baniwa, local populations maintain intimate knowledge of the environmental preferences of a large number of plant species (see also Instituto Socioambiental 2003; Milliken 1998; Shepard and Chicchon 2001).

Study Area

Research was carried out in Baniwa indigenous communities of the middle Içana River. Initial visits were made to the communities of São José, Juivitera, Tucumã Rupitá, Jandu Cachoeira, Mauá Cachoeira, Trindade, and Aracu Cachoeira. The communities of Juivitera, Jandu Cachoeira, and Aracu Cachoeira (Figure 4.1) were chosen for a more intensive study of *campinarana* habitats.

The Baniwa belong to the Arawakan cultural-linguistic family and inhabit tributaries of the upper Rio Negro in Brazil along the Colombia-Brazil-Venezuela border region. The Baniwa are closely related to the Coripaco of Colombia and the Wakuenai of Venezuela. In Brazil, the Baniwa have a population of about 5,500, divided among ninety small communities along the Içana, Ayari, Cuyari, Xié, and other upper Rio Negro tributaries (Cabalzar and Ricardo 2006). The Baniwa form part of a diverse group of societies that have inhabited the upper Rio Negro region for at least 2,000 years, maintaining complex networks of cultural and economic exchange (Ribeiro 1995). With a combined current population of nearly 35,000, some twenty-two indigenous ethnic groups speaking languages belonging to five distinct linguistic families (Arawakan, Tukanoan, Maku, and Yanomami plus the introduced *lingua geral* trade language of Tupi-Guarani origin) inhabit this region. In the late 1990s, the lands occupied by these groups

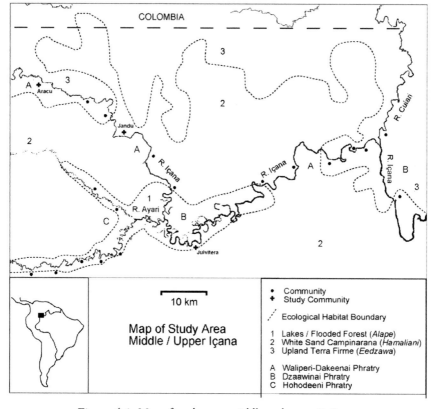

Figure 4.1 *Map of study area, middle and upper R. Içana, showing study communities and major habitat boundaries.*

were recognized and officially demarcated by the Brazilian government in a series of contiguous Indigenous Territories with a combined area of about 150,000 square kilometers (Cabalzar and Ricardo 1998).

Geologically, the upper Rio Negro comprises the southwest limit of the Guiana Shield, composed of ancient, eroded granite formations covered in soil depositions of rather recent origin. Annual rainfall, at 2,500 to 3,000 mm, is among the highest in Amazonia, while annual temperatures average 24° C with negligible seasonal variation. As its name implies, the Rio Negro and most of its major tributaries are black water rivers with little sediment, low biomass, and a dark, tea-like coloration (Sioli 1984).

Three major classes of soils and edaphic conditions define the dominant vegetation types in the region (Figure 4.2): podzols, hydrosols, and latosols (Ribeiro et al. 1999). Acidic, white-sand soils low in nutrients (podzols) are the most widespread, giving rise to a low-productivity (oligotrophic) vegetation with a structure ranging from closed-canopy forest to savanna-like woodland, which is known variably as *caatinga, caatinga amazônica* (Anderson 1981) or *campinarana* (Veloso et al. 1991) in the scientific literature, and *hamáliani* in Baniwa. White-sand vegetation contains many endemic species and is most common in the region centered on the Rio Negro and Rio Branco in Brazil, extending into Colombia, Venezuela, and northern Peru. Seasonally inundated soils (hydrosols) along the floodplains of black water rivers give rise to a specific flooded forest type known regionally as *igapó*, or *alápe* in Baniwa. Finally, clay-containing latosols, practically the only soils suitable for agriculture in the region, are exposed in larger stream valleys and hilly terrains and support upland forest, or *terra firme* vegetation known as *éedzawa* in Baniwa. The low overall productivity of fish and game in the region and the scarcity and patchiness of land suitable for agriculture have contributed to unique cultural-ecological adaptations among Rio Negro indigenous societies (Moran 1991).

Methods

Verification and Expansion of Preliminary Habitat Data

Anthropologist Geraldo Andrello (1998) carried out pioneering research on Baniwa forest classification while working with Instituto Socioambiental (ISA) to document indigenous occupation, knowledge, and resource use of the upper Rio Negro region. Using his preliminary list of Baniwa vegetation types as a starting point, in November–December 2003 authors Abraão and Shepard conducted semi-structured interviews and free-listing exercises with Baniwa informants from seven communities in different ecological settings along the middle and upper Içana. Informants were first asked if they recognized the naturally occurring vegetation types noted by Andrello (1998), and then were asked to

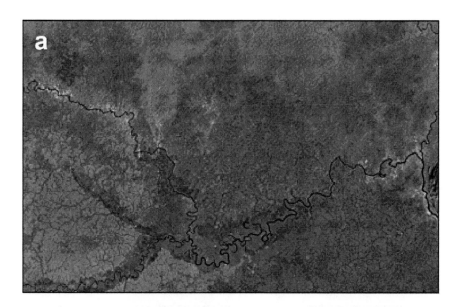

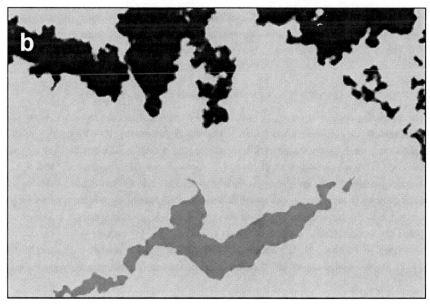

Figure 4.2 *Remote sensing images of study area. (a) Landsat Thematic mapper image, August 1993. (b) Visual classification of Landsat image showing major habitat areas:* terra firme *(black),* igapó *flooded forest (medium gray), and* campinarana *white-sand forests (light gray). Source: Endo (2005).*

list any additional vegetation types for each general habitat category: *hamáliani* (white-sand *campinarana*), *éedzawa* (upland *terra firme*), and *alápe* (*igapó* black water–flooded forest). Anthropogenic *beñame,* secondary forest in various stages of regeneration, for which Andrello (1998) noted fifteen named types, were excluded from the current study. Based on this additional free-listing with a larger group of informants from ecologically diverse communities, Andrello's preliminary list of fifteen *hamáliani* (white-sand *campinarana*) vegetation types was expanded to twenty-four, his list of twenty-one *éedzawa* (*terra firme*) types was expanded to twenty-five, and his list of seventeen *alápe* (flooded *igapó*) types was expanded to twenty-eight (see Tables 4.1 through 4.3; note that Andrello's [1998] Portuguese-based orthography has been updated to reflect the Baniwa orthography developed by Ramirez [2001]). More knowledgeable informants were asked to accompany the research team on walks through different vegetation types in the vicinity of each community. Waypoints for vegetation types and transition points between different types were registered with Garmin GPS 45 and GPS 12 units. These verification walks permitted a preliminary characterization of many common vegetation types, including the field identification of important indicator species. Botanical collections were not initially possible due to bureaucratic delays in the permit process, but collection and identification proceeded in 2008 after permits were finally acquired; also some collections from prior research (Hoffman 2001; Shepard et al. 2004b; Silva 2004) proved relevant to this study.

Structured Interviews Focusing on **hamáliani** Habitat Category

The *hamáliani* habitat category (white-sand *campinarana*), the most widespread, was chosen for more detailed study. Three focal communities—Juivitera, Jandu Cachoeira, and Aracu Cachoeira—were selected along a 100 km section of the middle Içana where white-sand forests predominated (see Figures 4.1 and 4.2). Abraão (initially with guidance from Shepard) carried out structured interviews with Baniwa informants concerning *hamáliani* vegetation types. Interviews were carried out in the villages (i.e., not *in situ* in the vegetation types), in order to assess the cognitive dimension of forest classification. Questions were posed and answered in Baniwa, with assistance from author João Claudio, a Baniwa high school student who served as Abraão's local collaborator. Informants were asked:

1) whether they recognized each *hamáliani* vegetation type (as compiled in the previous free-listing exercises), and whether each type occurred in the vicinity of their community. These questions referred to each vegetation type only by name (no photos or field visits);
2) which soils were associated with each vegetation type;
3) what activities (agriculture, hunting, gathering) were associated with each vegetation type;

4) which plant or tree species were found in each vegetation type. Answers were recorded in the form of a free list of Baniwa plant names.

Question 1 was posed to ten informants in the three focal villages at an early stage of the study. They were asked about all twenty-four *hamáliani* vegetation types elicited in the prior free-listing exercise. This initial list was reduced to fourteen types that were consistently recognized and confirmed by these ten informants as being distinctive and common near their communities (Table 4.4). The remaining types were either not recognized by many informants, were assigned consistently to other general habitat categories (*terra firme, igapó*), or were considered to be synonyms. The reduction from twenty-four to fourteen types was partly a result of more localized knowledge in the three study communities, and partly a culling of unreliable classes obtained at the early, free-listing stage of study. This result highlights the fact that studying diversity of vegetation types by simple free-listing can lead to overestimation unless informant variation and consensus are critically assessed. Questions 2, 3, and 4 were posed to the ten informants only for the fourteen consensus *hamáliani* types.

In the community of Aracu Cachoeira, informants mentioned a number of *hamáliani* localities or types that had distinctive mythological or cultural associations as well as specific vegetative characteristics (see Table 4.5). For the Baniwa, certain named white-sand localities—many of which may constitute ancient village sites—represent prominent cultural landmarks where significant historical or mythological events took place (Wright 1998). Indeed, sacred place naming seems a central element of environmental and social history for the Baniwa and other Arawakan societies (Hill 1989; Wright 1998). More generally, indigenous ecological knowledge is often bound up with cosmology, mythology, and spiritual conceptions (Basso 1996; Berlin 1977; Conklin 1957; Reichel-Dolmatoff 1976; Shepard 1999; Shepard et al. 2001; Strang 1997; Toledo 1992). However, Aracu was the only community where mythological information was volunteered spontaneously during interviews. Given time constraints and lack of comparable data from the other study communities, mythological aspects of ethnoecological classification were not explored in greater detail in this study. Baniwa landscape elements with mythological, historical, and other cultural significance were documented in a regional survey of indigenous ecological knowledge in which these authors participated (Instituto Socioambiental 2003), and have been explored in greater detail by Baniwa research collaborators in a larger project currently underway that grew out of this preliminary study (Instituto Socioambiental 2007).

Quantitative Analysis of Floristic Patterns

In response to Question 4 (above), a total of 150 Baniwa plant names were listed by the ten informants as present in one or more of the fourteen *hamáliani*

Table 4.1 Baniwa Vegetation Types for **hamáliani** (*campinarana*)

Asterisk (*) indicates vegetation types included under more than one general habitat category.

Vegetation type	Andrello (1998) term	Aspect	Habitat use	Indicator species	Preliminary identification†	Indicator species use
Anerima	anerimã	low, open *campinarana*	extractivism	***Ane***	*Eperua duckeana* (LEG)	no direct use
Aiholima	ahiurimã	high, closed canopy forest	extractivism	***Aiho***	*Micrandra spruceana* (EUP)	no direct use (inedible fruits)
Dzakoirima	zakoirimã	high, closed canopy forest	extractivism	***Dzakoi***	*Mezilarus synandra* (LAU)	edible fruit; trunk used to make dugout canoes
Dzaawakalima	dzaawakarimã	low, open *campinarana*	extractivism	***Dzaawaka***	*Eperua leucantha* (LEG)	no direct use; monkey fruit
Dzeekalima	—	high, closed canopy forest	extractivism	***Dzeeka***	*Hevea guianensis* (EUP)	edible fruit (siringa)
Dzokorodaapolima	—	low, open *campinarana*	extractivism	***Dzokoroda***	cf. *Calyptranthes* (MYT)	edible fruit; wood for construction
***Heridzorolima*[a]**	rerridzurrorimã	high, closed canopy forest on moist soils near streams	extractivism	***Heridzoro***	*Anaxagorea, Pseudoxandra* (ANN)	wood for construction
***Itamalerima*[c]**	itanalirimã	high, closed canopy forest in disturbed areas or near streams	extractivism	***Itanale***	*Vochysia* spp. (VCY)	wood for construction
Itewipalima	—	low, open *campinarana*	—	***Itewipa***	*Aspidosperma* sp. (APC)	—
Iitsaapolima	—	high, closed canopy forest in secondary vegetation of old garden sites	agriculture, extractivism	***Iitsaapo***	*Duguetia* spp. (ANN)	stems used as fishing rods
***Iiteuirima*[a]**	iteuirimã	open, swampy areas	extractivism	***Iitewi***	*Mauritia flexuosa* (ARE)	edible fruit (buriti); useful fibers
Kadaapolima	—	?	extractivism	***Kadaapo***	cf. *Duguetia* (ANN)	stems used as fishing rods

Koliwaipalima	kuriuaiparimã	low, open *campinarana*, dense understory of *Asplundia, Bromelia*	extractivism	*Koliwaipa*	*Asplundia vaupesiana* (CYC)	edible fruiting body
Koyapberima	kuiaperimã	high, closed canopy forest with understory palm stands	extractivism, agriculture	*Koyapbe*	*Attalea macrocarpa* (ARE)	edible fruit; leaves used for temporary thatching
Maarolima	—	high, closed canopy forest	extractivism	*Maaro*	*Hapoclabra paniculata* (CLU)	wood for construction
Maawiririma	—	?	extractivism	*Maawiri*	*Iriartella setigera* (ARE)	thin, hollow trunks used to make blowguns
Maporottirima	mapurutirimã	low, open *campinarana*, indicator of past human disturbance	extractivism, low-productivity agriculture	*Maporotti*	*Humiria balsamifera* var. 1 (HUM)	edible fruit (umiri)
Ponamalima	punamarimã	high, closed canopy forest	extractivism	*Ponama*	*Oenocarpus bataua* (ARE)	edible fruit (patahuá)
Poramolima	—	high, closed canopy forest near streams	extractivism	*Poramo*	*Euterpe catingae* (ARE)	edible fruit (açai chumbinho)
Ttiñalima	tinharimã	low, open swampy *campinarana*; shrub and herb vegetation, few trees	extractivism	*Ttiña*	*Mauritiella carana* (ARE)	palm leaves for roof thatch
Waalialima	uariarimãita	low, open *campinarana*	extractivism	*Waalia*	*Humiria balsamifera* var. 2 (HUM)	edible fruit (umiri)
*Waapalima[e]	uaparimada	high, closed canopy forest, transition to *terra firme*;	agriculture, extractivism	*Waapa*	*Eperua purpurea* (LEG)	no direct use; indicator of *terra firme* transition and good agricultural soils
Waittirima	uaiterimã	low, open *campinarana*; white, *waitti* trunks covered in epiphytes	extractivism	*Waitti*	*Aldina heterophylla* (LEG)	no direct use
Yaalirima	iaririmã	stream edges	extractivism	*Yaali*	(ANN)	useful fibers; fruit eaten by fish and birds

[a] Vegetation type listed also for alápe habitat category (Table 4.3).

[e] Vegetation type listed also for éedzawa habitat category (Table 4.2).

[†] Ongoing work has led to revision of some botanical identifications, leading to a few discrepancies with results presented in Abraão et al. (2008)

Table 4.2 Baniwa Vegetation Types for *éedzawa* (upland, *terra firme* forests)

Asterisk (*) indicates vegetation types included under more than one general habitat category.

Vegetation Type	Andrello (1998) term	Soil type (Andrello 1998)	Indicator species	Preliminary identification	Indicator species use
Awiñalima	auinharimã	yellow clay	*Awiña*	*Monopteryx uacu* (LEG)	no direct use
Axiropalima	achirruparimã	yellow	*Axiropa*	*Compsoneura* cf. *ulei* (MYS)	fruit edible, but not very delicious
Doohepalima	—	—	*Doohepa*	*Cyclanthus bipartitus* (CYL)	leaves used for wrapping, improvised basketry
Hipolilima	ripolirimã	yellow	*Hipoli*	*Alchornea triplinervi* (EUP)	no direct use (fruits not edible)
Hitawalima	ritauarimã	dark, thick soil	*Hitawa*	*Swartzia schomburgkii* (LEG)	no direct use
Idzepolima	—	—	*Idzepoli*	*Pourouma cucura* (MOR)	edible fruit (*cucura*)
Itanalerima[b]	itanalirimã	(not noted)	*Itanale*	*Vochysia* spp. (VCY)	wood for house beams
Kadaapolima	kadaporirimã	yellow or black	*Kadaapo*	cf. *Duguetia* (ANN)	thin stems used for fishing rods
Kawitrerima	kauitirririmã	yellow	*Kawittiri*	*Aniba panurensis; Licaria, Endlicheria* (LAU)	light, strong wood for canoes, boards
Kerererima	kererrririmã	(note noted)	*Kereri*	*Brosimum rubescens* (MOR)	fine hardwood (*pau brasil*)
Koonolirima	—	—	*Koonoli*	*Micrandra* cf. *siphonoides* (EUP)	—
Mainhitilima	máinhitirimã	(not noted)	*Mainhiti*	?	no direct use (fruits not edible)
Maipanablilima	maipanarririmã	yellow	*Maipanabli*	*Tapirira guianensis* (ANA)	edible fruit

Maliponenerima	maripunenirimã	yellow or black	**Maliponene**	*Geonoma deversa* (ARE)	palm leaves for roof thatch
Maperima	—	—	**Mape**	*Pourouma minor* (MOR)	edible fruit (*-cucura*)
Mookolilima	mukulirimã	best soil for agriculture, will produce bananas, maize, sweet potato, chilis	**Mookoli**	*Suartzia argentea* (LEG)	no direct use; fruit eaten by paca (*Agouti paca*); indicator of best agricultural soils
Moolirima	muririmã	yellow or black	**Mooli**	*Cedrelinga caateniformis* (LEG)	light, strong wood for canoes
Oowadalima	uadarimã	yellow	**Oowada**	*Dacryodes* sp. (BUR)	edible fruit (*uapixuna*)
Orokailima	urrukairimã	(not noted)	**Orokai**	*Protium* sp. (BUR)	—
Padzomalima	—	—	**Padzoma**	*Abarema* sp. (LEG)	—
Pooperima	puparimã	(not noted)	**Pooperi**	*Socratea salazarii* (ARE)	wood staves for internal support of roof thatch
Tioophirima	thopirimã	yellow or black	**Ttophi**	*Minquartia guianensis* (OLA)	rot-resistant trunk used for house posts
Teepalima	—	—	**Teepa**	*Clathrotropis macrocarpa* (LEG)	—
Waapalima[h]	uaparimada	—	**Waapa**	*Eperua purpurea* (LEG)	no direct use; sandy soils but still good for agriculture
Wiritaarima	—	—	**Wiritaa**	*Inga bicoloriflora, Inga* spp. (LEG)	dye fixer for basketry

Table 4.3 Baniwa Vegetation Types for *alápe* (black water–flooded forest, *igapó*)

Asterisk (*) indicates vegetation types included within more than one general habitat category.

Vegetation Ttype	Andrello (1998) term	Indicator species	Preliminary identification	Indicator species use
Adaphenalima	adapenarimã	**Adaphenali**	(APC?)	strong, light wood for boards, manioc graters
Adarokonalilima	adarrokunalerimã	**Adarokonali**	cf. *Alchorneopsis* (EUP)	strong, light wood for boards, manioc graters
Attaattilima	—	**Attaatti**	cf. *Trymatococcus amazonicus* (MOR)	no direct use (spiny palm with noninedible fruit)
Doowirilima	duirridaka	**Doowiri.**	*Bactris* cf. *concinnea*	wood for construction
Heridzorolima[h]	rerridzurrorimã	**Heridzoro**	*Anaxagorea, Pseudoxandra* (ANN)	wood for making oars
Ikoli-domalilima	ikulidumarerimã	**Ikoli-domali**	(ICA?; "big-headed turtle's *Poraqueiba*")	edible palm fruits (*buriti*); useful fibers
Iitewirimã[h]	iteuirimã	**Iitewi**	*Mauritia flexuosa* (ARE)	—
Kaamalima	—	**Kaama**	cf. *Hymenaea intermedia* (LEG)	light wood (molongó) for rafts, fishing bobs
Kadzalirima	kadzalida	**Kadzali**	*Tabernaemontana* sp. (APC)	worm is fishing bait
Katanapililima	katanapiririmã	**Katanapili**	(type of worm)	pineapple-like fruit eaten by turtles
Ketepanelilima	ketepanererimã	**Ketepaneli**	cf. *Ananas* (BRO)	the tallest tree in deeply flooded *igapó*, branches serve as hiding place for fish; excellent spot for fishing during flood season
Kodopilima	kodupirimã	**Kodopili**	?	

Komeerima	kumerimã	*Komee*	(MYT)	edible fruits (*-araçá*), also eaten by fish and game
Mainilima	mainirimã	*Maini*	*Symphonia globulifera* (CLU)	latex used to make sticky, tar-like resin (*breu*)
Makopaolima	—	*Makopao*	?	—
Manakbelima	manakerimã	*Manakbe*	*Euterpe precatoria* (ARE)	edible palm fruit (*açaí*)
Mapolima	maporimã	*Mapoli*	?	good wood for boards
Marakbelima	marrakerimã	*Marakbe*	*Moronobea pulchra* (CLU)	sweet fruits eaten by fish and game animals (*Agouti paca*)
Matboalima	—	*Matboa*	(LEG)	—
Padzwalima	—	*Padzawa*	*Pouteria torta* (SPT)	—
Pitiri-yolepapidalima	pitirriurepapidarimã	*Pitiri-yolepapi*	cf. *Uncaria* (RUB)	liana along stream and lake edges; provides shelter for fish
Poopalima	puparimã	*Poopa*	*Socratea exorrhiza* (ARE)	construction material
Poronarilima	—	*Poronari*	*Xylopia amazonica, Xylopia* spp. (ANN)	construction material
Towanbelima	—	*Towanbe*	*Mauritiella aculeata* (ARE)	leaves used for improvised shelter
Waarbelima	uacherimã	*Waarbe*	*Licania* sp. (CHY)	edible nut (*uchí*); very resistant wood
Yaamolima	—	*Yaamo*	*Parkia* sp. (LEG)	—
Yawalirima	iauaririmã	*Yawali*	*Astrocaryum jauari* (ARE)	edible palm fruits and seeds (*jawarí*)

[h] Vegetation type listed also for *hamãliani* habitat category (Table 4.1).

Table 4.4 Summary information (interview data) on fourteen consensus *hamáliani* vegetation types selected for further study.

Note that vegetation types assigned numbered codes (1–6) were the subject of quantitative ethnobotanical inventories and structural-ecological measurements (but not botanical collections), as detailed in Abraão et al. 2008. Remaining types were assigned alphabetical codes.

Code	Vegetation Type	Indicator Species	Associated Species
1	*Añholima*[†]	*Micrandra spruceana* (EUP)	dzeeka, móokoli, maaro, heridzoro, attale, towanhe, dzakoi, píikoli (edible fruit), hoopima, waapa
2	*Heridzorolima*[†]	*Anaxagorea, Pseudoxandra* (ANN)	dzeeka, poramo, waapa, maaro, tarawiña, itanale, dzaawaka, pimiwidzoli, añho, phíitsikawaapa
3	*Waapalima*[†]	*Eperua purpurea* (LEG)	mokoli, kodama, teepa, kaiwali, konoli, hitawa, taapo, hopima, dzaawaka, aropa
4	*Maarolima*	*Haploclathra paniculata* (CLU)	dzaawaka, attaale, píikoli, waapa, ettipa, pimiwidzoli, heridzoro, hopima, añho, kettere
5A 5B	*Anerima*[†] *2 subtypes:* dzenonipe kaawa - "high" madoape kaawa - "low"	*Eperua duckeana* (LEG)	waitti, dzakoi, ettipa, aropa, pottowarhe (edible fruit), waidane, ttiña, dowiriri, koliwaipa, nolono
6A 6B	*Waittirima*[†] *2 subtypes:* dzenonipe kaawa - "high" madoape kaawa - "low"	*Aldina heterophylla* (LEG)	ane, talawiña, ttiña, koliwaipa, dowiriíri, waalia, ettipa, towanhe, keterre, maporotti
a	*Waalialima*[†]	*Humiria balsamifera* var. 2 (HUM)	waitti, dowiriíri, koliwaipa, ane, píikoli, itewipa, kopií, hamale, nolono, kettere
b	*Koliwaipalima*[†]	*Asplundia vaupesiana* (CYC)	ane, móokoli, añho, waitti, itewipa, dowiriíri, pimiwidzoli, nolonolo, pottowarhe, aropa
c	*Ttiñalima*[†]	*Mauritiella carana* (ARE)	dowiriíri, píikoli, hopima, pimiwidzoli, nolono, waidani, waalia, ane, itewipa
d	*Maporottirima*[†]	*Humiria balsamifera* var. 1 (HUM)	ane, píikoli, waitti, kopií, mokoli, pimiwidzoli, wirari, kettere, dowiriíri, kawiripali
e	*Poramolima*	*Euterpe catingae* (ARE)	añho, heridzoro, hopima, píikoli, dzaawaka, attaatti, awiña, dzeeka, towanhe, kaiwali

Code	Vegetation Type	Indicator Species	Associated Species
f	*Dzeekalima*	*Hevea guianensis* (EUP)	waapa, dzakoi, awiña, añho, heridzoro, ettipa, atatti, towanhe, ponama
g	*Iitsaapolima*	*Duguetia* spp. (ANN)	waapa, dzawaka, añho, tiwali, heridzoro, poramo, ettipa, karoine, pottowarhe, itanale
h	*Ponamalima*[†]	*Oenocarpus bataua* (ARE)	waapa, awiña, heridzoro, maaro, teepa, taapo, itanale, añho, dzawaka, dzeeka

[†] Vegetation type mentioned by Andrello (1998) for *hamáliani*, correcting for updated orthography of Ramirez (2001).

Table 4.5 Special *hamáliani* Types or Localities in Aracu Cachoeira

Type	Observations	Vegetation Types
Iñaimimadza	"Place of the Earth Demon **Iñaimi**"; Mythological origin: house of mythological figure Kaali	*Anerima* *Waittirima*
Hipadakoa	"Place of Rocks"; Mythological origin: house of illness-causing Yopinai spirits	*Heridzorolima* *Añholima*
Nerittiapa	"Place where the deer lies down"	*Maarolima* *Ttiñalima*

vegetation types. These data were used to generate a matrix of fourteen columns by 150 rows, where the value for each cell is the number of informants who mentioned a given Baniwa plant name present in a vegetation type. Thus we actually registered Baniwa cognitive salience of each plant for each vegetation type, rather than relative abundance. We assume that those species for which there was the greatest consensus regarding their presence in a given vegetation type would also tend to be the most abundant species in that type, but we nonetheless acknowledge that cognitive salience may not always coincide with relative abundance.

Seeking to detect meaningful groupings and gradients based on reported floristic similarities, data in the matrix of species by vegetation type were analyzed using several different features of the PC-ORD (v. 4) ecological community analysis software package (see McCune and Grace 2002). First, Baniwa vegetation types were subjected to cluster analysis using Ward's method, generating a hierarchical tree of floristic groups. Plant communities (or vegetation types in this case) can be conceptualized as objects, plotted in an n-dimensional space (i.e., along n different axes), where n is the number of species. In our study there are 150 axes and the position of each vegetation type in this hyperdimensional "species space" is defined by the informed abundance of each of the 150 plant names. Pairs of vegetation types that are close together in the species space are floristically similar, while those that are far apart are less similar. Ward's method constructs a cluster dendrogram (branched, tree-like diagram) that represents the relative similarity among (in this case) the fourteen different vegetation types, and of hierarchically more inclusive clusters among them (see Results, Figure 4.4).

Second, indicator species analysis (ISA) identifies plant species and vegetation-type groups that are consistently and exclusively associated (see Results, Figure 4.5). ISA requires an a priori assignment of the hierarchical groups under consideration, provided in this case by the cluster analysis above. For each branch on the cluster diagram the procedure provides a list of species accompanied by an indicator score and a probability value, indicating which are the most reliable "indicator species" that discriminate a given branch from others. When the p-value for a given species exceeds a threshold, it is not a reliable indicator species for that branch; when a cluster branch has no reliable indicator species it is taken to be a non-significant grouping.

Third, the ordination technique known as non-metric multidimensional scaling (NMDS) was used to apprehend gradients of more gradual change in floristic composition. NMDS attempts to plot the positions of the original n-dimensional matrix onto a Cartesian space of lesser dimension, typically just one dimension (in this case, fourteen points arrayed along a line) or two dimensions (fourteen points arrayed in an x-y plot). Thus for this data set, NMDS would attempt to find fourteen positions in one or two dimensions that best preserve the original distances between all pairs of vegetation types in the original 150-dimensional species space (see Results, Figure 4.6). Here, we used the Bray-Curtis quantitative similarity index B as a distance measure, which gives greater weight to the quantitatively more abundant species (i.e., those with high informant consensus) in the analysis:

$$B = \frac{\Sigma_i |x_{ij} - x_{ik}|}{\Sigma_i |(x_{ij} + x_{ik})|}, \text{ over } i \text{ species and communities } j \text{ and } k.$$

If many species tend to change their abundances in concert as we advance progressively through neighboring vegetation types, then there is a main gradient of floristic change that can be detected and captured mostly in a single dimension (see Results, Figure 4.6). Such a floristic gradient may be associated with some real-world structural or environmental gradient, such as a progressive change in elevation, vegetation height, or soil fertility. When the fourteen vegetation types are plotted in one or two dimensions, it also becomes possible to visualize clusters of similar vegetation types, like those detected in the cluster dendrogram. Clusters of vegetation types detected with these methods were also compared with a simple grouping of vegetation types reported by the Baniwa based on vegetation structure: low, open understory savannas vs. high, closed canopy forest. (Note that such qualitative vegetation structure is fully independent from the interview-based floristic information used for the analyses).

Results and Discussion

Baniwa Landscape Classification: General Features

As described previously by Andrello (1998), the Baniwa recognize four general habitat categories, three defined by the predominant edaphic regimes (Tables 4.1 through 4.3) and the fourth (*heñame*) defined by anthropogenic disturbance and secondary regeneration associated with Baniwa agricultural activities (Andrello 1998; Hoffman 2001; Silva 2004). Baniwa landscape classification is thus similar to that described for other indigenous Amazonian societies, where most broad habitat categories are defined by overall geomorphology, hydrology, and soil (edaphic) conditions: e.g., uplands, lowlands, grasslands or savannas, mountains, etc. (Fleck and Harder 2000; Parker et al. 1983; Shepard et al. 2001). Secondary forest resulting mostly from human agricultural activity is likewise treated as a salient habitat category by the Baniwa, much as it is in the other systems of Amazonian ethnoecological nomenclature studied to date (Balée and Gély 1989; Carneiro 1978; Fleck and Harder 2000; Fleck 1997; Parker et al. 1983; Halme and Bodmer 2007; Shepard et al. 2001). In the domain of landscape classification, these general habitat terms would appear to occupy a position somewhat analogous to that of "life form" in Berlin's (1992) scheme for folk plant and animal classification (see Figure 4.3). As in Berlin's life form rank, general Baniwa habitat categories are few in number, are formed mostly by simple, monolexemic linguistic expressions, and include most vegetation types of lower rank. The Baniwa landscape term *awakada*,[1] referring to 'forest' or 'natural vegetation cover' generally, appears to be analogous to Berlin's "unique beginner" rank for ethnobiological classification (note that the same root, *awaka-*, also appears in the term *awakalona*, a capricious and magical being who confuses and misleads people in the forest, similar to the *curupira* of Brazilian folklore or *chuyachaki* of Peru).

Baniwa names for specific vegetation types within these broad habitat categories consist mostly of Baniwa plant names plus the suffix *-lima* or *-rima*, a collective expression meaning "grouping" or "abundance of." Thus *ttiña-lima* is a vegetation type containing an abundance of *ttiña*, the palm *Mauritiella carana*, known locally as *caraná* and used to make roof thatching. (Note that regional

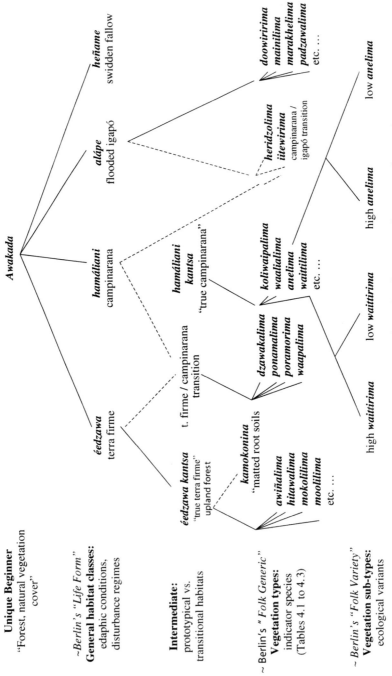

Figure 4.3 *Schematic diagram of Baniwa forest classification, with corresponding rank in Berlin's (1992) universal principles of folk biological classification.*

Brazilian folk landscape classification uses analogous terms of apparently mixed indigenous [Tupian] and Portuguese origin, employing the suffix *-al* or *-zal,* e.g., *caranázal* for stands of the thatch palm *caraná*). This level of habitat classification is analogous with Berlin's "generic rank" for plant and animal classification (Figure 4.3). Indeed, the Baniwa plant names used to label vegetation types are almost exclusively generic-level plant names: for example, **wiritaalima** (Table 4.2) is a *terra firme* (and also secondary) forest type containing **wiritaa,** a Baniwa generic name for a group of similar species of the genus *Inga* (Leg: Mimosaceae) used as dye fixers for decorative basketry (Hoffman 2001; Leoni 2005; Shepard et al. 2004b).

The plant species used by the Baniwa as vegetation type indicators include a wide range of botanical families and growth habits: herbs, palms, shrubs or small trees, and emergent canopy trees. Many, though not all, Baniwa indicator species are used for various purposes, from provision of edible fruits, to house and canoe construction, fishing or hunting technologies, and utility as ecological indicators of game species or good soils for agriculture (see Tables 4.1 through 4.3). Indicator species are not necessarily dominant species within the vegetation type they indicate. Rather, they appear to be species that are both perceptively salient (notably palms) and sensitive to environmental variation, and hence good indicators of transitions along environmental gradients. Most Baniwa vegetation types have trees as indicator species, but herbaceous species are sometimes noted. For example **koliwaipalima** is a vegetation type where **koliwaipa,** an herbaceous Cyclanthaceae (apparently *Asplundia vaupesiana*) with edible fruits, abounds. Only one habitat type referring to a non-plant indicator species was noted: **katanapililima,** a flooded *igapó* habitat where a large earthworm (**katanapili**), used as fishing bait, is found in abundance (see Table 4.3).

Our study was able to detect a number of Baniwa vegetation types not cited in Andrello's (1998) more limited preliminary study (see Tables 4.1 through 4.3). Our study also revealed intermediate and subordinate levels of habitat classification. For example, during both interviews and field visits to different **hamáliani** (white-sand) vegetation types, it became clear that the Baniwa recognized three subcategories of this habitat determined by soil characteristics. Pure white-sand soils (**halapokole**) are associated with low canopy, open understory, savanna-like vegetation, considered to be "true **hamáliani.**" Darker, clay-containing soils (**keramapere**) are associated with high, closed canopy forests in areas of transition to *terra firme,* especially **waapalima** (*Eperua purpurea* forest), the preferred vegetation type for swidden agriculture in communities where white-sand forest dominates. While people in predominately white-sand communities consider **waapa** to be an indicator of transition to *terra firme,* people in predominately *terra firme* regions regard it as an indicator of transition to lower-fertility white-sand *campinarana* (see Tables 4.1, 4.2, Figure 4.3). Waterlogged, muddy soils (**patsapatsapale, potiapedale**) unsuitable for agriculture are associated with specific white-sand vegetation types such as **heridzorolima** (*Anaxagorea/Pseudoxandra* forest) and **ttiñalima** (*Mauritiella carana* palm vegetation), found along streams, in swampy depressions and transitional areas to flooded **alápe** habitats (see Tables 4.1, 4.3, Figure 4.3). Soils covered by a dense layer of spongy, matted roots are referred to as **kamoko;** areas containing these soils are referred to as **kamokonina** and are treated as a salient intermediate type within the *terra firme* habitat category (see Figure 4.3) because they are not suitable for agriculture, even though the vegetation cover may be similar to that found in "true *terra firme*" (*éedzawa kantsa*).

These examples are analogous to the "intermediate" level in Berlin's (1992) universal scheme of folk biological classification (see Figure 4.3). Berlin's (1992: 42) notion of focal or "prototypical" genera and species within broader groupings also applies here, with salient distinctions between "true" vs. transitional or atypical habitat types (see Table 4.3).

In the field, informants indicated two contrasting subtypes for the vegetation type **anerima** (see Table 4.4, vegetation codes 5A–5B): "high-canopy" (**dzenonipe kaawa**) vs. "low-canopy" (**madoape kaawa**). Both appear to contain the same botanical indicator species (**ane**, *Eperua duckeana*), but different historical or soil conditions result in qualitative habitat differences. The same applies to the vegetation type **waittirima** (indicator species **waitti**, *Aldina heterophylla*), for which "high-canopy" and "low-canopy" subtypes are likewise distinguished (Table 4.4, vegetation codes 6A–6B; see also Abraão et al. 2008). Such examples appear analogous with Berlin's (1992) "specific" or "varietal" level of folk biological classification (see Figure 4.3).

Consensus-Floristic Patterns among **hamáliani** *Vegetation Types*

The ten Baniwa informants interviewed showed detailed knowledge about the plant species and soil types associated with each of the fourteen consensus **hamáliani** vegetation types (Tables 4.1, 4.4). Ward's cluster analysis of the fourteen types, based on the abundance of 150 Baniwa plant names elicited during interviews, revealed two major groups and several smaller clusters (Figure 4.4). The Indicator Species Analysis (ISA) showed that the two major groupings were significantly distinct in their floristic composition, while the smaller clusters were not (Figure 4.5). These tools for quantitative floristic analysis confirmed the distinction made during interviews and field visits between the two Baniwa-recognized intermediate categories noted above and in Figure 4.3. "True **hamáliani**" (**hamáliani kantsa**; white circles in Figures 4.4 and 4.6) consists of open savanna with exposed white-sand soil (**halapokole**) whose woody elements form a low and sparse canopy, including typical white-sand species such as **ane** (*Eperua duckeana*), **waitti** (*Aldina heterophylla*), **waalia** (*Humiria balsamifera* var. 2), thatch palm **ttiña** (*Mauritiella carana*), and **koliwaipa** (*Asplundia vaupesiana*). By contrast, the higher, closed canopy white-sand forest (not given a habitual

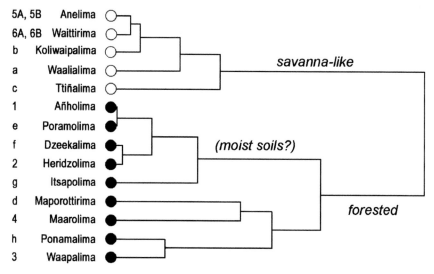

Figure 4.4 *Cluster analysis by Ward's method of fourteen* **hamáliani** *vegetation types based on informant reports of floristic composition.*

habitat name in Baniwa; black circles in Figures 4.4 and 4.6) is characterized by darker, more fertile soils (**keramapere**) and is marked especially by the presence of the tree **waapa** (*Eperua purpurea*), prime indicator of the transition to *terra firme* (**éedzawa**).

Two of the three **hamáliani** vegetation types (among the fourteen studied here) that were cited independently by Andrello (1998) as having agricultural potential (Table 4.1) were grouped together by cluster analysis (Figure 4.4): **waapalima** (*Eperua purpurea* forest), and **maporottilima** (*Humiria balsamifera* var. 1 forest[2]). The cluster analysis (Figure 4.4) also suggests an association among certain species and vegetation types with humid soils, especially **heridzoro** (*Anaxagorea/Pseudoxandra*) and the edible palm **poramo** (*Euterpe catingae*); however this relationship was not robustly supported in the indicator species analysis (Figure 4.5).

Using NMDS ordination, most of the variability in the floristic distances (Bray-Curtis similarity index) of all pairs of vegetation types could be preserved in few dimensions. Specifically, 91 percent of variability was preserved in the two-dimensional ordination (Figure 4.6), while 82 percent was preserved in just one dimension (see Figure 4.7, below), results indicating the presence of a single main floristic gradient. In accord with the cluster analysis and the Baniwa's independent structural classification, Figures 4.6 and 4.7 also clearly show a grouping into two major subhabitats: "true" white-sand savanna vs. forested habitats ultimately transitioning to *terra firme*. Furthermore, there appears to be a fairly tight cluster of vegetation types in the upper left of Figure 4.6 that includes the vegetation types with humid soils, notably (1) **anholima,** (2) **heridzorolima** and

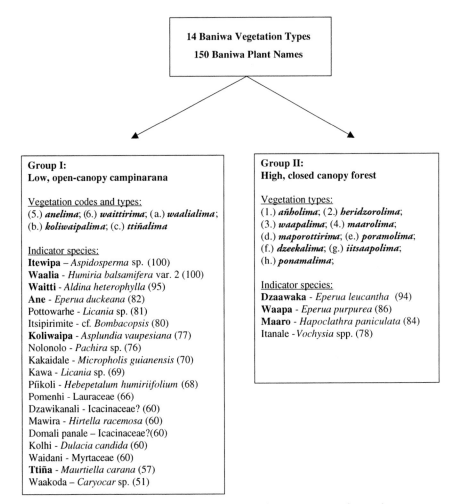

Figure 4.5 *Indicator species analysis based on informant reports of 150 plant names for the fourteen* **hamáliani** *vegetation types.*

(e) *poramolima,* with a more dispersed arrangement for the remaining types. Though the results of the three analyses (cluster analysis, indicator species analysis, ordination) are similar—based, after all, on the same data set of informant-reported floristic associations—the ordination in two dimensions presents a more visually apparent representation for inferring possible environmental and structural correlates with the floristic clusters and gradients.

Figure 4.7 shows how the most-reported species change in abundance (i.e., informant consensus regarding presence) along the main one-dimensional floristic gradient. The fourteen Baniwa vegetation types are ordered along the horizontal

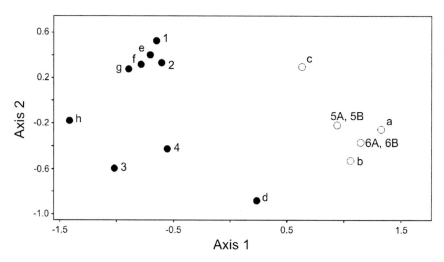

Figure 4.6 *Two-dimensional ordination by non-metric multidimensional scaling (MDS), of fourteen Baniwa-defined **hámaliani** vegetation types, (see Table 4.4) based on ten informants' reports of presence for 150 predominant plant species.*

dimension according to their one-dimensional ordination score (at the bottom of the figure are fourteen dots showing the positions of the fourteen **hamáliani** types along this ordination). Each histogram—one for each Baniwa plant species, labeled on the right—indicates how many informants (0 to 10) stated a species was present in each vegetation type. Each vertical stack of bars above a dot (i.e., one vegetation type) shows the informant-reported floristic composition for that vegetation type. Two main groups of vegetation types can be seen with very different compositions—one group with negative (left side) and the other with positive (right side) ordination scores.[3] Again, these correspond to the two structurally defined classes of **hamáliani** described by the Baniwa: low, open-canopy *campinarana* ("true **hamáliani**") and high, closed-canopy vegetation types transitioning to *terra firme*. However, the ordination appears to support a finer division of **hamáliani** types than did the indicator species analysis (ISA), highlighting the distinctiveness of vegetation types associated with humid soils, suggested by cluster analysis but not detected in ISA. There is a crowding of five vegetation types with very similar compositions in a narrow range of ordination scores 0.7 to 1.0 (Figure 4.7, bottom). These five types may be overclassified by the Baniwa, or this result may represent a deficiency in our use of informant consensus as a proxy for abundance. Since these are the tall, forested **hamáliani** vegetation types with presumably higher species diversity, these types could share many species while nonetheless differing considerably in true abundances of some species if the types were inventoried in the field (cf. Abraão et al. 2008).

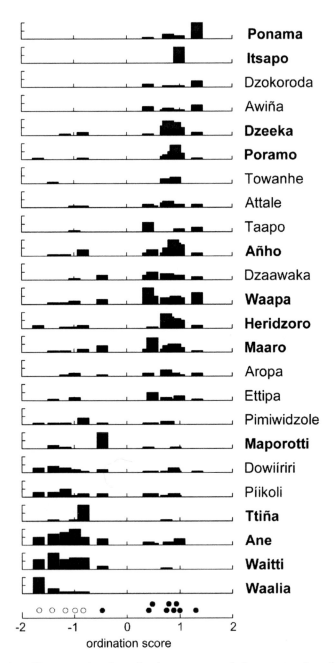

Figure 4.7 *Change in abundance for the most reported plant species along the main floristic gradient detected by a 1-dimensional ordination.*

In addition to these groups, one notices a fairly orderly replacement of species as one moves diagonally up from lower left to upper right of Figure 4.7; a similar visual effect is seen in Table 4.6, which organizes the raw data in tabular form. Unlike the other forms of data presentation, here the reader becomes acutely aware that vegetation types are not sharply circumscribed objects arranged in a hierarchical set of relationships (as, for example, in cluster analysis; Figure 4.4). Instead, the Baniwa vegetation types are associations of plant species that change gradually along a gradient, though of course certain qualitative or "gestalt" shifts are also noted. As such, there is a greater degree of overlap in species composition between vegetation types that are close to one another along that gradient. As one moves further along the **hamáliani** gradient, species composition changes until there is almost complete replacement of one set of species by another. In other words, the species composition at one end of the gradient is not merely a quantitatively different subset of (for example) a richer flora at the other end, but rather a qualitatively different flora. The actual floristic gradient may prove to show strong turnover with little overlap between types, in which case the overall plant community analysis would corroborate the finely divided Baniwa typology. On the other hand, there may be weak turnover and high overlap, so that the Baniwa typology—which is largely based on a single indicator species for each type, rather than the entire flora—could represent overclassification from a community analysis perspective. These patterns and judgments are relevant to an evaluation of the value of indigenous typology for broader applications such as conservation planning and landscape management (see Abraão et al. 2008; Instituto Socioambiental 2003).

In sum, an ordination as portrayed in Figures 4.6 and 4.7 allows researchers to use indigenous knowledge of species composition and habitat preferences to perceive gradual patterns of change in vegetation types that might not be apparent from more traditional ethnoecological typologies consisting of distinctive, hierarchically arranged types. Such groupings of vegetation types revealed through ordination can be compared with Baniwa cognitive groups based on independent criteria, like vegetation structure. Moreover, ordination reveals subtle patterns of knowledge structure within Baniwa landscape classification and might have broader applicability to other kinds of ethnoecological data.

Conclusions

The Baniwa have a detailed system of habitat classification that recognizes multiple vegetation types subsumed within four broad habitat categories: **éedzawa,** upland *terra firme* forest; **alápe,** flooded *igapó* forest; **hamáliani,** white-sand *campinarana* (or *caatinga amazônica*) vegetation; and **heñame,** secondary forest in various stages of regeneration following human agricultural activities. Preliminary free-listing exercises with Baniwa informants about the naturally-occurring

Table 4.6 *Consensus among ten Baniwa informants concerning predominant species in fourteen* **hamáliani** *vegetation types.*

Codes follow Table 4.4. Only plants mentioned by at least four informants per vegetation type are included, for a total of twenty-five Baniwa plant names. Bold text (column headers) and numbers indicate plants that serve as indicator species for the fourteen vegetation types. Vegetation types and indicator species are organized following the results of single axis ordination (see Figure 4.7).

Baniwa plant names* mentioned by four or more informants:

Code (Table 4.4)	Vegetation type	**waalia**	**waitti**	dowiriri	**koliwaipa**	piikoli	**ane**	**ttiña**	pimiwidzoli	**maporotti**	**maaro**	ettipa	**añho**	dzaawaka	**Heridzoro**	aropa	**poramo**	towanhe	**dzeeka**	attaale	**iitsaapo**	**waapa**	taapo	**ponama**	awiña	dzokoloda
a.	*waalialima*	10	6				5																			
6.	*waittirima*		10	4			7																			
b.	*koliwaipalima*		6		9	5	8																			
5.	*anelima*		7				10																			
c.	*ttiñalima*		7				5	10	4				4													
d.	*maporottirima*						4	4		10	4															
4.	*maarolima*									5	9	5	4	5	6				7			5				
1.	*añholima*										5		10				5									
2.	*heridzolima*													4	9	4			6			4				
e.	*poramolima*										5		8		6		9	4	8			5				
f.	*dzeekalima*										4		5	4	7		4		9	4		4				
g.	*itsapolima*						4						7		5				5		10	4				
3.	*waapalima*																					10	6			
h.	*ponamalima*																					7		10	4	4

* Preliminary botanical identifications of Baniwa plant names:

ane: *Eperua duckeana* (LEG)
anho: *Micrandra spruceana* (EUP)
aropa: ?
attaale: *Couma utilis* (APC)
awiña: *Monopteryx uacu* (LEG)
dowiriri: *Pithecellobium* sp.
dzaawaka: *Eperua leucantha* (LEG)
dzeeka: *Hevea guianensis* (EUP)
dzokoroda: cf. *Calyptranthes* (MYT)
ettipa: *Guatteria* spp., others (ANN)
heridzoro: *Anaxagorea/Pseudoxandra* (ANN)
itsapoli: *Duguetia* spp. (ANN)
koliwaipa: *Asplundia vaupesiana* (CYL)

maaro: *Haploclathra paniculata* (CLU)
maporotti: *Humiria balsamifera* var. 1 (HUM)
piikoli: *Hebepetallum humiriifolium* (HUG)
pimiwidzoli: *Couma* sp. (APC)
ponama: *Oenocarpus bataua* (ARE)
poramo: *Euterpe catingae* (ARE)
taapo: ?
towanhe: *Mauritiella aculeata* (ARE)
ttiña: *Mauritiella carana* (ARE)
wadia: *Humiria balsamifera* var. 2 (HUM)
waapa: *Eperua purpurea* (LEG)
waitti: *Aldina heterophylla* (LEG)

habitats revealed twenty-five forest types for *éedzawa,* twenty-eight types for *alápe,* and twenty-four vegetation types for *hamáliani.* Andrello (1998) had previously documented fifteen secondary *heñame* forest types, not included in the current study.

Further interviews in three focal communities reduced the original number of twenty-four to fourteen predominant and salient categories for *hamáliani,* confirmed through inter-informant consensus. This result suggests that free-listing of indigenous habitat typologies without considering consensus may likely exaggerate actual beta-diversity of the botanical landscape. Consensus from diverse informants and incisive questions directed at all informants to reveal the criteria underlying the cognitive classes are necessary if indigenous knowledge is to be used as a proxy for botanical inventories and applied to landscape management or conservation planning.

The Baniwa have an extensive repertoire of plant names—150 names mentioned by a consensus of ten informants for the *hamáliani* category alone—as well as knowledge of the habitat and soil preferences and ethnobotanical uses of many species. Given this detailed knowledge about floristic and habitat associations, informant consensus regarding the presence of each species is here used as a proxy for its actual abundance in each vegetation type. This quantitative information was then subjected to standard plant community analyses to search for floristically defined clusters and gradients of change in composition.

Ordination of reported abundances of plant species for the fourteen study types proved to be a useful tool for detecting floristic gradients, providing substantiation for more vegetation types than did the cluster and ISA analyses that were also applied. A single-axis gradient explained 82 percent of overall variability in reported abundances between the vegetation types. Baniwa perceptions of habitat diversity are likewise independently confirmed by ordination of floristic data from quantitative ethnobotanical plots and structural measures (Abraão et al. 2008). Together, these results suggest that Baniwa perceptions of habitat diversity are well correlated with changes in overall species composition and may have application for rapid assessment of conservation priorities (Instituto Socioambiental 2003). Moreover, the wealth of mythological, historical, and cultural information imbedded within landscape knowledge, although not explored in detail in this study, is fundamental to an ongoing collaborative research program including these authors (Instituto Socioambiental 2007) that aims to document and validate indigenous environmental knowledge while affirming the Baniwa's historical and cultural rights to their ancestral territory.

The use of ordination techniques allows the researcher to apprehend vegetation "types" (both in indigenous and botanical terms) not as distinct and separate categories, but rather as variable assemblages along continuous gradients. Habitats are composed of associations of multiple species, the composition of which shifts gradually in response to external variables such as soils, water regime, and disturbance regimes as well as geographical distance more generally. Long periods of observation over relatively large areas are required to perceive these subtle variations, and to develop a scheme for identifying important transitional points along the gradient. For these reasons, indigenous peoples with generations of accumulated, daily knowledge of forest habitats have distinct advantages over tropical biologists in apprehending and classifying habitat diversity in local landscapes. Beyond such pragmatic concerns, indigenous people are deeply vested culturally and spiritually in their territories, which serve as "bastions not only of biodiversity but also of indigenous culture, history and spirit" where people like the Baniwa and others have "breathed their history into a landscape which echoes with the triumphs and tragedies of the past" (Shepard and Chicchon 2001: 173).

Acknowledgments

The Amazonas State Foundation for Research (FAPEAM) provided field support funds, a master's student stipend to Abraão and a Young Amazonian Scientist stipend to João Claudio Baniwa. The study was carried out within a broader collaborative project evaluating the socio-environmental sustainability of Baniwa commercial and subsistence practices, funded by both FAPEAM and Brazil's National Science Council (CNPq). Participating institutions are Instituto Nacional de Pesquisas da Amazônia (INPA), Instituto Socioambiental (ISA), and the Baniwa indigenous organization Organização Indígena da Bacia do Içana (OIBI). The Baniwa study communities, represented through OIBI, provided their informed consent for this research. The project was evaluated by Brazil's national council on genetic patrimony and associated traditional knowledge (CGEN), and was deemed exempt from the CGEN permit process for access to components of genetic patrimony. We thank Paulo Assunção Apóstolo for field verification and correction of several botanical identifications.

Notes

1. This term is often heard in the locative expression *awakadaliko,* 'in the forest' or 'into the forest', used to indicate any excursion away from the inhabited village area to the surrounding forest-savanna-garden fallow matrix. *Awakada* most often refers to primary forest, in contrast to agricultural fallows (*heñame*), but depending on the context *awakada* can also refer to any kind of non-agricultural vegetation cover, natural or anthropogenic. The suffix *-da* (e.g., *awaka-da*) indicates a delimited, roughly circular area or patch of vegetation, and is sometimes used as an added suffix on some vegetation type names, for example *waapalimada* (see Table 4.1, Andrello's terminology), namely a small, defined patch of *waapa* (*Eperua purpurea*).
2. The Baniwa consider *maporotti* and *waalia* to be similar but nonetheless distinct species. They occupy different ecological zones, *waalia* occurring only in open *campinarana*, *maporotti* in more forested environments such as agricultural fallows. They have slightly different leaves, though their fruits and flowers are nearly identical. In current botanical classification, the two taxa appear to refer to distinctive varieties of a single species, *Humiria balsamifera*, of which there are at least two taxonomically recognized varieties: var.

balsamifera and var. *floribunda*. Though voucher specimens of *waalia* were not examined, *maporotti* has been identified positively as *Humiria balsamifera*. Association of the two Baniwa-recognized names with the recognized botanical varieties (*balsamifera* vs. *floribunda*) is still uncertain.

3. In Figure 4.7 (lower part of graph) the higher-canopy forests (as opposed to savanna-like vegetation) have positive scores, while in Figure 4.6 axis 1, they have negative scores. Rotation and inversions of the cloud of data points in successive iterations of an ordination has no significance, since the analysis attempts to establish proper relative distances between data points, not their absolute position. In this case, rotation of the data cloud was prevented by the "varimax" option available in the PC-ORD software, but inversion from positive to negative scores (e.g., between Figures 4.6 and 4.7) could not be prevented.

References

Abraáo, Marcia Barbosa. 2005. *Conhecimento Indígena, Atributos Florísticos, Estruturais e Espectrais como Subsídio para Inventariar Diferentes Tipos de Florestas de Campinarana no rio Içana, Alto Rio Negro*. Master's thesis, Instituto Nacional de Pesquisas da Amazônia.

Abraáo, Marcia Barbosa, Bruce W. Nelson, Joáo Claudio Baniwa, Douglas W. Yu, and Glenn H. Shepard Jr. 2008. "Ethnobotanical Ground-Truthing: Indigenous Knowledge, Floristic Inventories and Satellite Imagery in the Upper Rio Negro, Brazil." *Journal of Biogeography* 35(12): 2237–2248.

Anderson, A. B. 1981. "White-Sand Vegetation of Brazilian Amazonia." *Biotropica* 13(3): 199–210.

Andrello, Geraldo. 1998. "O ambiente natural e a ocupação tradicional dos povos indígenas." In *Povos Indígenas do Alto e Médio Rio Negro: Uma Introdução à Diversidade Cultural e Ambiental do Noroeste da Amazônia Brasileira*, ed. A. Cabalzar and C. A. Ricardo. Sáo Paulo: Instituto Socioambiental (ISA)/ Federação das Organizações Indígenas do Rio Negro (FOIRN).

Balée, William, and A. Gély. 1989. "Managed Forest Succession in Amazonia: The Kaapor Case." *Advances in Economic Botany* 7: 129–158.

Basso, Keith H. 1996. *Wisdom Sits in Places: Landscape and Language among the Western Apache*. Albuquerque: University of New Mexico Press.

Berlin, Brent. 1977. *Bases empíricas de la cosmología Aguaruna Jívaro, Amazónas, Perú*. Volume report no. 3. Berkeley: Behavior Research Laboratory, University of California at Berkeley.

———.1984. "Contributions of Native American Collectors to the Ethnobotany of the Neotropics." *Advances in Economic Botany* 1: 24–33.

———. 1992. *Ethnobiological Classification: Principles of Categorization of Plants and Animals in Traditional Societies*. Princeton, NJ: Princeton University Press.

Berlin, Brent, Dennis Breedlove, and Peter Raven. 1974. *Principles of Tzeltal Plant Classification: An Introduction to the Botanical Ethnography of a Mayan-Speaking People of Highland Chiapas*. New York: Academic Press.

Bulmer, Ralph. 1974. "Folk Biology in the New Guinea Highlands." *Social Science Information* 13(4–5): 9–28.

Cabalzar, Aloísio, and Carlos Alberto Ricardo, eds. 1998. *Povos Indígenas do Alto e Médio Rio Negro: Uma introdução à diversidade cultural e ambiental do noroeste da Amazônia Brasileira*. Sáo Paulo and Sáo Gabriel da Cachoeira: Instituto Socioambiental (ISA) & Federação das Organizações Indígenas do Rio Negro (FOIRN).

————, eds. 2006. *Mapa-Livro Povos Indígenas do Rio Negro.* São Paulo and São Gabriel da Cachoeira: Instituto Socioambiental (ISA) & Federação das Organizações Indígenas do Rio Negro (FOIRN).

Carneiro, Robert. 1978. "The Knowledge and Use of Rainforest Trees by the Kuikuru Indians of Central Brazil." In *The Nature and Status of Ethnobotany,* ed. R. I. Ford. Anthropological Papers, Vol. 67. Ann Arbor: Museum of Anthropology, University of Michigan.

Chapin, Mac. 2004. "A Challenge to Conservationists." *World Watch* (November/December): 17–31.

Conklin, H. C. 1954. *The Relation of Hanunóo Culture to the Plant World.* Ph.D. thesis, School of Forestry, Yale University, New Haven, CT.

————. 1957. *Hanunóo Agriculture: A Report on an Integral System of Shifting Cultivation in the Philippines.* Volume 12. Rome: FAO.

Diamond, J. M. 1966. "Zoological Classification System of a Primitive People. *Science* 151: 1102–1104.

Ellen, R. 1979. "Omniscience and Ignorance: Variation in Nuaulu Knowledge, Identification, and Classification of Animals." *Language in Society* 8: 337–364.

Encarnación, Filomeno. 1993. "El bosque y las formaciones vegetales en la llanura amazónica del Perú." *Alma Máter* (Universidad Nacional de San Marcos, Lima) 6: 95–114.

Endo, Whaldener. 2005. *Campinarana e Índios Baniwa: Influências Ambientais e Culturais sobre a Comunidade de Vertebrados Terrestres no Alto Rio Negro, AM.* Master's thesis, Instituto Nacional de Pesquisas da Amazônia.

Fleck, David William. 1997. *Mammalian Diversity in Rainforest Habitats as Recognized by Matses Indians in the Peruvian Amazon.* M.S. thesis, Ohio State University.

Fleck, David William, and John D. Harder. 2000. "Matses Indian Rainforest Habitat Classification and Mammalian Diversity in Amazonian Peru." *Journal of Ethnobiology* 20(1): 1–36.

Gentry, Alwyn H. 1993. *A Field Guide to the Families and Genera of Woody Plants of Northwest South America (Colombia, Ecuador, Peru).* Volume 85. Washington, D.C.: Conservation International.

Halme, K. J., and R. E. Bodmer. 2007. "Correspondence Between Scientific and Traditional Ecological Knowledge: Rain Forest Classification by the Non-indigenous Ribereños in Peruvian Amazonia." *Biodiversity and Conservation* 16: 1785–1801.

Hill, Jonathan D. 1989. "Ritual Production of Environmental History among the Arawakan Wakuenai of Venezuela." *Human Ecology* 17: 1–25.

Hirsch, Eric, and Michael O'Hanlon, eds. 1995. *The Anthropology of Landscape: Perspectives on Place and Space.* Oxford: Clarendon Press.

Hoffman, Daniel J. 2001. *Arumã no alto Rio Içana: Perspectivas para o manejo.* São Paulo: Instituto Socioambiental (ISA).

Hunn, Eugene S. 1977. *Tzeltal Folk Zoology: The Classification of Discontinuities in Nature.* New York: Academic Press.

Instituto Socioambiental. 2003. *Terras Indígenas do Alto e Médio Rio Negro: Resultados do macrozoneamento etno-ambiental.* São Paulo: Instituto Socioambiental.

————. 2007. *Paisagens Baniwa do Içana: Etnoecologia de unidades de paisagem como base para a gestão socioambiental.* Report (22 pgs.) to Conselho de Gestão do Patrimônio Genético (CGEN), Ministério do Meio Ambiente, Brasília. São Paulo: Instituto Socioambiental.

Krohmer, Julia. 2004. "Was Fulbe bewegt: Umweltkonzepte und handlungsmotive agropastoraler Fulbe in Burkina Faso, Benin und Nigeria." In *Mensch und Natur in Westafrika: Ergebnisse aus dem Sonderforschungsbereich* 268, "Kulturentwicklung und Sprachgeschichte im Naturraum Westafrikanische Savanne," ed. K.-D. Albert, D. Löhr, and K. Neumann. Weinheim: Wiley-VCH Verlag.

Leoni, Juliana Menegassi. 2005. *Ecologia e Extrativismo de Plantas Utilizadas como Fixadoras de Corantes no Artesanato Baniwa, Alto Rio Negro.* Master's thesis, Instituto Nacional de Pesquisa da Amazônia.

Martin, Gary J., Agnes L. Agama, John H. Beaman, and Jamili Nais. 2002. *'Projek Ethnobotani Kinabalu': The Making of a Dusun Ethnoflora (Sabah, Malaysia).* People and Plants Working Paper No. 9. Paris: UNESCO.

McCune, Bruce, and James B. Grace. 2002. *Analysis of Ecological Communities.* Gleneden Beach, OR: MJM Software Design.

Milliken, William. 1998. *Levantamentos Etnoecológicos em Reservas Indígenas na Amazônia Brasileira: Uma metodologia.* Edinburgh: Edinburgh Development Consultants.

Moran, Emilio F. 1991. "Human Adaptive Strategies in Amazonian Blackwater Ecosystems." *American Anthropologist* 93: 361–82.

Parker, Eugene, Darrell A. Posey, John Frechione, and Luiz Francelino da Silva. 1983. "Resource exploitation in Amazonia: Ethnoecological examples from four populations." *Annals of the Carnegie Museum of Natural History* 52(8): 163-203.

Phillips, Oliver L. 1996. "Some Quantitative Methods for Analyzing Ethnobotanical Knowledge." In *Selected Guidelines for Ethnobotanical Research: A Field Manual,* ed. M. N. Alexiades. Bronx: New York Botanical Gardens.

Pires, João Murça, and Ghillean T. Prance. 1985. "The Vegetation Types of the Brazilian Amazon." In *Key Environments: Amazonia,* ed. G. T. Prance and T. E. Lovejoy. Oxford and New York: Pergamon Press/IUCN.

Posey, Darrell A., John Frechione, and John Eddins. 1984. "Ethnoecology as Applied Anthropology in Amazonian Development." *Human Organization* 43: 95–107.

Ramirez, Henri. 2001. *Dicionário Baniwa-Portugués.* Manaus: Editora da Universidade do Amazonas (EDUA).

Reichel-Dolmatoff, Gerardo. 1976. "Cosmology as Ecological Analysis: A View from the Rain Forest." *Man* 11(3): 307–318.

Ribeiro, Berta R. 1995. *Os Índios das Águas Pretas.* São Paulo: Companhia das Letras, Universidade de São Paulo.

Ribeiro, José Eduardo L., M. J. G. Hopkins, A. Vicentini, C. A. Sothers, M. A. S. Costa, J. M. de Brito, M. A. A. D. de Souza, L. H. P. Martins, L. G. Lohmann, P. A. C. L. Assunção, E. da C. Pereira, C. F. da Silva, M. R. Mesquita, and L. Procópio. 1999. *Flora da Reserva Ducke: Guia de identificação das plantas vasculares de uma floresta de terra-firme na Amazônia Central.* Manaus: Instituto Nacional de Pesquisas da Amazônia (INPA).

Santos-Granero, Fernando. 1998. "Writing History into the Landscape: Space, Myth and Ritual in Contemporary Amazonia." *American Ethnologist* 25(2): 128–148.

Sheil, Douglas, and Anna Lawrence. 2004. "Tropical Biologists, Local People and Conservation: New Opportunities for Collaboration." *Trends in Ecology and Evolution* 19(12): 635–638.

Shepard, Glenn H., Jr. 1999. "Shamanism and Diversity: A Matsigenka Perspective." In *Cultural and Spiritual Values of Biodiversity,* ed. D. A. Posey. UNEP Global Biodiversity Assessment, Vol. Supplement 1. London: United Nations Environmental Programme and Intermediate Technology Publications.

Shepard, Glenn H., Jr., and Avecita Chicchon. 2001. "Resource Use and Ecology of the Matsigenka of the Eastern Slopes of the Cordillera Vilcabamba." In *Biological and Social Assessments of the Cordillera de Vilcabamba, Peru*, ed. L. E. Alonso, A. Alonso, T. S. Schulenberg, and F. Dallmeier. RAP Working Papers No. 12 and SI/MAB Series 6. Washington, D.C.: Conservation International.

Shepard, Glenn H., Jr., Douglas W. Yu, Manuel Lizarralde, and Mateo Italiano. 2001. "Rainforest Habitat Classification Among the Matsigenka of the Peruvian Amazon." *Journal of Ethnobiology* 21(1): 1–38.

Shepard, Glenn H., Douglas W. Yu, and Bruce Nelson. 2004a. "Ethnobotanical Ground-Truthing and Forest Diversity in the Western Amazon." In *Ethnobotany and Conservation of Biocultural Diversity*, ed. T. Carlson, and L. Maffi. Advances in Economic Botany 15.

Shepard, Glenn H., Jr., Maria N. F. da Silva, Armindo F. Brazão, and Pieter van der Veld. 2004b. "Arte Baniwa: Sustentabilidade socioambiental de arumã no Alto Rio Negro." In *Terras Indígenas e Unidades de Conservação da Natureza: O desafio das sobreposições*, ed. F. Ricardo. São Paulo: Instituto Socioambiental.

Sillitoe, Paul. 1996. *A Place Against Time: Land and Environment in the Papua New Guinea Highlands*. Amsterdam: Harwood Academic Publishers.

Silva, Adeilson Lopes da. 2004. *No Rastro da Roça: Ecologia, Extrativismo e Manejo de Arumã (Ischnosiphon spp. Marantaceae) nas Capoeiras dos Índios Baniwa do Içana, Alto Rio Negro*. M.S. thesis, Instituto Nacional de Pesquisas da Amazônia/Universidade Federal do Amazonas, Manaus.

Sioli, Harald. 1984. "The Amazon and Its Main Affluents: Hydrography, Morphology of the River Courses, and River Types." In *The Amazon: Limnology and Landscape Ecology of a Mighty Tropical River and its Basin*, ed. H. Sioli. Monographiae Biologicae, vol. 56. Boston: Dr. W. Junk Publishers.

Strang, Veronica. 1997. *Uncommon Ground: Cultural Landscapes and Environmental Values*. New York: New York University Press.

Toledo, Marisol, and Jan Salick. 2006. "Secondary Succession and Indigenous Management in Semideciduous Forest Fallows of the Amazon Basin." *Biotropica* 38(2): 161–170.

Toledo, Victor M. 1992. "What Is Ethnoecology? Origins, Scope, and Implications of a Rising Discipline." *Etnoecología* 1(1): 5–21.

Tuomisto, Hanna. 1998. "What Satellite Imagery and Large Scale Field Studies Can Tell About Biodiversity Patterns in Amazonian Forests." *Annals of the Missouri Botanical Gardens* 85: 48–62.

Veloso, H. P., A. L. R. Rangel Filho, and J. C. A. Lima. 1991. *Classificação da Vegetação Brasileira, Adaptada a Sistema Universal*. Rio de Janeiro: Instituto Brasileiro de Geografia e Estatística, Departamento de Recursos Naturais e Estudos Ambientais.

Wright, Robin M. 1998. *Cosmos, Self and History in Baniwa Religion: For Those Unborn*. Austin: University of Texas Press.

Zent, Stanford. 1996. "Behavioral Orientations toward Ethnobotanical Quantification." In *Selected Guidelines for Ethnobotanical Research: A Field Manual*, ed. M. N. Alexiades. Advances in Economic Botany, vol. 10. Bronx: New York Botanical Gardens.

Chapter 5

Why Aren't the Nuaulu
Like the Matsigenka?
Knowledge and Categorization of Forest Diversity
on Seram, Eastern Indonesia

Roy Ellen

Theoretical Background

Ethnobotanical studies of the knowledge of tropical forest peoples have demonstrated an extensive local knowledge of trees, local recognition of forest diversity, and the existence of coherent vernacular classifications of forest types. While folk classifications of habitats, biotopes, and landscapes more generally have received much less attention than folk systematics (Sillitoe 1998: 104; Meilleur 1986: 54–90; Martin 1974; Torre-Cuadros and Ross 2003), the data that are available on indigenous forest classifications in particular suggest significant variation in the extent to which recognition of compositional diversity actually translates into complex, fixed, and labeled categories for different types of forest.

Although there are some early references to the importance of establishing ethnoecological categories in the Asian tropics (e.g., Bartlett 1936; Conklin 1954), the pioneer work on this subject was conducted in the Amazon, and has since extended elsewhere. Carneiro (1978) reports four forest types for the Kuikuru, as do Parker, Posey, Frechione, and da Silva (1983: 170–171) for the Brazilian Kayapó, with several subdivisions of gallery forest, as well as ten forest types for the Brazilian Yekuana around Lake Coari. Balée and Gély (1989: 131–132) report four Ka'apor forest categories (old swidden, fallow forest, mature forest, and swamp forest), plus a number of unspecified named ecotones. For New Guinea, Sillitoe (1998) lists nine basic Wola terms for vegetation types, with a further four subtypes, but this is not all forest, and it covers an extensive altitudinal range, 1600–2000 m asl. Dependable data for island southeast Asia are more difficult to obtain, but Puri (1997: 104), in his work on Penan forest knowledge, elicited just eight categories that map onto what we might generally accept as "forest," despite systematic attempts to generate more detailed labeled categories. By contrast, a number of researchers working in the Amazon region have recently reported folk classifications of forest evidently more terminologically refined and extensive than these reports suggest. Thus, Andrello (cited in Shepard, Yu, Lizarralde, and Italiano 2001) has reported 53 natural habitat types. Fleck and Harder (2000: 1–3), working amongst the Peruvian Matses, list 47 labeled vegetation types overall, and claim that by combining vegetational and geomorphologic designations, Matses distinguish 178 habitats. Of these, 104 are described as types of primary rainforest and 74 as types of successional forest. Shepard, Yu, Lizarralde, and Italiano (2001) suggest great similarities between Matses and Matsigenka, also in Peru, who operate with about 40 categories for lowland forest.

That local peoples have a profound knowledge of forest diversity is now hardly doubted, but how can we account for such discrepancies in the lexicalization of knowledge for people living in ecologically very similar environments? In order to address this problem we might consider a number of hypotheses:[1]

- That not all knowledge, everywhere, is equally lexicalized or that some research is less thorough;
- That ecological and subsistence differences influence the extent to which people categorize and lexicalize;
- That models based on the structure of folk taxonomies generated in studies of folk systematics bias our methodologies.

One approach to studying local knowledge of forest or more general habitat variation is to use a model based on the structure of folk taxonomies generated in studies of folk systematics. Unfortunately, such an approach treats forest types and habitats as kinds of folk species, and as functioning, organizational, entities they are quite different: at the level of organism classification there are more terms and a greater fixity in the terms used. In other words, habitats are not "things" in the sense, I think, implied in, say, Meilleur's multileveled, lexically defined ranks of "ethnoecosystem" and "folk biotope," a system reminiscent of Berlin's (e.g., Berlin, Breedlove, and Raven 1973) ranks for natural kind classification (Meilleur 1986: 54–90).

By comparison with the morphological discontinuousness of individual species, where unidimensional characteristics, salient prototypes, and contrasting segregates abound, the ecological variation of forest is multidimensional and continuous, and such variation does not encode well into simple taxonomic models. For example, many folk classifications of forest terminologically encode the phases of swidden cultivation and forest fallow regrowth, and the categories so established on the basis of these phases inevitably tend to merge with those on either side. Multidimensional models work better, especially as we now know that the tropical rainforest varies spatially much more in compositional terms than was at one time supposed. Even in the world of plant ecology, the basis for distinguishing habitats and vegetation types in rainforests is notoriously difficult (Ellen 2007).

In the light of this, therefore, it is perhaps hardly surprising that Shepard, Yu, Lizarralde, and Italiano (2001: 7) provide Matsigenka data that they interpret as indicating no single hierarchy of habitat categories, but rather multiple systems of habitat description, intersecting to define forest types, and with habitat definitions overlapping. Nevertheless, they are sufficiently confident to report that these dimensions generate seventy-six biotically defined habitats, including fifty lowland primary forest types defined by indicator species. The categories of lexically recognized habitats are based on various kinds of criteria: abiotic features, disturbance regimes, soil vocabulary, indicator species (palms, bamboos, ferns, and herbs, as well as trees), characteristic secondary and weedy vegetation, typical montane vegetation types, those defined by overall vegetative aspect, and habitats defined by faunal association. Allan (2002), in her work on the Makushi of Guyana, provides further data that suggest similar intersecting "frameworks," but without any evidence for extensive lexical complexity or, indeed, for the routine expression of such ecological differences in lexical terms. Shepard, Yu, Lizarralde, and Italiano (2001: 27) also suggest that there are a relatively small number of perceptual features underlying this otherwise elaborate scheme: flatland : montane, river mouth : headwaters, river edge : forest interior, weedy secondary growth : primary forest, terrestrial : arboreal, male : female, diurnal : nocturnal, natural : domesticated, and native : introduced. Allan (2002) too notes the small number of perceptual regularities underlying the forest classifications of her informants, and shows how Makushi forest classification can be understood through the intersection of four 'forest frameworks': disturbance history (high bush : low bush), land form (hill : swamp), soil category (sand : clay), and a small number of special associations with indicator species. These (mostly oppositional) criteria are seen to generate the twenty-four categories recorded for characterizing forest: four based on landform, nine based on soil, two to three based on human disturbance history, two based on growth form, and six based on species association or dominance.

What is interesting about these two studies from tropical south America is that between them they indicate a number of organizational features of folk classifications of forest that I suspect are very widespread. Additionally, despite the extensive evidence of multidimensionality and flexibility, they also display continuing underlying assumptions of fixed hierarchical organization, lexical fixity, and shared knowledge, rather than it being suggested that the categories are the outcome of knowledge organized along a small number of dimensions to generate extensive, but essentially ad hoc, terminologies and classifications. I wish here to show that the features highlighted are also evident in Nuaulu classifications of forest variation, but that unlike the Matsigenka, and more like the Makushi, Nuaulu are much more flexible in the way in which limited perceptual characters (such as toponyms) are used to generate organizational frameworks, variable in the classifications that become operative, and that the rather small fixed lexical repertoire of forest types systematically underestimates what people actually "know." It does so because the amount of variation is so great that it cannot, and need not, be handled in a single static classification, which is anyway arguably less cognitively efficient.

Background

The study reported in this paper compares local knowledge of tree vegetation and forest variation in a series of plots widely distributed spatially and in ecologically distinct areas (Ellen 2007). The plot surveys were conducted as part of long-term qualitative and quantitative ethnoecological fieldwork that I have been engaged in since 1970 among the Nuaulu, a people who in 1996 constituted a group of intermarrying clans with a total population of about 2,000, located in south central Seram, Maluku, eastern Indonesia (Figure 5.1, inset). Nuaulu subsist largely through sago extraction, hunting, swidden cultivation, and forest gathering, with some fishing. Historically and conceptually they perceive their subsistence and general cultural orientation as one of forest and mountain rather than of coast and sea. Until the late nineteenth century most Nuaulu lived in dispersed inland settlements (Figure 5.1). Between about 1880 and the 1980s they concentrated in a small number of settlements surrounding the Muslim village of Sepa, though continuing to extract from a wide inland area. However, since the 1980s, the impact of transmigration, road-building, and more recently, civil disturbance, has led once again to a more dispersed pattern of settlement, though one constrained by the new political realities.

The ecological differences between the Nuaulu plots selected reflect altitude, substrate, species composition, and anthropic influence. One of the objectives in surveying them has been to measure the extent to which knowledge varies according to different kinds of forest and geographic area, as well as between different adult male informants, and why. The analysis undertaken demonstrates a high ability to name trees consistently, irrespective of locality and ecology, and a high degree of shared knowledge between male informants. It also illustrates the extent to which Nuaulu understanding of forest diversity and patterning matches recent scientific ecological modeling of rainforest in terms of a complex mosaic rather than as an aggregation of discrete types, and how much of that variation is culturally informed. The data support the claim that simple folk classifications of large numbers of forest types based on the analogy of folk taxonomies of individual plants do not reflect accurately how Nuaulu perceive and encode forest

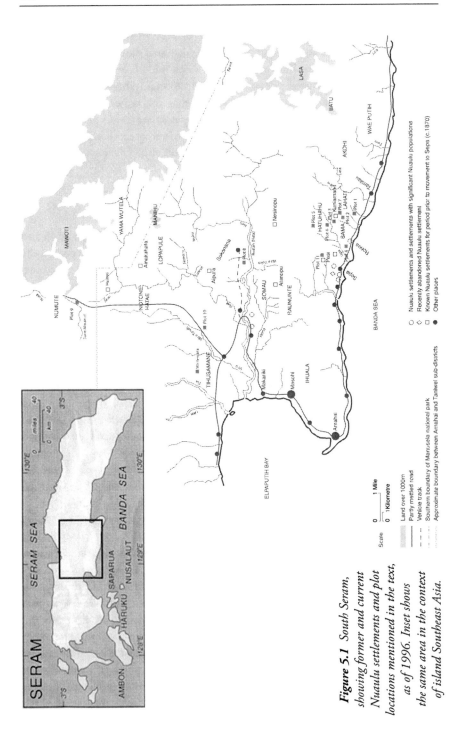

Figure 5.1 South Seram,
showing former and current
Nuaulu settlements and plot
locations mentioned in the text,
as of 1996. Inset shows
the same area in the context
of island Southeast Asia.

differences. Rather, it is suggested that Nuaulu representation of forest is nontaxonomic, constructed on the basis of the intersection of a number of classificatory dimensions based on different criteria. While these multiple dimensions might be represented etically as interacting to generate a large number of discrete 'types', the evidence suggests that they are actually deployed in a more flexible and less mechanical way. Terminologies arising from classificatory stimuli are more likely to be ad hoc descriptions of difference rather than indicating the presence of widely shared and fixed categories. What encodes much of the difference is not recognition of ecological difference per se, but rather its cultural encoding. This relates back to the observation that what underlie many of the differences in lowland tropical forest are impacts made by humans that have short- and long-term ecological consequences.

Nuaulu Ethnoecological Classification of Forest

The structural and compositional complexity of forest, as understood by Nuaulu, is evident from: (1) structured interview data and more discursive qualitative data obtained from both forest hikes and informal village group discussions, (2) patterns of differential use and extraction, and (3) linguistic evidence. The qualitative evidence is referred to in this chapter but not presented in detail. The behavioral data on uses consist of both a cumulative data set for a period of some thirty years in which species are the units of record (the Nuaulu Ethnobotanical Database or NED), and household surveys (over a period of four months) of work inputs and consumption, based on work conducted in 1970–71.

Let us first consider the behavioral data. The entire forested area from which the Rohua Nuaulu (180 persons) extracted during the period in which the 1970–71 fieldwork took place was some 900 square kilometers (Ellen 1978: 61, Map 7). However, Nuaulu temporo-spatial interaction with this vast area of forest is very patchy. Work diary data indicate that more time was spent in two toponymically designated localities than in all other localities put together (some 62 percent of the total number of man days) (1978: 79, Table 12a). Together, these areas occupy about sixty-two square kilometers. Compared with the total area from which Nuaulu extract in, say, one year, this is very small, and the productivity of such areas is accordingly high, though as these are quite large areas, other, smaller terminologically designated localities have a higher productivity when area is divided by the number of days spent extracting (1978: 63, Table 12b). Since the 1970–71 survey, in-migration, population growth and movement, logging, market-driven patterns in forest product extraction, civil disturbance, and the introduction of surfaced roads and therefore improved access have all modified this picture somewhat, particularly through increased levels of extraction in more distant localities, where Nuaulu have either resettled or been allocated space and facilities within transmigration zones created by the provincial government during the 1980s. However, I have no reason to suspect that the general pattern revealed in the 1970–71 data—that extraction varies quite considerably from area to area according to vegetational cover, faunal composition, topography, and relative accessibility—has altered over the intervening thirty years. This patchiness in human extraction patterns matches the patchiness in the ecology of the forest that has been described above, and underpins the way in which Nuaulu perceive forest.

The linguistic evidence includes terminological data on individual forest plants, but also data suggesting the existence of categories grouping different plant species and other distinctive components of forest into broader categories based on ecological, social, and cultural criteria. However, there is a marked contrast between the large number of folk species named and the small number of forest types indicated terminologically. Thus, there are in excess of 214 distinguishable Nuaulu terms for different nondomesticated tree species associated with the most inclusive forest category, *wesie,* compared with a total of 339 Nuaulu terms for trees as a whole coded in the NED. Approximately 156 tree species are listed for the 1996 plots, translating into about 153 distinguishable vernacular terms. Of course, compositional complexity is evident not only in woody tree species that might be considered as nondomesticated, but also in the presence of various domesticates and semidomesticates that are either introduced or self-seeded, plus lianas, epiphytes and other species of the canopy, and ground cover. Overall, in excess of 234 distinguishable Nuaulu terminal plant categories are associated with *wesie.*

Wesie is the most general term applied to forest by Nuaulu.[2] At a symbolic level it connotes vastness, wildness, and uncontrollability. It is opposed to *niane,* village, which is associated with the contrasting qualities of discreteness, tameness and controllability. Indeed, Nuaulu explicitly describe villages as "islands" (*nusa*) within a sea of forest. The symbolic significance of this apparent contrast between the realms of nature and culture has been explored by me elsewhere (Ellen 1996b, 1999a). However, such an apparently rigid symbolic contrast belies a more fluid characterization of forest that embodies cross-cutting distinctions and acknowledges the merging of categories. This too is reflected at a symbolic level, though from this point on my primary concern will be only with the recognition of distinctions of material and practical significance.

Most of what Nuaulu describe as *wesie* covers both lowland rain forest, which stretches from sea level mountainwards, and montane rainforest. Although the Nuaulu recognize zonal variation within *wesie* areas, it is not systematically labeled. The only occasions on which montane zones are traversed are during journeys to North Seram, or on the longer hunting trips to the headwaters of the Nua, Ruatan, Kawa or Lata (Figure 5.1), or in collecting resin from conifer *Agathis dammara,* a particularly prominent feature of the higher forest zones (plot 9). For the Nuaulu the main distinguishing features of *wesie* (largely unowned and uncultivated forest land) and *wasi* (land that is owned and has been cultivated) are the trees from which they are comprised, in particular their size. Thus, almost synonymous with the contrast set *wesie* : *wasi* is the set *ai ia onata* : *ai ia ana wasi*, 'great trees': 'trees of the garden', trees of mature and secondary forest associations respectively. *Wesie* has all the characteristics of *ai ia onata* and more: it is experienced as a source of useful plant materials and of game; it has a mystical dimension, particular visual, olfactory and acoustical qualities (Ellen 1978: 65; c.f. Feld 1996), and its overall ecology is experienced synesthetically.

Here I describe the much more limited Nuaulu lexicon and array of categories applied to 'forest' in a broad etic sense, that for Nuaulu represents a fuzzy conceptual domain with *wesie* at its core, but necessarily includes peripheral transitional vegetational types that overlap cultivated land and patches of nonforest vegetation within areas overall designated *wesie*. In summary, Nuaulu categorization and general cognition of forest reflect

- disturbance history
- land form and substrate
- salient species associations

nuanced in terms of

- land ownership and
- toponyms

I argue that the broadness and flexibility of such a framework, involving a limited set of fixed, shared, lexically labeled categories, do not reflect lack of knowledge of wide-ranging ecological differences in forest, but rather constitute a pragmatic response to the recognition of its complexity.

Terms and Categories Reflecting Disturbance History

Wesie broadly indicates all forest, but narrowly and prototypically is understood as mature forest, far away from human settlement, that has not been modified in recent times, what is sometimes called *wesie huie,* or 'empty forest' (Ellen 1978: 23–4). Once forest is cut by a known individual or household it becomes *wasi,* until such time as the rights which became so established have lapsed or been forgotten about. Clearly such distinctions are fundamental to an understanding of Nuaulu land tenure, and indeed conceptions of tenure have a bearing on Nuaulu ecological perceptions of forest, but for the time being we shall confine ourselves to those distinctions that indicate different kinds and degrees of disturbance. Once cut, individual areas of cultivation are known as *nisi,* 'garden', and these in turn can be divided into three basic types:

(1) *nisi honue,* a recently cleared garden plot up until the end of the first year;

(2) *nisi monai,* a garden after the first year, including sago, clove and coconut groves, and that may (as in the case of coconut) be up to 60 years old;

(3) *nisi ahue,* secondary growth of various kinds.

Of all new gardens created in a single year, on the basis of data for 1969–70, and 1970–71, between 20 and 30 percent are cut from mature forest growth, or rather from what Nuaulu describe as *wesie.* As such gardens tend to be larger than those cut from secondary growth (*nisi ahue*) the percentage area is some-what greater: around 4–5 hectares, or 30 to 36 percent (Ellen 1978: table 21, p. 109). Neither *nisi honue* nor *nisi monai* are in any sense forest in terms of Nuaulu perceptions, though from an external perspective some *nisi monai* might be regarded as agroforest. For the Nuaulu themselves, however, the status of *ahue* is ambiguous, ranging from plots that have been cleared within the last three years to plots that are twenty-five years old or more. But not only is it a temporally wide category it is spatially extensive. Of all land within a four kilometer radius of Rohua village in 1971, totalling some 2430 hectares, 15 percent was *ahue* and 58 percent *wesie.* Of the 102 hectares designated *nisi* and cultivated by Rohua - Nuaulu in 1970–71 23 percent was classified as *nisi ahue* (ibid., p. 185).While a stand of young sago (2–10 years) might be considered *nisi monai,* the same sago grove a few years later might be *ahue.* Despite the ecological distinctiveness and economic importance of depleted forest, and the reasonable assumption that its marginality and ambiguousness might create the kind of blur between the natural and cultural, *wesie* and *wasi,* that elsewhere might be considered symbolically salient, it is not terminological distinct, or even categorically identifiable.

One special category based on disturbance history is indicated by the term *nia monai* (literally, 'old village'). This term may refer to an old settlement that is still inhabited, but also to old settlement sites at different stages following abandonment. Figure 5.1 shows all existing Nuaulu settlements for 1996, and the plots surveyed in the same year. In 1996, *nia monai* was used to describe the site at Aihisuru six years following its abandonment as a village (marked as an open diamond in Figure 5.1). The term was also applied to two areas surveyed: plot 5 at Pesi and plot 11 at Amatene (Figure 5.2)—both about 200 years old—as well as at least seven more old Nuaulu settlement sites (marked as empty squares in Figure 5.1) dated variously within a period of 120 to 500 years since abandonment, and many additional sites associated with other ethnic and political groupings. For example, Amahoru is an old Sepa clan site now mainly *Imperata* grassland, while the site of the old village of Ouh (which at least 500 years ago moved to the nearby island of Saparua) is now entirely covered in bamboo. Amatene (plot 11) was presumably vacated in the 1880s and may have been periodically visited since; its active use today encourages perpetuation of the surrounding resources. The old Numanaeta clan site at Mamokoni, further inland than Amatene, is renowned for its density of sago. As we shall see, the category *nia monai,* while clearly flagging an ecologically distinctive habitat characterized mainly by a high

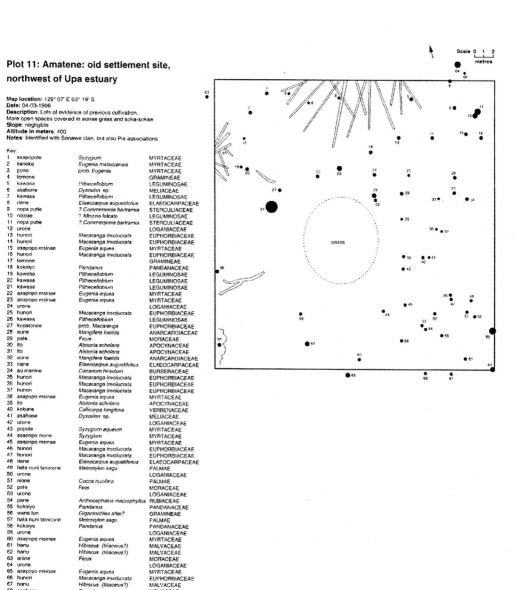

Plot 11: Amatene: old settlement site, northwest of Upa estuary

Map location: 129° 07' E 03° 19' S
Date: 04-03-1996
Description: Lots of evidence of previous cultivation.
More open spaces covered in sonae grass and soka-sokae
Slope: negligible
Altitude in meters: 400
Notes: Identified with Sonawe clan, but also Pia associations

Key:

1	asapopote	Syzygium	MYRTACEAE
2	kanoke	Eugenia malaccensis	MYRTACEAE
3	pono	prob. Eugenia	MYRTACEAE
4	tomone		GRAMINEAE
5	kawasa	Pithecellobium	LEGUMINOSAE
6	asahune	Dysoxion sp.	MELIACEAE
7	kawasa	Pithecellobium	LEGUMINOSAE
8	riane	Elaeocarpus augustifolius	ELAEOCARPACEAE
9	nopa putie	? Commersonia bartramia	STERCULIACEAE
10	nisoae	? Altizzia falcata	LEGUMINOSAE
11	nopa putie	? Commersonia bartramia	STERCULIACEAE
12	urone		LOGANIACEAE
13	hunori	Macaranga involucrata	EUPHORBIACEAE
14	hunori	Macaranga involucrata	EUPHORBIACEAE
15	asapopo msinae	Eugenia aquea	MYRTACEAE
16	hunori	Macaranga involucrata	EUPHORBIACEAE
17	tomone		GRAMINEAE
18	kokoiyo	Pandanus	PANDANACEAE
19	kawasa	Pithecellobium	LEGUMINOSAE
20	kawasa	Pithecellobium	LEGUMINOSAE
21	kawasa	Pithecellobium	LEGUMINOSAE
22	asapopo msinae	Eugenia aquea	MYRTACEAE
23	asapopo msinae	Eugenia aquea	MYRTACEAE
24	urone		LOGANIACEAE
25	hunori	Macaranga involucrata	EUPHORBIACEAE
26	kawasa	Pithecellobium	LEGUMINOSAE
27	kupatonae	prob. Macaranga	EUPHORBIACEAE
28	aune	Mangifera foetida	ANARCARDIACEAE
29	pate	Ficus	MORACEAE
30	ito	Alstonia scholaris	APOCYNACEAE
31	ito	Alstonia scholaris	APOCYNACEAE
32	aune	Mangifera foetida	ANARCARDIACEAE
33	riane	Elaeocarpus augustifolius	ELAEOCARPACEAE
34	au marine	Canarium hirsutum	BURSERACEAE
35	hunori	Macaranga involucrata	EUPHORBIACEAE
36	hunori	Macaranga involucrata	EUPHORBIACEAE
37	hunori	Macaranga involucrata	EUPHORBIACEAE
38	asapopo msinae	Eugenia aquea	MYRTACEAE
39	ito	Alstonia scholaris	APOCYNACEAE
40	kokune	Callicarpa longifolia	VERBENACEAE
41	asahune	Dysoxion sp.	MELIACEAE
42	urone		LOGANIACEAE
43	popole	Syzygium aqueum	MYRTACEAE
44	asapopo mane	Syzygium	MYRTACEAE
45	asapopo msinae	Eugenia aquea	MYRTACEAE
46	hunori	Macaranga involucrata	EUPHORBIACEAE
47	hunori	Macaranga involucrata	EUPHORBIACEAE
48	riane	Elaeocarpus augustifolius	ELAEOCARPACEAE
49	hata nuni taminone	Metroxylon sagu	PALMAE
50	urone		LOGANIACEAE
51	nione	Cocos nucifera	PALMAE
52	pate	Ficus	MORACEAE
53	urone		LOGANIACEAE
54	pane	Anthocephalus macrophyllus	RUBIACEAE
55	kokoiyo	Pandanus	PANDANACEAE
56	wana tun	Gigantochloa atter?	GRAMINEAE
57	hata nuni taminone	Metroxylon sagu	PALMAE
58	kokoiyo	Pandanus	PANDANACEAE
59	urone		LOGANIACEAE
60	asapopo msinae	Eugenia aquea	MYRTACEAE
61	hanu	Hibiscus (tiliaceus?)	MALVACEAE
62	hanu	Hibiscus (tiliaceus?)	MALVACEAE
63	anine	Ficus	MORACEAE
64	urone		LOGANIACEAE
65	asapopo msinae	Eugenia aquea	MYRTACEAE
66	hunori	Macaranga involucrata	EUPHORBIACEAE
67	hanu	Hibiscus (tiliaceus?)	MALVACEAE
68	asahune	Dysoxion sp.	MELIACEAE

Figure 5.2 *Plot 11. Amatene: old village site, northwest of Upa estuary. Much evidence of previous cultivation. Open patches covered in grass and* Selaginella.

density of useful resources, also potently encodes history and a cultural reconstruction of the landscape.

Topography and Substrate

In addition to disturbance history, Nuaulu may also describe forest locations in terms of four categories based on topography and altitude (Ellen 1978: 114, Map 10):

- ***watane:*** flat (areas, the coastal margins, valley floors, alluvium);
- ***sanene:*** valley sides;
- ***pupue:*** ridge land, crests, the higher reaches of valley walls;
- ***tinete, pupue tinete:*** mountains, peaks.

Wesi pupue, therefore, may refer to ridge forest, and **wesi tinete,** to montane forest.

The categorization of forest in terms of its underlying substrate is even more basic. The soil underlying forest in the Nuaulu area, by and large favorable dark-red and reddish brown latosols with little variation, is associated with an alluvial geology containing conglomerates and the characteristic outcrops of coralline limestone ("karang"), or with phyllite formations (Ellen 1978: 27, Map 14). There are some elevated coral reefs and conglomerates in the area, but none of the plots were located on these. The neogenous soft white porous karang is called **nokase** by Nuaulu, the iron stained weathered muscovite schists **ina inate,** the laminated schists **maia,** and iron stained quartzose sandstone **tapune** (**msinae**) (ibid., p. 216, table 41). Soils are either hard or soft, though they may be nuanced by using adjectival qualifiers referring to parent material (e.g. **tapu msinae**). The plots were distributed as follows: coralline limestone (1, 4, 5, 7, 10), sandstone (2), alluvium (6, 8), schists (9), limestone/phyllite (11). Soil in itself is not used to characterize forest, though the presence of **nokase** may be indicative; otherwise topographic terms are a significant encoder of substrate.

Categories Indicating Species-Specific Types

Shared labeled categories indicating species dominance are rare in Nuaulu, though the more salient species in particular areas are certainly recognized. Nevertheless, there is plenty of evidence to suggest that Nuaulu are well aware that the dominance of certain species or groups of plants in a particular area will indicate a particular micro-ecology and compositional pattern. For example, once while traveling to a plot, Nuaulu voluntarily remarked clearly and explicitly on how much **wesie** still existed on the north side of the Upa River, above Nepinama's extensive **nisi ahue,** which was noted as containing much **paune** (a kind of *Pandanus*) and, on another trek, on the density of **mukune** (*Cyathea* spp.) at the place they call Sama. Additionally, Nuaulu are well aware that all tree species have a particular autoecology that includes the presence of other indicative species (see Ellen 1996a). This knowledge can be elicited easily enough via queries based on the interrogative logic of 'if species x, then species y'. For example, in forest dominated by **onie** (*Shorea selanica*) (e.g., plot 3) Nuaulu are aware that unusually, relatively few other species are found, but that those other species, often disproportionately represented, include **niotune** (*Duabanga moluccana*) and **ai bunara** (*Octomeles sumatrana*), and that **kinekane,** the polypod fern *Phymatodes nigrescens,* forms a characteristic understory ground cover. Species or groups of species that generate networks of ecological knowledge of this kind and would seem to justify distinct terminological recognition include *Metroxylon sagu, Canarium,* the large bamboos, the grass *Imperata exaltata,* rattans, *Agathis dammara,* and *Cyathea.* Indeed, it is the case that any patch where a particularly salient species is dominant may be described with a term such as **wesi mukune** (tree-fern forest), **wesi iane** (*Canarium indicum* forest), or **oni-oni** (*Cylindrica exaltata*). Still, we should not mistake these for fixed terms, even though their use evokes widely shared meanings and bodies of knowledge. Interestingly, where there are special terms, these tend to be for deliberate anthropogenic patches, groves, or plantations. Thus, strictly ecological criteria elide with cultivation and ownership categories.

For the most part forest variation is fairly continuous, and the discrete patches that could be described in this way are fairly small in size. However, on some well-trodden trails, the passage from one patch "type" to another may be quite salient and encoded in shared memory as part of the landscape description used to evoke particular places and routes. Thus, in descending from plot 11, the old village site of Amatene, the traveler passes through fourteen zones that are distinguished by Nuaulu according to a diverse selection of indicators: mixed orchard arboriflora, highly indicative of an old habitation site (including *Eugenia, Durio, Bambusa, Gigantochlea, Cocos, Metroxylon*); open bamboo scrub; a belt of **iane** (*Canariun,* with some **meu munate** rattan); depleted **wesie**; **nisi ahue** with **oni** (*Imperata,* 'alang-alang'); a new Rohua Nuaulu garden (**nisi honue**); **nisi ahue** with **kokoiyo** (*Pandanus* sp.); bamboo **ahue**; riparian vegetation along the banks of the small Wae Ia Ana Ikine River; a **tepine** (*Caryota rumphiana*) and rattan belt; an old Bunara Nuaulu garden (**nisi monai**); mixed secondary forest; old garden land; and coastal coconut grove. Some of this diversity is systematically labeled using ethnoecological categories, but most is not.

The Categories of Land Tenure

As we have seen, **wasi** is the local jural term for any patch of land that is owned by a particular clan, and it is subject to the ultimate authority of the Matoke clan, personified in the **ia onate Matoke,** the 'Lord of the Land'. Through its status as a senior and autochthonous clan to which all other clans owe their existence, Matoke is guardian of all land that the Nuaulu consider their original territory. This territory stretches as far as the upland divide between north and south of the island and around the drainage basins of the major Ruatan and Nua rivers (Ellen 1978: 86, maps 2 and 3). The **ia onate Matoke** is regarded as having ultimate jurisdiction over all Nuaulu cultivated lnd and tenurial arrangements, land-people and people-forest relationships. Swiddens cut from forest in the general sense are described as **nisi.** I shall not say anything further here about **nisi honue** (first year gardens) or **nisi monai** (gardens of approximately 2–15 years in age). Only *nisi ahue* (gardens acknowledged as fallow) are considered to be simultaneously gardens and forest, but even here land that might otherwise be described as forest will sometimes be identified using the name of the resource and the clan that owns it: e.g., *hata* Soumori ('Soumori sago'), or generically *hata ipan* or *ipane hatana* ('sago belonging to the clan'). Similarly, sago, or some other resource belonging to the village, becomes *hata niane* ('village sago'). Categories designating particular forest types with the name of the principle floristic component overlap with the terminology used for specialized gardens: thus we have *nisi* (or now sometimes, *rusun,* from Ambonese Malay 'dusun'), *hatane* ('sago garden'), *nisi kanai* ('areca garden'), etc.

A special case of tenure that intrudes into the lexicon to describe different kinds of forest is *sin wesie. Sin(e) wesie* are areas of sacred, protected forest (e.g., plot 7) that are resource-rich and whose composition may reflect age and successional stages of long fallow (cf. Balée and Gély 1989; Fernandez-Gimenez 1993). They are owned at the clan level, though at any one time not all clans in a village will have one. The *sin wesie* sampled in plot 7 at the headwaters of the Mon River belongs to the Peinisa clan. Resources in such areas cannot be harvested without the express permission of a clan chief, and even sago that has reached fruition must be allowed to rot.

Toponyms

Finally, Nuaulu perceive forest through a detailed toponymic grid. The main components of this grid are named rivers, even small creeks, supplemented by names of peaks, hills, prominent rock outcrops, stones, waterfalls, lakes, swamps, caves, and the like. In addition, transient features such as large trees, paths, log bridges, burned patches, and patches of grassland are named, along with the recorded evidence of human activity, such as gardens belonging to particular individuals, old gardens, abandoned gardens and—most importantly—old village sites, and sites of some locational specificity such as sites within the sago forest at Somau (*Kamnanai ukune* or *Nusi ukune*).

The extensive character of participatory mapping exercises elsewhere has shown just how detailed this knowledge can be. In the Nuaulu case, evidence of the geographic extensiveness of such toponymic knowledge comes not only from the mapping exercises that I have conducted with Nuaulu over a period of some thirty years, but from maps produced by the Dutch Topographische Dienst in 1917. A large swathe of Central Seram, approximately corresponding to the area that Nuaulu clans claim as their territory, is annotated with river and mountain names that are clearly, and for the most part, Nuaulu. These names were presumably generated by the mapmakers surveying with Nuaulu guides in the first decade of the twentieth century. Working with Nuaulu from 1970 onward, and particularly in 1996, I was struck—as were the Nuaulu themselves—by the congruence between current Nuaulu knowledge as indicated in culturally annotated sketch maps that Nuaulu produced for me (cf. Fernandez-Gimenez 1993) and the toponyms provided in the 1917 Dutch map. On the whole, as a reference system, these toponyms begin with the names of particular mountains on the one hand and large rivers on the other. However, the mountains or hills are fixed points that also give their names to large areas surrounding them. Similarly, the large rivers indicate extensive riparian and valley areas rather than simply the rivers themselves. Linking a river name with a mountain name provides general coordinates, which are then refined by referring to tributaries of the main rivers and to tributaries of tributaries. Only when this set of coordinates is insufficient to locate places will other indicators be introduced. This set of toponyms serves to identify particular patches of forest and to some extent bypasses the need to identify forest in terms of its distinctive floristic composition or habitat structure. The toponymic references manifestly indicate the investment of history in the description of a particular landscape, and they are no better revealed than through the narrative associations of old village sites. A stretch of vegetation is never seen as an example of some generic ahistorized type, but rather as a place whose character must be understood only through its particular historical associations and the overall "cultural density" of the landscape (Brosius 1986: 175).

Plot Analysis in Relation to Local Knowledge and Use of Forest

Table 5.1 summarizes selected 1996 plot data reflecting Nuaulu botanical knowledge in relation to levels of identification. The first point to note is the completeness of Nuaulu terminological data. Of all the trees over 10 cm dbh in all the plots there is a mean level of identification of 97.5 percent, with seven of the eleven plots yielding 100 percent agreed identifications. As might be expected, identifications in the more remote plots proved more difficult, including the most distant plot 9 in a submontane area, which yielded a 94 percent identification rate. Some species were unfamiliar to Nuaulu assistants in this plot, but a more practical problem proved to be tall trees with little leaf visible below canopy level.

Table 5.2 summarizes 1996 plot data relating to tree uses. The measure employed is a crude one that scored trees in each of the plots in terms of whether they had a volunteered use at all and then whether uses could be allocated to one of more of three aggregate categories: food, medicinals, and physical equipment. How some of the more marginal specific uses of trees were allocated between these three categories is explained in the key accompanying the table.

However, when we look at the plot descriptions using the general words employed by Nuaulu to describe their overall character, five are described as *wesie,* two as *nisi ahue,* two as *niamonai* and one as *sin wesie* (Table 5.3). In describing the vegetation in these plots no terms other than those applied to

Table 5.1 Summary of Selected 1996 Plot Data: Levels of Identification

Plot	1	2	3	4	5	6	7	8	9	10	11
1	N	34	30	37	26	28	54	124	116	80	35	68
2	N with Nuaulu name[a]	34[b]	28	37	26	28	54	124[b]	115	74	35	68[b]
3	Percentage N identified with agreed Nuaulu name	100	93	100	100	100	100	100	99	94	100	100
4	N Identified to family level	30	25	35	25	16	50	108	98	54	25	68
5	N identified to generic level	20	21	32	24	16	49	103	90	37	25	58
6	N identified to species level	5	12	19	24	1	38	46	41	29	12	40

Key: N = number of standing trees 10 dbh or above; a = name agreed by minimum of three adult male field assistants; b = one doubtful

Table 5.2 Summary of Selected 1996 Plot Data: Selected and Aggregated Uses

Plot	1	2	3	4	5	6	7	8	9	10	11
N trees above 10 dbh	34	30	37	26	28	54	124	116	80	35	68
N species	20	21	16	10	9	22	47	33	28	21	26
N species with indicated uses	14	17	12	9	4	18	28	17	15	10	19
N trees with indicated uses	22	23	31	25	5	45	78	56	39	20	55
N species used for food	1	3	4	5	0	10	6	5	1	2	9
N trees used for food	1	4	19	21	0	15	16	18	1	2	20
N species used medicinally	1	2	0	2	0	2	2	0	0	1	1
N trees used medicinally	1	2	0	2	0	4	2	0	0	1	1
N species used for equipment	11	12	11	6	3	14	21	15	11	9	14
N trees used for equipment	19	14	29	6	4	36	64	42	32	19	34

Note: Fuel is excluded as a use. Food includes additives such as spices. Medicinal uses include stimulants and cosmetics. Equipment includes manufactured items of all sizes, clothing and building timber.

Table 5.3 General Characteristics of Nuaulu Forest Patch Categories Based on 1996 Plot Data

Nuaulu description	Plots indicated	N plots	Mean S	Mean % N useful trees	Mean % S useful trees
1. *wesie*	2, 3, 8, 9, 10	5	25 (16, 22, 27, 28, 32)	57	61
2. *nisi ahue*	1, 4, 6	2	18 (10, 20, 23)	81	77
3. *nia monai*	5, 11	2	48 (28–68)	26	66
4. *sin wesie*	7	1	50	63	60

individual kinds of plant were elicited, despite demonstrably articulate descriptions provided by Nuaulu co-researchers of the relations between species found and the general ecological characteristics of the plots. Table 5.3 attempts to show the aggregate "usefulness" of four Nuaulu categories applied to forest at the plot sites by measuring the number of trees listed with uses in the Nuaulu Ethnobotanical Database. The data are indicative rather than definitive, and more plots, over a wider spectrum of forest variation and of larger size, would provide a test of the plausibility of these patterns. For index of species richness (S), *sin wesie* (protected sacred forest: e.g., Figure 5.3) was found to have the highest value, followed by *nia monai* (old village sites), and then *wesie* (general mature forest) and *nisi ahue* (long-term fallow). These data provide evidence to support the hypothesis that long-term anthropic influences increase local tree diversity. If, however, we measure mean percentage of trees listed as "useful" against the total number of trees in plots of that category, and against the index of species richness, we find that recent long fallow, *nisi ahue,* provides the greatest concentration of resources in areas most influenced by humans by whatever measure. While old village sites score well when total numbers of useful species are determined, and the old Sonaue village site of Amatene (plot 11) was still being regularly visited for its *Durio, Canarium,* and *Artocarpus* in 1996, the percentage of actual trees indicated as useful is markedly the lowest of all the categories plotted. And in terms of species diversity and number of useful trees, protected sacred forest (*sin wesie*) is much like any *wesie,* which suggests that forest may not always be selected for special protection because it has a larger concentration of resources than any other area of forest, but rather because of its accessibility—physically and socially—to a particular clan. The usefulness figures for *sin wesie* also suggest that forest is not particularly likely to be protected because of its higher degree of anthropic influence. Finally, the data show wide variation within the category *wesie,* which is most unlikely to be maximally reflected in the plots selected in the 1996 survey.

Plot 7: Part of sin wesie Peinisa (protected forest) at Mon sanae

Map location: 129° 10' E 03° 20' S
Date: 21-02-1996
Description: Edge of protected area
Slope: 20°
Altitude in meters: 300
Notes:
The boundaries of this *sin wesie* (protected sacred forest) go many kilometres inland and include large areas of sago (*Metroxylon sago*).

—— = rock edges
..... = stream bed

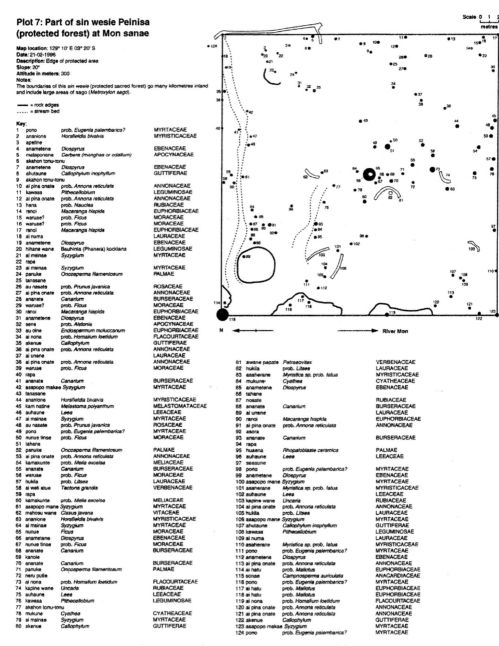

Key:
#	Name	Identification	Family
1	pono	prob. *Eugenia palembarica*?	MYRTACEAE
2	ananione	*Horsfieldia bivalvis*	MYRISTICACEAE
3	apetine		
4	anametene	*Diospyrus*	EBENACEAE
5	mataponane	*Gerbera (manghas or odallum)*	APOCYNACEAE
6	akahon tonu-tonu		
7	anametene	*Diospyrus*	EBENACEAE
8	ahutaune	*Callophylum inophyllum*	GUTTIFERAE
9	akahon tonu-tonu		
10	ai pina onate	prob. *Annona reticulata*	ANNONACEAE
11	kawasa	*Pithecellobium*	LEGUMINOSAE
12	ai pina onate	prob. *Annona reticulata*	ANNONACEAE
13	hana	*Nauclea*	RUBIACEAE
14	ranoi	*Macaranga hispida*	EUPHORBIACEAE
15	waruae?	prob. *Ficus*	MORACEAE
16	waruae?	prob. *Ficus*	MORACEAE
17	ranoi	*Macaranga hispida*	EUPHORBIACEAE
18	ai numa		LAURACEAE
19	anametene	*Diospyrus*	EBENACEAE
20	hihane wane	*Bauhinia (Phanera) kockiana*	LEGUMINOSAE
21	ai msinae	*Syzygium*	MYRTACEAE
22	rapa		
23	ai msinae	*Syzygium*	MYRTACEAE
24	panuke	*Oncosperma filamentosum*	PALMAE
25	tanasane		
26	au nasate	prob. *Prunus javanica*	ROSACEAE
27	ai pina onate	prob. *Annona reticulata*	ANNONACEAE
28	ananate	*Canarium*	BURSERACEAE
29	waruae?	prob. *Ficus*	MORACEAE
30	ranoi	*Macaranga hispida*	EUPHORBIACEAE
31	anametene	*Diospyrus*	EBENACEAE
32	sene	prob. *Alstonia*	APOCYNACEAE
33	au oine	*Endospermum moluccanum*	EUPHORBIACEAE
34	ai nona	prob. *Homalium foetidum*	FLACOURTIACEAE
35	akenue	*Callophylum*	GUTTIFERAE
36	ai pina onate	prob. *Annona reticulata*	ANNONACEAE
37	ai unene		LAURACEAE
38	ai pina onate	prob. *Annona reticulata*	ANNONACEAE
39	waruae	prob. *Ficus*	MORACEAE
40	rapa		
41	ananate	*Canarium*	BURSERACEAE
42	asapopo makae	*Syzygium*	MYRTACEAE
43	tanasane		
44	ananione	*Horsfieldia bivalvis*	MYRISTICACEAE
45	kam natine	*Melastoma polyanthum*	MELASTOMATACEAE
46	auhaune	*Leea*	LEEACEAE
47	ai msinae	*Syzygium*	MYRTACEAE
48	au nasate	prob. *Prunus javanica*	ROSACEAE
49	pono	prob. *Eugenia palembarica*?	MYRTACEAE
50	nunue tinse	prob. *Ficus*	MORACEAE
51	lahane		
52	panuke	*Oncosperma filamentosum*	PALMAE
53	ai pina onate	prob. *Annona reticulata*	ANNONACEAE
54	kamakunte	prob. *Melia excelsa*	MELIACEAE
55	ananate	*Canarium*	BURSERACEAE
56	waruae	prob. *Ficus*	MORACEAE
57	hukila	prob. *Litsea*	LAURACEAE
58	ai weti atue	*Tectona grandis*	VERBENACEAE
59	rapa		
60	kamakunte	prob. *Melia excelsa*	MELIACEAE
61	asapopo mane	*Syzygium*	MYRTACEAE
62	mahosu wane	*Cissus javana*	VITACEAE
63	ananione	*Horsfieldia bivalvis*	MYRISTICACEAE
64	ai msinae	*Syzygium*	MYRTACEAE
65	nunue	*Ficus*	MORACEAE
66	anametene	*Diospyrus*	EBENACEAE
67	nunue tinse	prob. *Ficus*	MORACEAE
68	ananate	*Canarium*	BURSERACEAE
69	kanole		
70	ananate	*Canarium*	BURSERACEAE
71	panuke	*Oncosperma filamentosum*	PALMAE
72	neru putie		
73	ai nona	prob. *Homalium foetidum*	FLACOURTIACEAE
74	kapine wane	*Uncaria*	RUBIACEAE
75	auhaune	*Leea*	LEEACEAE
76	kawasa	*Pithecellobium*	LEGUMINOSAE
77	akahon tonu-tonu		
78	mukune	*Cyathea*	CYATHEACEAE
79	ai msinae	*Syzygium*	MYRTACEAE
80	akenue	*Callophylum*	GUTTIFERAE
81	awane papate	*Petraeovitex*	VERBENACEAE
82	hukila	prob. *Litsea*	LAURACEAE
83	asaherane	*Myristica sp.* prob. *fatua*	MYRISTICACEAE
84	mukune-	*Cyathea*	CYATHEACEAE
85	anametene	*Diospyrus*	EBENACEAE
86	tahane		
87	nosate		RUBIACEAE
88	ananate	*Canarium*	BURSERACEAE
89	ai unene		LAURACEAE
90	ranoi	*Macaranga hispida*	EUPHORBIACEAE
91	ai pina onate	prob. *Annona reticulata*	ANNONACEAE
92	asora		
93	ananate	*Canarium*	BURSERACEAE
94	rapa		
95	huaena	*Rhopaloblaste ceramica*	PALMAE
96	auhaune	*Leea*	LEEACEAE
97	seaaune		
98	pono	prob. *Eugenia palembarica*?	MYRTACEAE
99	anametene	*Diospyrus*	EBENACEAE
100	asapopo mane	*Syzygium*	MYRTACEAE
101	asaherane	*Myristica sp.* prob. *fatua*	MYRISTICACEAE
102	auhaune	*Leea*	LEEACEAE
103	kapine wane	*Uncaria*	RUBIACEAE
104	ai pina onate	prob. *Annona reticulata*	ANNONACEAE
105	hukila	prob. *Litsea*	LAURACEAE
106	asapopo mane	*Syzygium*	MYRTACEAE
107	ahutaune	*Callophylum inophyllum*	GUTTIFERAE
108	kawasa	*Pithecellobium*	LEGUMINOSAE
109	ai numa		LAURACEAE
110	asaherane	*Myristica sp.* prob. *fatua*	MYRISTICACEAE
111	pono	prob. *Eugenia palembarica*?	MYRTACEAE
112	anametene	*Diospyrus*	EBENACEAE
113	ai pina onate	prob. *Annona reticulata*	ANNONACEAE
114	ai hatu	prob. *Mallotus*	EUPHORBIACEAE
115	sonae	*Camponosperma auriculata*	ANACARDIACEAE
116	pono	prob. *Eugenia palembarica*?	MYRTACEAE
117	ai hatu	prob. *Mallotus*	EUPHORBIACEAE
118	ai hatu	prob. *Mallotus*	EUPHORBIACEAE
119	ai nona	prob. *Homalium foetidum*	FLACOURTIACEAE
120	ai pina onate	prob. *Annona reticulata*	ANNONACEAE
121	ai pina onate	prob. *Annona reticulata*	ANNONACEAE
122	akenue	*Callophylum*	GUTTIFERAE
123	asapopo makae	*Syzygium*	MYRTACEAE
124	pono	prob. *Eugenia palembarica*?	MYRTACEAE

N ◄———————————————► River Mon

Figure 5.3 Plot 7. Part of **sin wesie** Peinisa (*protected forest*) at Mon sanae.

Discussion: Why Aren't the Nuaulu Like the Matsigenka?

On the basis of the data provided in this chapter we can conclude that Nuaulu forest classification conforms to a model in which there is evidence for subtle and extensive understanding of variation, and of the ecological properties of different vegetative associations. However, only a small number of categories are shared and consistently organized, and there is a low degree of formal lexicalization. Systematic data on ecological knowledge and linguistic evidence indicate that:

- The categories *wesie* (forest) and *nisi ahue* (long fallow) absorb a great deal of variation.
- Disturbance history is the main and unifying basis for local understandings of variation, modified by occasional considerations of topography and substrate. Nuaulu perceive all forest as in a state of flux, and recognize that this flux is in large part due to interaction with humans.
- Some stable categories are associated with specific species, but named categories of this kind are rare.
- Perception and classification of forest are inseparable from categories dividing forest in terms of patterns of ownership and the cultural reconstruction of landscape reflected in the use of toponyms.

Overall, what is remarkable about the classification of forest types is the *small* number of formal categories, the *low* degree of lexicalization that could be elicited despite plot data indicating the objective ecological complexity and patchiness of forest, and the flexibility of content locally permitted with respect to the Nuaulu categories *wesie* (forest) and *nisi ahue* (long fallow) that describe them. While there are few widely shared terms systematically glossing types of forest composition and structure, there is good evidence from participant observation that this does not prevent Nuaulu from being aware of, and fluently talking about, often quite subtle variation in floristics and in the ecological properties of different vegetative associations. Some of these conclusions now need to be explored in a bit more detail.

I do not think we should be surprised by the small number of labeled categories given the ecological and social conditions under which Nuaulu live, and given the body of existing contemporary evidence in cognitive anthropology that challenges some of the more simplistic versions of the Sapir-Whorf and Nida-Conklin hypotheses. In this context, I wish to argue that no useful purpose is served by a detailed classification of types, as opposed to recognition of some loose, intersecting dimensions of variation. This finding is consistent with the observation that where peoples' folk knowledge is extensive and profound, and where there is a high degree of "semantic" contact between people and a nuanced, complex perceptual environment, that is experienced intricately and intimately on a daily basis, extensive, systematic, and formal lexical codifications are often the least satisfactory way of organizing that knowledge (Ellen 1999b). Indeed, they may be too inflexible and cumbersome to be a particularly productive way of retrieving and sharing information. In such situations, and especially where, as in this case, the subject of classification is highly variable and patterns of extraction are very individualized (rather than collectivized), and there is evidence of much intracultural variation, local categories need to be sufficiently flexible to incorporate personal experience, which may be more important than transmission of a formalized body of knowledge (e.g., Bulmer 1974; Allan 2002: 179; Sillitoe 2002).

The data presented here indicate the difficulty of separating categories based on ecology, resource use, and social and cultural criteria, and thus challenge the idea of an "ethnoecological classification" independent of other factors relating to land and vegetation that people see as being important. All forest is perceived as through a social grid: moving through the forest constantly reinforces particular parts of its history as they have impinged on Nuaulu lives: the location of past villages, the territories of clans, or the patterns of extraction of particular individuals. In the same way that it is difficult to isolate a separate "technical" lexicon of forest types based solely on ecological criteria, so it is difficult to separate a classification of "forest types" from general vegetation types and other land use categories. Forest, in the sense of *wesie,* is a fundamental cultural category with profound symbolic resonance, but in terms of people's ecological understanding it is continuous with other vegetative associations, including those created or modified by the Nuaulu themselves. However, none of this is to deny the rich understandings of forest ecology embedded and encoded in everyday experience and in superficially non-ecological language.

The analysis also demonstrates that ethnoecological categories are probably best not approached as analogues of folk classifications of individual organisms. There is little evidence of conventional taxonomic organization; categories are only weakly contrastive, seldom easily identifiable as cognitive prototypes, and much better seen as ad hoc categories in which criteria clustering may better describe their mode of cognition. There are particular dangers in deliberately seeking to find complex classifications of vegetation types that resemble the conventions of scientific ecology in their mode of organization, at the very least because these latter conform to a narrow linguistic logic derived from the conventions of written language as these have been absorbed into Western science. To assume a priori that the underlying mode of organization is such is to fall into the trap of the linear-sentential fallacy (Bloch 1991). How categories based on radically different kinds of criteria interact, and are presented in a field situation, will in the end be reflected by the way and context in which questions are asked that yield the data, which is basically the way the categories work and are shared amongst local decision-makers, a process I have elsewhere called *prehension* (Ellen 1993: 229–234).

While we can agree that the frameworks of categories used to describe forest are based on essentially simple, sometimes binary, distinctions, we must be wary of rejecting a conventional "ethnotaxonomic" mode of representation only to replace it with a model of intersecting dimensions that is imagined as a matrix-like structure, capable of generating a larger number of categories simply by combining the dimensions of the matrix. Thus, when Allan (2002) derives twenty-four different categories for Makushi characterizations of forest, by linking terms for landform, soil, human disturbance history, growth form, and species association or dominance, we have to interpret such data with care. We should neither assume that we can generate additional "local" categories by the simple logical manipulation of other categories that appear to cross-cut them, nor give undue, or even spurious, arithmetical or linguistic fixity to particular numbers of labeled categories discovered. While Nuaulu data reveal numerous simple contrasts of this type, I am skeptical as to the extent to which they can be formally constituted into a grid. There is not much cognitive evidence for this. Methodologically, we should operate with caution when it comes to cognitive models of dichotomous frameworks, treating them (until sufficient evidence to the contrary) as nothing other than a useful technique for yielding the range of criteria that might be involved in local classificatory behavior. Such models are a useful preliminary analytic device to get some measure of the data, rather than a definitive tool for yielding certain knowledge of how people cognize their environment.

I want to suggest that the general model of understanding forest variation that I have elucidated here may apply more widely, and is—if you will—the default way in which tropical forest peoples comprehend and describe forest variation more generally. Because of the strong ecological and cultural similarities, a good starting point for any comparison is other material relating to the island of Seram. Conveniently, two studies exist of Alune (West Seram) vegetation classification conducted in two different areas, that of Suharno (1997: 153, 173) in the Eti-Lumoli area, and that by Florey and her associates in Lohiasapalewa and Lohiatala (Wolff and Florey 1998; Florey, personal communication). We find here that the mixture of criteria, the difficulty of distinguishing forest from other kinds of vegetation, and the primacy accorded disturbance regime are the same as for the Nuaulu. However, on the face of it we also find a greater terminological specification of different types of forest. This is in part due to more routine resource extraction from higher altitudes than is the case with the Nuaulu (an ecological difference), but it is also partly an artifact of the way in which the lexical data are presented. Thus the prefix *lusun* can apparently be attached to any number of different cultivated trees grown in groves: as the number of tree species incorporated as adjectival qualifiers varies, so does the recorded number of 'vegetation types'. It is also partly accounted for by including specific ecological descriptors for features which are not of themselves forest types (e.g. *tape ial uwei,* 'rich soil found around *Canarium commune*', or *lasa porole,* 'infertile ground around *Intsia bijuga*'). It is impossible to say, for example, to what extent the categories *mosole* (lower altitude montane forest) and *lasa porole,* are commensurable.

However, it is with respect to categories for forest regeneration—the dimension of disturbance history—where we can see some really interesting differences.

Thus, the Alune term for the first year of regeneration is **kwesie buini,** that for forest of 4–8 years regeneration, **kwesie,** and (in contrast) for dense forest, **a ela uei** (**ela** = big, **uei** = base of tree) (Suharno 1997: 153, 173). Thus, although the Alune term **kwesie** is clearly cognate with Nuaulu **wesie,** its meaning is subtly different, and whereas Nuaulu **wesie** is all forest where specific garden histories are no longer traceable, **kwesie** refers specifically to regrowth where specific histories are traceable. Indeed, Nuaulu forest that may not have been cultivated for twenty years, but whose history is still traceable, is still a kind of garden, **nisi ahue.** The Alune term for garden, **ndinu,** refers only to the few years that a garden is being actively cultivated.

Other terminologies from other parts of island Southeast Asia are more consistent in having a single overarching term for forest, and in dividing this up into three to four stages of regrowth. We have already noted that, in his Penan Benalui work, Puri (1997: 104) elicited just eight categories for types of forest, despite systematic attempts to delineate more. However, he notes that further discriminations are possible by plotting these against a riverine (**la' bai**) : inland (**la' daya**) contrast. The geographically close Lundayeh in Sabah divide up the semantic domain forest (**fulung**) into three categories of regrowth: **amog darii** (less than ten years), **amog karar** (10–20 years), and **amog balui** (20–50 years). Forest where histories of cultivation are no longer traceable is known as **fulung karar** (Hoare 2002). Thus, some peoples have a generic term for cultivated land that includes productive and traceable regrowth (like the Nuaulu), while others have a separate term for regrowth that encompasses periods of different regrowth duration: eight years in the case of the Alune and fifty years in the case of the Lundayeh. These differences may reflect the local importance attached to different kinds of extraction, pressure on resources, and general subsistence orientation. Different peoples terminologically distinguish "true" forest from "ancient garden" in different ways, presumably because there are local differences in the correlation between usefulness and the age of a plot that was previously a swidden (Grenand 1992: 32).

Common Themes and Emerging Patterns

In looking more generally at the ethnoecological classifications used by tropical forest peoples to describe forest diversity, Shepard et al. (2001: 31–32) have noted that "several common themes and patterns emerge ... [and suggest] an overall pattern of extraordinary concordance between habitat classification by culturally distinct and geographically separate groups." Their comparative data were largely Amazonian, but some of their core generalizations are upheld against the island Southeast Asian data examined here. They suggest that abiotic factors (topography, flooding and disturbance regimes, soils) are used to distinguish a small number of general categories. The distinction between river/coast and uplands is found in all indigenous systems. This is borne out in Allan's work and is reflected in the Nuaulu data presented here. The suggestion by Shepard and his co-authors that the distinction between primary forest and secondary forest, including various stages of swidden fallow regeneration, also appears as a salient category in all systems is confirmed, and indeed disturbance history turns out to be the single most important dimension in classifying forest for people engaged in swidden cultivation. Indeed, within the range of variation described we detect a pan-cultural and geographically widespread conceptual model based on a limited number of dimensions of perceived experience.

Where our own data do not conform with the picture that emerges from these Amazonian studies, it is in terms of overall category differentiation, degree of lexicalization, and, most specifically, in relation to the claim of the extent to which biotic features—mostly indicator plant species—are used to define more specific habitat types. Even in tropical South America the pattern is far from uniform. For example, frameworks based on species association were described by only 4 percent of Allan's Makushi participants (2002: 154). While we do not disagree that the concept of indicator species is widely understood by tropical forest peoples, and that substantive knowledge relating to recognized differences based on these is often extensive, we are not convinced that this is universally translated into systematic lists of labeled categories. Our own data also show the difficulties of eliciting ethnoecological classifications that are independent of distinctions based on use strategies, land tenure, and cultural significance. We are concerned about the dangers of treating complex multidimensional landscape categories in the same way that many have analyzed folk classifications of species. We suggest that in many cases we should not expect the degree of shared systematic categorization implied in the Matsigenka data, and indeed expect people to lexicalize their environment more flexibly and with more limited shared encoding.

Acknowledgments

The fieldwork reported here was undertaken mainly in 1996 but also draws on long-term fieldwork conducted between 1969 and 1971, and for shorter periods in 1973, 1975, 1981, 1986, and 1992. On all occasions it has been sponsored by the Indonesian Institute of Sciences (LIPI: Lembaga llmu Pengetahuan Indonesia) and, more recently, also by Pattimura University Ambon and the Maluku Study Centre attached to the university. Financial support between 1969 and 1992 came from a combination of the former UK Social Science Research Council, the London-Cornell Project for East and Southeast Asian Studies, the Central Research Fund of the University of London, the Galton Foundation, and the Hayter Travel Awards Scheme. Most recently, directly and substantially, it relies on the support of ESRC (Economic and Social Research Council) grant R000 236082 for work on "Deforestation and forest knowledge in south central Seram, eastern Indonesia." Additionally, for this work I need to acknowledge the support of the Herbarium Bogoriense and in particular Dr. Johannis Mogea, and of Dr. Johan Iskandar of the Institute of Ecology, Padjadjaran University, Bandung. It also draws on research conducted for the European Commission DG-8–funded program entitled "L'Avenir des Peuples des Fôrets Tropicales," of which UKC was a consortium partner, and on an earlier ESRC grant (R000 23 3088) for work on "The ecology and ethnobiology of human-rainforest interaction in Brunei (a Dusun case study)." I am indebted to Christine Eagle and Lesley Farr for help with the maps, production of the plot diagrams, and work on the development of the ethnobotanical database on which this work depends. Simon Platten and Amy Warren have also provided valuable assistance in data processing. Christie Allan and Margaret Florey kindly permitted me to refer to unpublished material. Finally, I am grateful for the continuing and enthusiastic involvement and support of Nuaulu friends and co-researchers in Rohua. A version of this essay was presented at the Eighth International Congress of Ethnobiology held in Addis Ababa in 2002, and I am indebted to the British Academy for a grant which made attendance at this event possible.

Notes

1. Paul Sillitoe (2002, and personal communication) has suggested that large numbers of formal categories might also imply the presence of some overarching authority to define them and arbitrate in disputes, whereas a flexibly constituted understanding is more consistent with a decentralized sociopolitical system. I can think of one clear example, from a nonliterate animist population, of formal control over categories for certain kinds of primary biodiversity: the validation of the names for new landraces of rice by Kasepuhan ritual specialists in upland West Java (Soemarwoto 2007: 96–105). Something similar might well apply to categories of secondary biodiversity in populations where agricultural decision-making is linked to a complex divided landscape, such as for Balinese irrigation associations. However, I remain to be convinced that the distinction would be helpful in the context of the present discussion. If anything, Nuaulu political and ritual authority is more formal than that of the Matsigenka or the Wola. Nuaulu patrilineal clans are divided into complementary "houses," each with ascribed leadership in a relation of formal diarchy with the other. In turn, clans are arranged in a mythically legitimated order of precedence, where the head of the most senior clan acts as a *primus inter pares*. Ritual and political authority does not necessarily or systematically translate into technical authority in matters relating to ethnoecological categories, except in literate societies.
2. *Apane* appears as a synonym for **wesie.** Thus the leguminous vine *Derris trifoliata*, used locally as a fish poison, is described variously as **awane munu apane,** **(awane) munu wesie,** and **munu apane.**

References

Allan, C. L. 2002. *Amerindian Ethnoecology, Resource use and Forest Management in Southwest Guyana.* Unpublished PhD thesis, University of Surrey (Roehampton).

Balée, W., and A. Gély. 1989. "Managed Forest Succession in Amazonia: The Ka'apor Case." In *Resource Management in Amazonia: Indigenous and Folk Strategies,* ed. D. A. Posey and W. Balée. *Advances in Economic Botany* 7: 129–158.

Bartlett, H. H. 1936. "A Point of View and a Method for Rapid Fieldwork in Tropical Phytogeography." *Botany of the Maya Area.* Miscellaneous Papers 1–13. Carnegie Institution.

Berlin, B., D. E. Breedlove, and P. H. Raven. 1973. "General Principles of Classification and Nomenclature in Folk Biology." *American Anthropologist* 75: 214–242.

Bloch, M. 1991. "Language, Anthropology and Cognitive Science." *Man* 26(2): 183–198.

Brosius, J. P. 1986. "River, Forest and Mountain: The Penan Gang Landscape." *Sarawak Museum Journal* 36(57) (New Series): 173–184.

Bulmer, R. N. H. 1974. "Folk Biology in the New Guinea Highlands." *Social Science Information* 13(4/5): 9–28.

Carneiro, R. 1978. "The Knowledge and Use of Rainforest Trees by the Kuikuru Indians of Central Brazil." In *The Nature and Status of Ethnobotany,* ed. R. I. Ford. Anthropological Papers, No. 67. Ann Arbor: Museum of Anthropology, University of Michigan.

Conklin, H. C. 1954. "An Ethnoecological Approach to Shifting Agriculture." *Transactions of the New York Academy of Sciences* 17: 133–142.

Ellen, R. F. 1978. *Nuaulu Settlement and Ecology: The Environmental Relations of an Eastern Indonesian Community.* Verhandelingen van het Koninklijk Instituut voor Taal-, Land- en Volkenkunde No. 83. The Hague: Martinus Nijhoff.

Ellen, R. F. 1993. *The Cultural Relations of Classification: An Analysis of Nuaulu Animal Categories from Central Seram.* Cambridge: Cambridge University Press.

Ellen, R. F. 1996a. "Putting Plants in Their Place: Anthropological Approaches to Understanding the Ethnobotanical Knowledge of Rainforest Populations." In *Tropical Rainforest Research - Current Issues,* ed. D. S. Edwards, W. E. Booth, and S. C. Choy. Dordrecht: Kluwer.

Ellen, R. F. 1996b. "The Cognitive Geometry of Nature: A Contextual Approach." In *Nature and Society: Anthropological Perspectives,* ed. P. Descola and G. Palsson. London: Routledge.

Ellen, R. F. 1999a. "Forest Knowledge, Forest Transformation: Political Contingency, Historical Ecology and the Renegotiation of Nature in Central Seram." In *Transforming the Indonesian Uplands: Marginality, Power and Production,* ed. T. Li. Amsterdam: Harwood.

Ellen, R. F. 1999b. "Modes of Subsistence and Ethnobiological Knowledge: Between Extraction and Cultivation in Southeast Asia." In *Folkbiology,* ed. D. L. Medin and S. Atran. MIT Press.

Ellen, R. F. 2007. "Local and Scientific Understandings of Forest Diversity on Seram, Eastern Indonesia." In *Local Science vs Global Science: Approaches to Indigenous Knowledge in International Development,* ed. P. Sillitoe. Oxford: Berghahn.

Feld, S. 1996. "A Poetics of Place: Ecological and Aesthetic Co-evolution in a Papua New Guinea Rainforest Community." In *Redefining Nature: Ecology, Culture and Domestication,* ed. R. Ellen and K. Fukui. London: Berg.

Fernandez-Gimenez, M. 1993. "The Role of Ecological Perception in Indigenous Resource Management: A Case Study from the Mongolian Forest-Steppe." *Nomadic Peoples* 33: 31–46.

Fleck, D. W. and J. D. Harder. 2000. "Matses Indian Rainforest Habitat Classification and Mammalian Diversity in Amazonian Peru." *Journal of Ethnobiology* 20(1): 1–36.

Grenand, P. 1992. "The Use and Cultural Significance of the Secondary Forest Among the Wayapi Indians." In *Sustainable Harvest and Marketing of Rain Forest Products,* ed. M. Plotkin and L. Famolare. Washington, D.C.: Island Press.

Hoare, A. L. 2002. *Cooking the Wild: The Role of the Lundayeh of the Ulu Padas (Sabah, Malaysia) in Managing Forest Foods and Shaping the Landscape.* Unpublished PhD dissertation, University of Kent at Canterbury.

Martin, M. 1974. "Essai d'ethnophytogeographie Khmere." *Journal D'Agriculture Tropicale et de Botanique Appliquée* 22(7–9): 219–238.

Meilleur, Brien A. 1986. "Alluetain Ethnoecology and Traditional Ecology: The Procurement and Production of Plant Resources in the Northern French Alps." PhD diss., University of Washington, Seattle.

Parker, E., D. Posey, J. Frechione, and L. F. da Silva. 1983. "Resource Exploitation in Amazonia: Ethnoecological Examples from Four Populations." *Annals of the Carnegie Museum of Natural History* 52(8): 163–203.

Puri, R. K. 1997. "Hunting Knowledge of the Penan Benalui of East Kalimantan Indonesia." PhD diss., University of Hawai'i, Honolulu.

Shepard, G. H., D. W. Yu, M. Lizarralde, and M. Italiano. 2001. "Rainforest Habitat Classification Among the Matsigenka of the Peruvian Amazon." *Journal of Ethnobiology* 21(1): 1–38.

Sillitoe, P. 1998. "An Ethnobotanical Account of the Vegetation Communities of the Wola Region, Southern Highlands Province, Papua New Guinea." *Journal of Ethnobiology* 18(1): 103–128.

Sillitoe, P. 2002. "Contested Knowledge, Contingent Classification: Animals in the Highlands of Papua New Guinea." *American Anthropologist* 104(4): 1162–1171.

Soemarwoto, R. 2007. "Kasepuhan Rice Landrace Diversity, Risk Management and Agricultural Modernization." In *Modern Crises and Traditional Strategies: Local Ecological Knowledge in Island Southeast Asia,* ed. R. Ellen. Oxford: Berghahn.

Suharno, D. M. 1997. *Representation de l'Environment Vegetal et Pratiques Agricoles chez les Alune de Lumoli, Seram de l'ouest (Moluques Centrales, Indonesie de l'Est.* PhD thesis, l'Université Paris VI.

Torre-Cuadros, M. de L. A. I., and N. Ross. 2003. "Secondary Biodiversity: Local Perceptions of Forest Habitats: The Case of Soferino, Quintana Roo, Mexico." *Journal of Ethnobiology* 23(2): 287–308.

Wolff, X. Y., and M. Florey. 1998. "Foraging, Agricultural and Culinary Practices Among the Alune of West Seram, with Implications for the Changing Significance of Cultivated Plants as Foodstuffs." In *Old World Places, New World Problems: Exploring Issues of Resource Management in Eastern Indonesia,* ed. S. Pannell and F. von Benda-Beckman. Canberra: Australian National University.

Chapter 6

The Cultural Significance of the Habitat *Mañaco Taco* to the Maijuna of the Peruvian Amazon

Michael P. Gilmore, Sebastián Ríos Ochoa, and Samuel Ríos Flores

Introduction

The Maijuna Indians of the Peruvian Amazon have a complex and detailed habitat classification system for both the forest and nonforest habitats found within the Sucusari River basin (Gilmore 2005). Their habitat classification system is not a perfectly hierarchical system; instead it is composed of multiple, separate overlapping subsystems that they use to classify habitat types. The Maijuna classify over seventy different habitats within the Sucusari River basin based on geomorphology, physiognomy, indicator plant species, indicator animal species, and disturbance.

Geomorphologically defined habitat types recognized by the Maijuna within the Sucusari River basin are identified and classified based mainly on topography and hydrology (Gilmore 2005). Two main categories can be distinguished among the geomorphologically defined habitats recognized by the Maijuna, upland forest (*imi t̪t̪i*[1] or *imi coti*) and floodplain forest (*yiaya coti*); a dichotomy that is also fundamental to current Western scientific classification systems (Shepard et al. 2001). Within the general categories of upland and floodplain forest, the Maijuna recognize and classify a variety of more specific habitat types based on geomorphology. For example, within floodplain forest (*yiaya coti*) the Maijuna identify and classify levee islands (*yiaya coti t̪t̪i* or *cuedaca t̪t̪i*), swamps (*cuadubi*), and the banks and/or edges of rivers and streams (*yiaya unu*), among other habitats. It is also important to note that the geomorphologically defined habitats identified by the Maijuna can describe and classify any terrestrial area encountered within the Sucusari River basin, and that all other habitats classified by physiognomy, indicator plant species, indicator animal species, and disturbance overlie these geomorphologically defined habitats.

Two habitats defined by physiognomy or outer appearance are recognized by the Maijuna within the Sucusari River basin: *aquibi* ('place with ugly forest') and *deo bese* ('good clarity') or *deo dadi* ('good place') (Gilmore 2005: 44–45). In addition, over forty different habitats types are identified and classified by the Maijuna based on indicator plant species or plant life forms (Gilmore 2005: 45–63). Three main categories can be identified within this general group of habitats, including those located in areas with 'soft earth' (*cuadu*) (swamp habitats), habitats located in areas with 'ugly forest' (*aqui*), and habitats located in areas that do not have 'soft earth' or 'ugly forest'. For example, the Maijuna recognize *Mauritia flexuosa* L. f. palm swamps (*ne cuadu*), liana forests (*bichi aqui*), and forests with an understory dominated by the palm *Lepidocaryum tenue* Mart. (*miibi* or *mii nui nicadadi*), respectively. Palms are an especially important group of indicator plant species to the Maijuna as they are used to classify and identify a large number of habitats.

Animal species are used by the Maijuna to identify and classify an additional eight habitats within the Sucusari River basin (Gilmore 2005). The majority of these habitats are identified and classified based on indicator ant and mammal species. The Maijuna also identify a number of different types of disturbed or successional habitats (Gilmore 2005). These habitats may be human induced, such as swiddens (*yio*) and cemeteries (*mai tate taco*), or naturally occurring like natural tree fall gaps (*tutu badu yio*). The Maijuna do not have a general named category for secondary forest, as they do for primary forest (*maca*). Instead, secondary forest can be considered a covert category within the Maijuna habitat classification system, since the Maijuna recognize several characteristics that make all secondary forest habitats different from primary forest.

In addition to providing a description of the Maijuna habitat classification system, the use, importance, and significance of the different habitat types was also investigated in order to gain a better appreciation of how the Maijuna ultimately perceive these areas and interact with them (Gilmore 2005). We here use a case study approach to assess in detail the cultural significance of one habitat, called *mañaco taco* by the Maijuna. This habitat is one of the many different habitat types defined by indicator plant species by the Maijuna of the Sucusari River basin. *Mañaco taco* was chosen for an in-depth study because the Maijuna have well-defined and constructed traditional supernatural beliefs associated with these forests. Investigation of this habitat provides insights into cosmological and spiritual beliefs about the forest and provides a window into the emic construction of ecological relationships.

Mañaco taco are found in upland forest and are dominated by the small myrmecophytic tree or shrub *Durioa hirsuta* (Poepp.) K. Schum. (Rubiaceae). The most striking feature of a ***mañaco taco*** is its very open understory, contrasting sharply with the normally dense Amazonian rain forest (Figure 6.1). The name ***mañaco taco*** literally translates as an 'open area with *D. hirsuta*', which provides an appropriate description of this habitat. *Duroia hirsuta* forests have been reported elsewhere in the literature (e.g., Aquino et al. 1999; Campbell, Richardson, and Rosas 1989; Duke and Vasquez 1994; Encarnación 1985, 1993; Fleck and Harder 2000; Frederickson, Greene, and Gordon 2005; Page, Madriñán, and Towers 1994; Pfannes and Baier 2002; Schultes 1969, 1987; Schultes and Raffauf 1992). These forests are generally called *jardín del diablo* ('Devil's garden') in the Colombian Amazon (Schultes 1987), *limpo de canelo de vehlo* ('clearing of the shinbone of an old man') in the western Brazilian Amazon (Campbell et al. 1989), and *supay chacra* ('Devil's field', 'Devil's swidden', or 'Devil's garden') in the Peruvian Amazon (Duke and Vasquez 1994; Gilmore pers. obs.).

The Maijuna

The Maijuna (Mai huna), also known as the Orejón or Coto (Koto), are a Western Tucanoan people (Bellier 1993a, 1994; Steward 1946) presently found along the Sucusari, Yanayacu, and Algodón rivers of the northeastern Peruvian Amazon

Figure 6.1 **Mañaco taco** *(forest dominated by* D. hirsuta*) located in upland forest in the Sucusari River basin. Note the very open understory of this forest.*

(Bellier 1993a, 1994). There are approximately 300 Maijuna individuals living in a total of four communities located along the rivers mentioned above (Bellier 1993a, 1994).[2] All four Maijuna communities have received parcels of legally titled land from the Peruvian government (Brack-Egg 1998). Unfortunately, the titled land that the Maijuna have received represents a very small portion of their ancestral lands.

Traditionally, the Maijuna are organized into patrilineal clans named after both plants and animals (Bellier 1993a, 1994). Clans practiced exogamy and uxorilocal residence upon marriage, ultimately dispersing the men of each clan (Bellier 1993a, 1994). As Bellier (1994) notes, clans did not have independent ancestors, stories, leaders, or territories. Today, many Maijuna traditions and cultural practices are no longer practiced by the Maijuna or have been significantly altered due to the impact of missionaries, the *patrón* system, governmental policies, mestizos,[3] the regional society, and the formal education system, among other things (Bellier 1993a, 1994; Gilmore pers. obs.).

Study Site

All field research was conducted in the Maijuna community of Sucusari. Sucusari is located along the Sucusari River, a tributary of the Napo River, in northeastern Peru. The Sucusari community is located approximately 126 kilometers by river from Iquitos, the largest city and commercial center of the Peruvian Amazon. This general region of Peru has a mean annual precipitation of almost 3,100 mm per year and a mean annual temperature of 26° C (Marengo 1998). The Sucusari River basin is dominated by upland tropical wet forest, yet seasonally inundated forest is also present (Gilmore pers. obs.). No other communities are located along the Sucusari River, although an ecotourism lodge (ExplorNapo Lodge), established in approximately 1983 (Castner 2000), is located about 4 to 4.5 kilometers downriver from the main community.

Sucusari is recognized as an official Native Community by the Peruvian government and has legal title to 4,771 hectares (Brack-Egg 1998), a small fraction of their traditional land. The Sucusari community contains twenty monofamilial or plurifamilial houses with ninety-seven residents in total. The majority of the residents of Sucusari are indigenous Maijuna; 71 percent are pure Maijuna and 12 percent are at least one-half Maijuna (all figures from July 2001). The Maijuna and other members of the Sucusari community participate in a variety of subsistence strategies, including hunting, fishing, swidden-fallow agriculture, and the gathering of various forest products. A variety of income generating strategies are also employed by members of the community. For example, community members sell game meat, domestic animals, agricultural produce, timber, and nontimber forest products to make money, and they occasionally participate in wage labor within the community (i.e., logging) and outside of it, among other things. At the beginning of this study, several families also occasionally sold tourist crafts to visitors from the ecotourism lodge downriver from the community; however, presently tourists very rarely visit the community.

Methods

All field research was conducted by Michael P. Gilmore and was completed over three field seasons totaling nine months during 2003–04. During this time period nine Maijuna individuals (seven men and two women), ranging in age from approximately thirty-eight to seventy-eight years old, were interviewed regarding the Maijuna habitat classification system. During these interviews, *mañaco taco* was identified as being a Maijuna-recognized and classified habitat type, and several consultants briefly and generally explained its significance. Intrigued by this information, Gilmore extensively interviewed his main consultant and co-author, Sebastián Ríos Ochoa (a fifty-year-old Maijuna man), using semistructured interviewing techniques (Cotton 1996) about the supernatural beliefs associated with *mañaco taco* and its overall significance to the Maijuna. Within the Sucusari community, Ríos Ochoa is recognized as one of the most deeply versed in Maijuna traditional knowledge, and most importantly, he is an excellent and patient teacher.

In addition to this information, a Maijuna traditional story regarding the supernatural residents of *mañaco taco* was told by S. Ríos Flores in the Maijuna language and recorded. Transcribed in Maijuna using the practical orthography established by Velie (1981), the story was translated into Spanish by S. Ríos Ochoa. In addition to being very knowledgeable about Maijuna cultural traditions, S. Ríos Ochoa is also perfectly bilingual and literate in both Maijuna and Spanish.

Data regarding ethnobotanical and ethnoecological knowledge and use practices associated with *mañaco taco* and *D. hirsuta* were collected via participant observation and open-ended interviews (Cotton 1996). Two groups of consultants, consisting of individuals previously interviewed about the Maijuna habitat classification system, were also interviewed using semistructured interviewing techniques (Cotton 1996) to verify and supplement this information. One of the groups interviewed consisted of two males and the other group consisted of two males and one female.

In addition, several *mañaco taco* were visited and qualitative ecological observations were made during the above mentioned nine months of field research. Voucher specimens of *D. hirsuta* were collected and are deposited in the Herbarium Amazonense (AMAZ), Universidad Nacional de la Amazonía Peruana, Iquitos, Peru, and the Willard Sherman Turrell Herbarium (MU), Miami University, Oxford, Ohio.[4] It should also be noted that all interviews during the course of this study were conducted in Spanish and, when necessary, translated into Maijuna by Ríos Ochoa. In addition, all data collected during the course of this study were coded, organized, and analyzed using a modified version of the methods described by Strauss and Corbin (1998).

Results and Discussion

As previously described, **mañaco taco** are anomalous open areas dominated by the small tree or shrub *D. hirsuta* in the normally dense and diverse western Amazonian forest. Western ecologists and scientists have come up with a variety of different explanations as to why these areas are open. It was hypothesized by some that the open understory of **mañaco taco** was primarily due to allelopathy (Aquino et al. 1999; Campbell et al. 1989; Page et al. 1994; Pfannes and Baier 2002), but it has recently been determined that *Myrmelachista* ants are responsible for clearing the vegetation around *D. hirsuta* (Frederickson 2005). Specifically, Fredrickson et al. (2005) report that the ant *Myrmelachista schumanni*, which nests in the hollow and swollen distal woody stems of *D. hirsuta*, keeps the understory of **mañaco taco** open by poisoning all plants except *D. hirsuta* with formic acid. *Myrmelachista schumanni* worker ants bite small holes in the leaf tissue of invading plants and inject formic acid from their abdomens into these holes, causing necrosis along primary veins within hours (Fredrickson et al. 2005).

Not surprisingly, the Maijuna ascribe a completely different cause to the strikingly open nature of **mañaco taco**. According to traditional Maijuna beliefs, the understory vegetation in these areas is kept clear by invisible supernatural beings called **Ma baji** that reside in these forests. As explained by Ríos Ochoa:

> … I asked my mom one day. "Well," she said to me, "here in this **mañaco taco** where you see that it is open the people (**Ma baji**) of these **mañaco taco** are workers, workers, and they can keep it open." She also told me, "In other swiddens (**mañaco taco**) (where) you see ugly places (i.e. understory plants) the residents [**Ma baji**] of these [**mañaco taco**] are lazy, they do not know, they do not know how to clear their places."

Therefore, although **Ma baji** are invisible supernatural beings, the physical manifestations of their work can be observed in the open appearance and structure of **mañaco taco**.

According to consultant testimony, **Ma baji** are male supernatural beings, not female, and they reside in all **mañaco taco** regardless of size. The components of the name **Ma baji** can be dissected and translated ultimately providing a good description of these supernatural beings. For example, the word **ma** can be literally translated as 'red' whereas the morpheme **baji** is the Maijuna designation for clan which bilingual Maijuna consultants generally translate as 'race' or 'group' (Bellier 1993a, 1994; Gilmore pers. obs.). Therefore, **Ma baji** can be translated as 'red group', 'red race', or 'red clan'. According to consultants, these supernatural beings are called **Ma bajɨ** because they paint their bodies red. As S. Ríos Ochoa explained:

> Because **ma** is red and **bajɨ** is its race, no, of this group. So, we say **ma** because it (**Ma bajɨ**) is painted, they are painted with annatto, no, and it makes them red and for this reason we call them **Ma bajɨ**. Even though you cannot see them, no, we call them **Ma bajɨ** because sometimes our ancestors told us that **Ma bajɨ** is this way, they are painted with annatto, and that is their race. For that reason we call [them] **Ma bajɨ.**

It is important to note that the species of annatto that is used by **Ma bajɨ** is not the normal cultivated variety *Bixa orellana* L. Instead, **Ma bajɨ** use their own wild relative of annatto, *Bixa platycarpa* Ruiz & Pav. ex G. Don, that the Maijuna call **bati bosa ñi** ('*B. orellana* of the spirit'). Interestingly, although **Ma bajɨ** reside in **mañaco taco** they also occasionally leave these areas to wander around other parts of the forest. During these trips the smell of annatto tips off the Maijuna that **Ma bajɨ** is close:

> **Ma bajɨ** not only lives inside of a **mañaco taco,** he also always leaves to walk (around), this is certain. Because when we were in a house, a hunting camp, in our hunting camp, where there is no **mañaco taco,** nothing, in the evening we smelled the scent of annatto, no, and my mom said, "well, children **Ma bajɨ** is passing by (and) for that reason there is the scent of annatto and we are smelling it."... So when they smelled (annatto) they said **Ma bajɨ** is passing by and for that reason the scent of annatto arrives. Because certainly he [**Ma bajɨ**] has to paint or put (annatto) on his body and have the scent of this, of annatto.

When asked to describe **Ma bajɨ** it was common for bilingual Maijuna consultants to use the words *madre* 'mother' (**aico**) and *dueño* 'owner' in regards to their relationship with **mañaco taco** and *demonio* 'demon' or 'devil' and *espíritu malo* 'bad spirit' in general. Bellier (1993b: 42) provides a very good general description of the Maijuna concept of *madres*:

> The forest is inhabited by "madres" that incarnate the vital essence of plant and animal species, and all of the places. Protectors of their creatures, the "madres" possess aggressive powers that they use against mankind, alone or mediated by shamans. Under the aspect of "vital essence," the concept of "madre" is linked to the notions or ideas of "the power to generate" and "animation" that are typically feminine.

It is interesting to note that even though **Ma bajɨ** are male supernatural beings, they are still described as "*madres*" by the Maijuna. Although most of the Maijuna "*madres*" that Bellier (1993a; 1994) introduces and describes are female, at least one is also a male supernatural being like **Ma bajɨ.** In addition to using the aforementioned words to describe **Ma bajɨ**, Ríos Ochoa also stated that **Ma bajɨ** are *brujos* (sorcerers) that possess some evil supernatural powers.

Ma bajï, as is evident in the words that Maijuna consultants use to describe them, are not benevolent beings. In addition to clearing the understory of *mañaco taco*, the Maijuna also attribute several other actions to *Ma bajï*. For example, according to Maijuna consultants, *Ma bajï* are malevolent beings that can rob the spirits or souls of babies and children up to approximately twelve years of age, eventually killing them. As S. Ríos Ochoa stated:

> *Ma bajï* are bad people ... bad spirits. Because my father said that when you walk in the forest there are *Ma bajï* and they also rob the [souls of] boys and girls and for that reason they [the boys and girls] die. Because that is what they do, that is his plan ... They take away the spirits or souls of boys and girls and the boys and girls die.

Due to the fact that *Ma bajï* can harm babies and children, Maijuna parents traditionally took several precautions to protect their children while in the presence of *mañaco taco*. For example, Maijuna parents did not enter *mañaco taco* with their children when traveling in the forest; instead they avoided entering these areas all together by walking around them. As a Maijuna male consultant explained:

> They say that in *mañaco taco* it is forbidden to walk with small children because the *dueño* [owner] [*Ma bajï*] of the swidden [*mañaco taco*] always robs the spirits or souls of the small ones and the small ones die. That is why a shaman, a shaman, forbids the mothers who have small children to enter the *mañaco taco* because if they enter it is certain that the small one could die, because in the swidden [*mañaco taco*] it is forbidden to go in with children. Those that have children always must make a curve [around the *mañaco taco*] they must go around.

To provide more protection while in the presence of these forests, Maijuna parents would also cover their babies and young children with leaves or a piece of bark cloth and/or burn dry leaves while walking in front of their children. Burning dried leaves while walking in front of their children ensured that they would inhale smoke from the burning leaves and therefore would "not be contaminated with the air of this *Ma bajï*." S. Ríos Ochoa recounted in detail what his mother did to protect him when he was approximately ten to twelve years old:

> If the trail entered within this *mañaco taco* my mom would always protect us by making a fire with *hungurahui* [local name for *Oenocarpus bataua* Mart.] leaves that she looked for. To prevent us from suddenly breathing in this air [of *Ma bajï*] ... we had to smell something from the fire or from the smoke. And she always led us alongside of the *mañaco taco*. My mom did this I saw her. This is not a story that my mom told me that she did, no, I saw what she did. She tied up a portion of dried leaves and she went in front of us [burning the leaves] and we went behind [her]. This is how she protected her children.

In the unfortunate event that a Maijuna child did become ill due to perceived soul-loss caused by *Ma bajɨ* the services of a Maijuna shaman were sought out. As Ríos Ochoa noted:

> A child six years old can become sick ... Then you can approach a shaman and say, "My son suffers from this, he has this problem." Then the shaman cures you ... Then the shaman is going to tell you what your son really has. Then he cures [your son] later at night [and] the following day [he will] say, "Well, this child suffers from this. He does not have his spirit or soul in his body. The *mañaco taco* already took it, the spirit of the *mañaco taco* [*Ma bajɨ*]." But if he cured him his spirit is going to return. And afterwards you should not go into the forest all of the time. It is forbidden to take small children because if you go all the time again you are going to have problems.

He also added that if the shaman is not successful in recovering the spirit or soul of the child, then that child will eventually die. Also, according to consultant testimony, shamans alone are able to cure soul-loss; Western medicine and other treatments are futile.

In addition to robbing the spirits and souls of babies and children, the Maijuna also traditionally believe that *Ma bajɨ* abduct young girls to raise them as their wives in *mañaco taco*. The abduction of a Maijuna girl by *Ma bajɨ* is recounted in the following traditional Maijuna story titled "Ma bajɨde quɨɨjá" (The story of *Ma bajɨ*)[5] as told by Samuel Ríos Flores:

[1]She was sweeping next to her house. [2]"What types of *chichibi* [South American coatis; *Nasua nasua*] are making a lot of noise?" [she said as] they bothered her. [*She mistook the noise created by the Ma bajɨ as chichibi.*] [3]After getting upset she returned to sweeping again. [4]Instantly, instantly the *Ma bajɨ* grabbed [her daughter] and left. [5]Taking her, taking her [the *Ma bajɨ*] raised her in a *mañaco taco*. [6]He always sang to her in a hammock. [7]Now she was grown-up and she felt hungry. [8][The *Ma bajɨ*] left to bring (her) several *bichi* [pineapples; *Ananas comosus* (L.) Merr.] but she did not want to eat [them] and she said that [they] were not good. [9]"Perhaps you want to eat a *toto aqui* [a type of armadillo; *Dasypus* sp.]?" [the *Ma bajɨ* asked her] and he brought it to her and she said that it was not good because it had a bad smell. [10]"What is it that you want now? I am going to bring you a *bichi toto aqui* [Southern naked-tailed armadillo; *Cabassous unicinctus*]," [the *Ma bajɨ* said] and he brought it to her and she did not want it. [11]"What will I do with you? I am going to bring you a fruit of *micabi* [Annonaceae?]" [the *Ma bajɨ* said] and he brought it to her and she did not want it and she said that it was not good to eat because it had a bad smell. [12]"If you talk that way, perhaps you want your mom," [the *Ma bajɨ* said]. [13]"Yes, I want my mom," [she said] [14]"Tomorrow I am going to take you so that you see [her] if you are sad, if you are sad," [he said] annoyed. [15]He took her [to see her mom] and he said, "Your mom lives here, enter. [16]I am going to wait for you here in this place, afterwards come to meet me." [17]She left and her mom saw her and it made her [mom] happy. [18]"My little daughter is coming, my little daughter is coming," [her mom said]. [19]When she was

going up the stairs [of the house], [her] mom said, "A person, a person can live [with my daughter] by giving her food, a person can live [with my daughter] by giving her food. [Traditionally, the giving or exchange of food was one of the steps in establishing a permanent union.] [20]The **Ma bajɨ** is not a person that you can live [with]." [21]"Did you raise her, did you raise her, why do you speak so quickly?" [her husband said as] he was annoyed with his wife. [22]She was there until the evening when they were sifting *chapo* [a type of thick drink] from ripe [plantains], already it was much too late. [23][Her mom] gave her a half-burnt stick from (the) fire, a clay pot, and a bunch of ripe [plantains] and she returned. [24]She returned, she returned … she returned and she carried it to her husband. [25]Now they were in the forest and [the **Ma bajɨ**] asked her a question, "What did your mom say, I heard [her talking]." [26]"She did not say anything, [she only said] my little daughter is coming, and it made her happy." [27]"I did not hear that. [28][I heard her say,] you can live with a person by giving them food, the **Ma bajɨ**, the **Ma bajɨ** is no good, it is not good that you have him [as your husband]. [29][He] is not a person that you live [with], she said that, I heard it," [the **Ma bajɨ** said]. [30]She annoyed him [and he said], "Return now, I do not want or love you." [31]Out of rage he took her spirit or soul, he took her spirit or soul, and immediately he disappeared. [32][She returned to her parent's house and] when her mom saw her [she said] "My little daughter is returning again." [33]"He heard what you said and he was annoyed with me and for that reason I am coming back [here]" she told her … [34]Grabbing her little brother she put him in a hammock and began to sing. [In another version of this story she sang her brother the songs that **Ma bajɨ** previously sang to her.] [35]She sang two times and then died. [36]That is everything.

This traditional Maijuna story ultimately highlights the danger that **Ma bajɨ** pose to young girls and reinforces the fact that Maijuna parents must remain vigilant in order to protect their children from **Ma bajɨ**.

Unlike Maijuna babies and children, Maijuna adults can enter **mañaco taco** without having their spirits or souls taken by **Ma bajɨ**; however, there are certain taboos that affect how they interact with these forests. For example, according to consultants, the Maijuna do not use *D. hirsuta* (**mañaco ñi**) in any way out of fear of reprisals from its *"dueño* ('owner') **Ma bajɨ**. For example, S. Ríos Flores stated the following when asked if he eats the fruits of *D. hirsuta*:

> The shamans say that it is forbidden to eat this fruit, you can die, it is forbidden … The old-timers did not eat this, they were scared … If you eat this you are not going to live, you are going to die … I have never tried it, never.

S. Ríos Ochoa also recounted what his mother told him about eating the fruits of *D. hirsuta*:

> She only said to me, "It is not good to eat the fruits [of *D. hirsuta*]; it is forbidden, the *"madre"* [**Ma bajɨ**] of **mañaco taco** can suddenly give, he can make you sick because he is a *brujo* [sorcerer] … he is a *brujo* and he can give you [hit you with] a *virote* [invisible "dart" in the arsenal of sorcerers] and you can die.

He also added his own thoughts:

> So far I have never eaten [the fruits of *D. hirsuta*], I see the fruits yes but I never touch them ... Because if you abuse [things] a lot he [*Ma bajɨ*] can do anything ... do not abuse [things] in the swiddens [*mañaco taco*]. It is the same if you have your [own] swiddens. If a person is going to abuse [things in your swidden] and you are missing something from your swidden you get upset as the owner. And this [*mañaco taco*] also has a *dueño* [owner] and he [*Ma bajɨ*] becomes bitter when you touch his things. That is it.

Given the attitudes expressed by Rios Ochoa above, it is not surprising that the Maijuna do not traditionally use *D. hirsuta*. Although the Maijuna do not make use of *D. hirsuta*, other Amazonian indigenous and local peoples use it in a variety of ways (Table 6.1). In addition to not collecting *D. hirsuta* in *mañaco taco*, the Maijuna also do not clear swiddens in these areas due to the same fears expressed above. Interestingly, there are no taboos against hunting in these forests. S. Ríos Ochoa suggested that this difference exists "... because the animal goes into that place [*mañaco taco*] for perhaps a moment. It does not live there all of the time."

It is also important to note that even though Maijuna adults can technically enter and pass through *mañaco taco* while in the forest, several Maijuna consultants interviewed during the course of this study expressed general fear and discomfort about doing so. As Ríos Ochoa explained:

> I am also afraid to enter [a *mañaco taco*]. Because I know that there is a, there is an *aico* [*madre*, mother] ... To repeat to you, I am older but I am afraid, I cannot readily enter [a *mañaco taco*] all the time. When I pass by or enter a [*mañaco taco*] I am always a little bit afraid, I am afraid ... I feel bad, I feel, I feel something in my body ... or (that something) is watching me, I don't know, but you feel it when you pass by or enter a *mañaco taco* ...

In addition, S. Ríos Flores recounted the fear that he felt one time while in a *mañaco taco*:

> When I was in a *mañaco taco* one time I heard "*fuing*" [a sound that he made with his mouth]. I was scared because they [*Ma bajɨ*] kidnap or abduct people. I ran ... This devil is named *Ma baji* ... he is the *dueño* ['owner'] of this swidden [*mañaco taco*], [he is] a type of devil.

These two quotes highlight the general fear and anxiety that some Maijuna adults currently experience when entering *D. hirsuta* forests.

As previously stated, the local name for *D. hirsuta* dominated forests is *supay chacra*, which can be translated as 'Devil's field', 'Devil's swidden', or 'Devil's garden' (Duke and Vasquez 1994; Gilmore pers. obs.). As suggested by the local

Table 6.1 Ethnobotanical information corresponding to the use of ***Duroia hirsuta*** within the Amazon Basin

Indigenous group or local population	Name for *D. hirsuta*	Use	Source
Quichua	***supai caspi***	ants associated with tree: eat (taste like lemon)	Bennett et al.(2002
Shuar	***iwiank*** (from *iwia* 'demon'); ***iwianki*** (from *iwia* 'demon')	trunk: used for firewood; fruits and bark: extract arrow poison	Bennett et al. (2002)
Waorani	*owekawe*	ants associated with tree: rub ant pheromones on inside of cheeks to relieve pain associated with excessive blowgun use or when mouth ulcers prevent blowgun use	Schultes and Raffauf (1990)
Kofán, Siona, Tikuna, Witoto and other indigenous groups	***sha-ka-ker-ná-sé*** (Kofán); ***solimán*** (Columbia)	bark: strips bound around arms and legs to make temporary tattoos	Schultes (1969, 1987); Schultes and Raffauf (1990, 1992)
Brazilian Amazon	not indicated	fruits: considered edible	Campbell et al. (1989)
Columbian Amazon	***solimán*** (*solymán*)	leaves: used as a fish poison	Schultes and Raffauf (1992)
Columbian Amazon	not indicated	fruits: occasionally chewed to prevent cavities	Duke and Vasquez (1994)
Peruvian Amazon	*huitillo del supay*	trunk: occasionally used in construction	Duke and Vasquez (1994)

name of this habitat, mestizos and other local inhabitants in this general part of the Peruvian Amazon also have interesting traditional beliefs associated with these forests. In their *Amazonian Ethnobotanical Dictionary*, Duke and Vasquez (1994) note, but do not clarify, that "[r]ural people (in the Peruvian Amazon), superstitious about the '*Supay chacra*', avoid walking nearby." In addition, Frederickson et al. (2005), from their work in the Peruvian Amazon, state that "'Devil's gardens' … according to local legend, are cultivated by an evil forest spirit." It is not clear exactly how local beliefs within the Peruvian Amazon specifically compare or contrast to the traditional beliefs of the Maijuna, as this was outside the scope of this study.

Several other brief references to the supernatural beliefs held by other local and indigenous peoples regarding *D. hirsuta* forests are also encountered in the literature. For example, according to Davidson and McKey (1993), Quechua-speaking peoples of the Amazon believe that *D. hirsuta* forests ('*Supay Chacras*') are cultivated by the devil or "Chuyachaqui" (*Chullachaqui*). Schultes and Raffauf (1992), in their discussion about "gardens of the Devil" in the Colombian Amazon, state that "[t]he Indian believes that it has a supernatural cause—the residence of invisible beings." In addition, Schultes (1969, 1987) and Schultes and Raffauf (1990) note that indigenous peoples of the Colombian Amazon also believe that the lack of vegetation around *D. hirsuta* is due to a poison released by this tree that kills the surrounding plants; certainly a more mundane explanation for this habitat. Fleck and Harder (2000) note that the Matses Indian name for this habitat is **mayanën sebad**, which translates as 'demon's swidden'. Although they do not discuss the cultural beliefs surrounding this habitat, the Matses name for *D. hirsuta* forests generally suggests that there may be supernatural beliefs associated with them.

Two general references were also found regarding supernatural beliefs associated with *D. hirsuta* in general. For example, Schultes and Raffauf (1992) state that in the Colombian Amazon, "The shrub (*D. hirsuta*) is also feared and respected as the sole arborescent inhabitant of strange, cleared areas in the forest called 'Gardens of the Devil'." Additionally, Bennett, Baker, and Gomez Andrade (2002), in their work with the Shuar of eastern Ecuador, note that the Shuar names of *D. hirsuta* are **iwiank** or **iwianki,** which are from the word **iwia,** 'demon'. Again, unfortunately no specific information is provided regarding the significance of these names and therefore one is left only to speculate on their origin and cultural importance.

It is clear from the examples provided that indigenous and local peoples from a wide geographical area generally associate *D. hirsuta* and, more specifically, *D. hirsuta* forests with supernatural beings and forces. Even though many of the specific details of these traditional beliefs may be different, the fact that they are generally similar is extremely interesting. Unfortunately, what is not known is if these traditional beliefs have arisen independently within each of these different cultures or if they represent an exchange of general beliefs amongst them. Another interesting but perplexing question regarding *D. hirsuta* forests is why supernatural beings are associated with these areas. With respect to the Maijuna, it is not known exactly why there is a supernatural origin and being associated with *mañaco taco*. The strikingly anomalous appearance and structure of *D. hirsuta* forests may have something to do with this association. The fact that there are no readily apparent natural origins of these forests may cause the Maijuna and other indigenous and local peoples to turn to the realm of the supernatural for answers.

In conclusion, the Maijuna have extensive traditional supernatural beliefs associated with *D. hirsuta* forests. As detailed, this habitat is traditionally perceived as an especially dangerous place for Maijuna babies and children due to their vulnerability and susceptibility to the malevolent deeds of *Ma baji*. Due to the fact that Maijuna parents completely avoid *mañaco taco* when accompanied by babies and children while in the forest, this habitat, in a traditional context, can reasonably be considered an *avoidance island* in certain situations. Avoidance islands are defined by Gilmore (2005) as "areas in primary or secondary forest that are generally avoided due to the plants or animals that are present and/or cultural beliefs (i.e. taboos, etc.) associated with them." Other examples of Maijuna recognized and classified habitats that can be considered avoidance islands are forests with high concentrations of an unidentified species of stinging ant (*jaiqui baidadi*) and low-growing, very dense thickets dominated by creeping vines and thorny plants found in floodplain forest along river margins (*aquibɨ*), among others (Gilmore 2005).

Avoidance islands contrast sharply with what Posey (1984) terms *resource islands*. According to Posey (1984: 117), resource islands are "areas in the primary forest where specific concentrations of useful plants or animals are found." It is important to note that most Maijuna recognized habitats can in fact be considered resource islands because they are dominated by ethnobotanically and/or economically important plant species (Gilmore 2005). For example, *M. flexuosa* palm swamps (*ne cuadu*) and forests with an understory dominated by the palm *L. tenue* (*mɨɨbɨ* or *mɨɨ nui nicadadi*) can be considered resource islands because they have high concentrations of ethnobotanically and economically important resources such as fruits and thatch leaves, respectively.

As detailed, not all Maijuna habitat types are of equal importance. Some Maijuna recognized and classified habitats are culturally important and useful while others are not, and some can be considered resource islands while others are avoidance islands. In order to obtain a more holistic and accurate understanding of how indigenous and traditional peoples actually perceive and interact with their environment and resources, it is critical to understand how and why they avoid certain areas and resources in addition to how they use them. Documenting solely how people use habitats and resources (as most ethnobiological and ethnoecological studies do) provides an incomplete and inaccurate understanding of how they actually perceive and interact with the environment as a whole.

It should also be noted that the information presented in this chapter represents traditional Maijuna knowledge and cultural beliefs. Today, most Maijuna individuals in the Sucusari community under the age of approximately thirty years old do not have extensive knowledge or understanding of the traditional supernatural beliefs associated with *mañaco taco* and its malevolent supernatural resident *Ma baji*. Due to the general degradation of this knowledge and cultural shifts in lifestyle, residence patterns, and subsistence strategies (resulting in Maijuna parents rarely, if ever, encountering *D. hirsuta* forests while accompanied by babies and young children), most of the cultural practices described in this paper as being associated with *mañaco taco* are no longer carried out by the Maijuna of Sucusari. This is one of many examples of the decline of traditional knowledge and beliefs in general amongst the Maijuna of the Sucusari community.

Acknowledgments

Many thanks are owed to the Maijuna of the Sucusari community for all of their help and support over the years. Without their assistance and interest in participating in this study, this research project would never have been possible. Special thanks to Isidora, Seberino, Victorino, Mamerto, Nancy, Felipe, and Nicolas of the Sucusari community for all of their help and patience throughout the many months of this research project. Research was conducted with the approval of the Sucusari community, the Miami University Committee on the Use of Human Subjects in Research, and the Instituto Nacional de Recursos Naturales (INRENA), Peru. Botanical specimens were collected under permit N° 71-2003-INRENA-IFFS-DCB issued by INRENA. The Herbarium Amazonense (AMAZ), Universidad Nacional de la Amazonía Peruana, Iquitos, Peru, provided general institutional support during the course of this research project. We would especially like to thank Cesar Grández Ríos, Meri Nancy Arévalo García, and Guillermo Criollo Díaz of AMAZ. Financial support for this project was provided by: the National Science Foundation, the Elizabeth Wakeman Henderson Charitable Foundation, Phipps Conservatory & Botanical Gardens (Botany in Action), The Society for Economic Botany, the Willard Sherman Turrell Herbarium (MU), the Garden Club of Ohio, the Garden Club of Allegheny County, and the Department of Botany Miami University. Thanks also to Hardy Eshbaugh, Adolph Greenberg, David Gorchov, Leslie Main Johnson, Susan Barnum, Susan Lamont, and Chris Myers for providing valuable and insightful comments on this manuscript. Very special thanks to Jyl Lapachin for all of her help, support, and encouragement throughout the entire course of this research project.

Appendix 6.1

The Maijuna Version of *Ma bajɨde quɨijạ* (The story of *Ma bajɨ*)

[1]Ue unu yuaco. [2]Ɨgue, chichibɨ junaca bejɨ botạta ani oco. [3]Na ode monɨde beco yuaco. [4]Ma bajɨ tea, tea ini etajọguɨ. [5]Sade, sade baquɨ mañaco taco debajọguɨ. [6]Jaɨdɨ beoquɨ ojɨ. [7]Ai ñi nede ạo oico. [8]Bichide jiyejani daquɨde ñameco bɨocona jạ ịco. [9]Queta, toto aquɨ anita caɨnade daquɨde bɨoquɨna jạ ịco. [10]Ɨguedeca, oico jana bichi toto aquɨde ɨjachi na daquɨde ñameco. [11]Ɨgueta, mɨde micabɨ jiyo jai cachi daquɨde ñameco bɨo biomɨade sa acueyo oco. [12]Mɨ cama ịco ani mɨaconade oida. [13]Ɨ jacode oiyi. [14]Ñata sayi ñiajɨ oico, oico ani ojɨ. [15]Saguɨ asade mɨacona ɨdadi baiyi cacama ɨjɨ. [16]Ɨdadi bequɨ ɨteyi, jete juadai yide. [17]Saicode bɨaco jɨaco chibaco. [18]Yibago ñi daico, yibago ñi daico. [19]Mɨi tada mɨicode da jɨcaco, mai, mai acode ba, mai acode ba. [20]Ma bajɨ mai aɨ bajaye. [21]Mɨ ñiaco baco debade, mɨ ñiaco baco ti jɨca anico eja ojɨ nɨjọde ojɨ. [22]O sujɨde beco, beco naijọco naijọquɨde. [23]Toa tou jạ, cuacodo jạ, o ojạ ɨchigode daico. [24]Daico, daico… daico ịco ɨjɨde. [25]Maca judu caca jaiquɨ jɨca asajɨ, mɨacona quima ɨjɨde asade yi. [26]Jɨcamago, yibago ñi daico chibago. [27]Yi cama asamadeca yi. [28]Mai nama acode bama, ma bajɨ, ma bajɨ oaquɨna jạ, oaquɨbɨ bayi. [29]Mai aɨ bajaye ɨjɨde asadeca yi. [30]O ñiaguɨ, jana mɨ monɨ oimayi mɨde. [31]Oquɨ ạbɨ naɨ utajọguɨ, naɨ utajọde tea beo bese nejọguɨ. [32]Bɨaco na jɨaco yibago najọ ñi daico. [33]Mɨ jɨcacode asade maitaquɨde daiyi ịco quɨaco… [34]Ñi doiquɨde neade jaɨdɨ tani ɨmede beco uja daico. [35]Tepe ñoa daico tɨni sanijọgo. [36]Casọa jạ.

Notes

1. All Maijuna terms are in bold italics. Transcription of Maijuna terms was accomplished with the help of S. Ríos Ochoa, a bilingual and literate Maijuna individual, using a practical orthography previously established by Velie (1981) consisting of twenty-seven letters that are pronounced as if reading Spanish, with the following exceptions: in a position between two vowels *d* is pronounced like the Spanish *r; ɨ* is pronounced like the Spanish *u* but without rounding or puckering the lips; and *ạ, ẹ, ị, ọ, ụ,* and *ɨ̠* are pronounced like *a, e, i, o, u,* and *ɨ* but nasalized. Also, the presence of an accent indicates an elevated tone of the voice; accents are used only when the tone is the only difference between two Maijuna words and the word's meaning is not clarified by its context. The twenty-seven letters that make up the Maijuna alphabet are: *a, ạ, b, c, ch, d, e, ẹ, g, h, i, ị, j, m, n, ñ, o, ọ, p, q, s, t, u, ụ, y, ɨ, ɨ̠.*

2. It is important to note that this population estimate does not include those Maijuna living in mestizo communities (see note 3), other indigenous communities, or Iquitos (Bellier 1994).

3. Mestizos, found throughout the Peruvian Amazon, are people of mixed Amerindian and Iberian descent who practice a mixture of traditional agriculture, hunting, fishing, and forest product extraction for their livelihoods (Chibnik 1994; Hiraoka 1985; Padoch 1988, as cited in Coomes and Ban 2004: 421).

4. Voucher specimens were collected by M. Gilmore under permit N° 71-2003-INRENA-IFFS-DCB issued by the Instituto Nacional de Recursos Naturales (INRENA), Peru. *D. hirsuta* collection numbers are 424 and 537.

5. A Maijuna version of this story is presented in Appendix 6.1. The numbered sentences in the English version of this story correspond exactly to the Maijuna version.

References

Aquino, R., N. De Tommasi, M. Tapia, M. R. Lauro, and L. Rastrelli. 1999. "New 3-methoxy-flavones, an Iridoid Lactone and a Flavanol from *Duroia hirsuta*." *Journal of Natural Products* 62: 560–562.

Bellier, I. 1993a. *Mai huna Tomo I: Los pueblos indios en sus mitos N° 7*. Quito: Abya-Yala.

———. 1993b. *Mai huna Tomo II: Los pueblos indios en sus mitos N° 8*. Quito: Abya-Yala.

———. 1994. "Los Mai huna." In *Guía Etnográfica de la Alta Amazonía*, ed. F. Santos and F. Barclay. Quito: FLACSO-SEDE.

Bennett, B. C., M. A. Baker, and P. Gomez Andrade. 2002. *Ethnobotany of the Shuar of Eastern Ecuador*. Advances in Economic Botany, Volume 14. Bronx: The New York Botanical Garden Press.

Brack-Egg, A. 1998. *Amazonia: biodiversidad, comunidades, y desarrollo* (CD-ROM). Lima, Peru: DESYCOM (GEF, PNUD, UNOPS, Proyectos RLA/92/G31, 32, 33, and FIDA).

Campbell, D. G., P. M. Richardson, and A. Rosas, Jr. 1989. "Field Screening for Allelopathy in Tropical Forest Trees, Particularly *Duroia hirsuta*, in the Brazilian Amazon." *Biochemical Systematics and Ecology* 17(5): 403–407.

Castner, J. L. 2000. *Explorama's Amazon: A Journey through the Rainforest of Peru*. Gainesville: Feline Press.

Chibnik, M. 1994. *Risky Rivers: The Economics and Politics of Floodplain Farming in Amazonia*. Tucson: University of Arizona Press.

Coomes, O. T., and N. Ban. 2004. "Cultivated Plant Species Diversity in Home Gardens of an Amazonian Peasant Village in Northeastern Peru." *Economic Botany* 58(3): 420–434.

Cotton, C. M. 1996. *Ethnobotany: Principles and Applications*. Chichester: John Wiley and Sons.

Davidson, D. W., and D. McKey. 1993. "Ant-Plant Symbioses: Stalking the Chuyachaqui." *TREE* 8(9): 326–332.

Duke, J. A., and R. Vasquez. 1994. *Amazonian Ethnobotanical Dictionary*. Boca Raton, FL: CRC Press, Inc.

Encarnación, F. 1985. "Introducción a la flora y vegetación de la Amazonia Peruana: estado actual de los estudios, medio natural y ensayo de una clave de determinación de las formaciones vegetales en la llanura amazónica." *Cadollea* 40: 237–252.

———. 1993. "El bosque y las formaciones vegetales en la llanura amazónica del Perú." *Alma Máter* 6: 95–114.

Fleck, D. W., and J. D. Harder. 2000. "Matses Indian Rainforest Habitat Classification and Mammalian Diversity in Amazonian Peru." *Journal of Ethnobiology* 20(1): 1–36.

Frederickson, M. E. 2005. "Ant Species Confer Different Partner Benefits on Two Neotropical Myrmecophytes." *Oecologia* 143: 387–395.

Frederickson, M. E., M. J. Greene, and D. M. Gordon. 2005. "'Devil's Gardens' Bedeviled by Ants." *Nature* 437: 495–496.

Gilmore, M. P. 2005. "An Ethnoecological and Ethnobotanical Study of the Maijuna Indians of the Peruvian Amazon." PhD diss., Miami University.

Hiraoka, M. 1985. "Changing Floodplain Livelihood Patterns in the Peruvian Amazon." *Tsukuba Studies in Human Geography* 9: 243–275.

Marengo, J. A. 1998. "Climatología de la zona de Iquitos, Perú." In *Geoecología y desarrollo Amazónico: estudio integrado en la zona de Iquitos, Perú,* ed. R. Kalliola and S. Flores Paitán. Annales Universitatis Turkuensis Ser. A II 114. University of Turku, Finland.

Padoch, C. 1988. "People of the Floodplain and Forest." In *People of the Tropical Rain Forest,* ed. J. S. Denslow and C. Padoch. Berkeley: University of California Press.

Page, J. E., S. Madriñán, and G. H. N. Towers. 1994. "Identification of a Plant Growth Inhibiting Iridoid Lactone from *Duroia hirsuta,* the Allelopathic Tree of the 'Devil's Garden'." *Experientia* 50: 840–842.

Pfannes, K. R., and A. C. Baier. 2002. "'Devil's Gardens' in the Ecuadorian Amazon: Association of the Allelopathic Tree *Duroia hirsuta* (Rubiaceae) and Its 'Gentle' Ants." *Revista de Biologia Tropical* 50(1): 293–301.

Posey, D. A. 1984. "A Preliminary Report on Diversified Management of Tropical Forest by the Kayapó Indians of the Brazilian Amazon." *Advances in Economic Botany* 1: 112–126.

Schultes, R. E. 1969. De plantis toxicariis e mundo novo tropicale commentationes IV. *Botanical Museum Leaflets: Harvard University* 22(4): 133–164.

———. 1987. "A Botanical Enigma in the Amazon." *Economic Botany* 41(3): 454–456.

Schultes, R. E., and R. F. Raffauf. 1990. *The Healing Forest: Medicinal and Toxic Plants of the Northwest Amazon.* Portland: Dioscorides Press.

———. 1992. *Vine of the Soul: Medicine Men, Their Plants and Rituals in the Colombian Amazonia.* Oracle, AZ: Synergetic Press, Inc.

Shepard, G. H., Jr., D. W. Yu, M. Lizerralde, and M. Italiano. 2001. "Rain Forest Habitat Classification Among the Matsigenka of the Peruvian Amazon." *Journal of Ethnobiology* 21(1): 1–38.

Steward, J. H. 1946. "Western Tucanoan Tribes." In *Handbook of South American Indians,* vol. 3, ed. J. H. Steward. Washington, D.C.: United States Government Printing Office.

Strauss A. L., and J. Corbin. 1998. *Basics of Qualitative Research: Techniques and Procedures for Developing Grounded Theory,* 2nd ed. Thousand Oaks, CA: SAGE Publications.

Velie, D. 1981. *Vocabulario Orejón.* Serie lingüística Peruana No. 16. Pucallpa, Perú: Instituto Lingüístico de Verano.

Chapter 7

The Structure and Role of Folk Ecological Knowledge in Les Allues, Savoie (France)

Brien A. Meilleur

Men have evidently been much the same in habits of thought and irrationalities during all of recorded history, so it must be recorded somewhere that men of cultures other than our own had the explicit idea of an ecosystem. Ideas are expressed in language. Man as an animal has been at least a part of the wild natural scene and many of the terms in languages succinctly encapsulate the idea of a particular kind of ecosystem. The English terms carr, moss, fen, and heath have very precise meanings in terms of kinds of plants concerned, habitat factors, and resulting landscapes. Siberian terms for regionally extensive ecosystems, such as tundra and taiga, are now universally used … Muskeg is a local Chippewa Indian term for a kind of ecosystem within the taiga … Every Old World Mediterranean country has a term for a shrubby, aromatic, usually sclerophyllous plant formation formed by man from forest vegetation … The names are different, but the aspect of the ecosystem, the dominant habitat factors, and often even many plants are the same.

—Jack Major, "Historical Development of the Ecosystem Concept," 10

Beginning with the pioneering work of Harold Conklin in the 1950s and Brent Berlin in the 1960s and 1970s, folk biological method and theory have attempted to account primarily for the internal logical structure of "native" or "folk" systems of biological classification (i.e., of flora and fauna). This effort resulted in cross-culturally comparative models of folk classificatory behavior, demonstrating what were often striking structural and intellectual similarities between the folk systems and "Western" scientific taxonomies. As such, folk biology continues to contribute to a better understanding of human psychology and cognition. In what might be called a second wave in folk biology, Hays (1982), Hunn (1982), and others drew attention to the implications of a utilitarian perspective in folk biological classification. As a consequence, ethnobiologists are now showing greater interest in what Drechsel has called the "product" of such classifications rather than primarily in their "substance" (Drechsel 1985: 55).

However, other than generally accepting that folk taxa are often economically significant, and that their lexicographical structure and perceptual content within classificatory systems are in some instances closely related to this functionality (especially at folk taxonomic levels above and below that of the "folk generic"; see Hays 1982: 90; Hunn 1982: 831; Posey 1984), few ethnobiologists have studied the cognitive mechanics through which these often-elaborate knowledge systems are practically applied in economic settings.

This chapter shifts the utilitarian focus of folk biological classification from the analysis of content and structure to that of folk ecological knowledge in use. To this end, the cognitive and functional relations between such knowledge and economic success in the alpine peasant community of Les Allues (Savoie, France) are briefly explored. The goal will be to demonstrate that only insofar as there is a continuous application of folk ecological knowledge in the resource discovery and acquisition processes can economic behavior be consistently successful and, thus, in the long term, adaptive. A more specific goal will be to show how folk knowledge of the plant world (represented by the set of folk botanical taxa or discrete, locally cognized units of the flora) is perceptually and practically linked to folk ecological knowledge, represented in Les Allues by a set of locally perceived higher-order units, called here "folk biotopes," analogous to the ecologist's combined concepts of biotic community and habitat. Terminology and concepts adapted from ecology are introduced, both to frame this analysis and as a means of appreciating the link between the cognized environment and practical action in the Alluetais economy.

The traditional Alluetais high mountain, agro-pastoral economy is composed of four major production sectors: "full field" agriculture, haymaking, stock-raising, and cheese-making. It also comprises eight secondary production sectors: fruit-tree arboriculture, viticulture, gardening, forestry, hemp cultivation, gathering, hunting/trapping/fishing, and beekeeping. The valley of Les Allues, a well-defined mountainous environment (600–3200 m asl) covering 90 square kilometers, has been occupied by humans since at least the Bronze Age (Meilleur 1986, 2008). Until the progressive but fairly rapid collapse of the agro-pastoral economy following the Second World War and a shift to ski tourism, the Alluetais had employed several classificatory schemes to order both natural and human-modified biotic and abiotic space within their high-altitude landscape. In the late 1970s and early 1980s I identified lexical sets that were still being used to refer to biological and ecological phenomena as well as to geographic-topographic, geologic, and hydrographic phenomena, sites, and formations. The combination of these lexical sets, along with those terms employed to refer to constructed space (for shelter, industry, intra-communal displacement, etc.) and an elaborate set of place names (Meilleur 1980), constituted a perceptual and semantic grid that permitted Alluetais to communicate about, and to direct productive effort toward, natural and domesticated resources within their valley.

I propose several new concepts here to help account for data that are pertinent to folk ecological ordering at levels above (or more inclusive than) the folk botanical and zoological domains in Les Allues and, as I believe is highly probable, to many small-scale traditional societies where economic activity brings people into direct contact with land and biotic resources. In the following section I describe the basic folk ecological unit—what I call the *folk biotope*—that is employed by Alluetais to structure biotic space in their alpine valley and, in all likelihood, in other small-scale traditional societies directly managing and exploiting natural and/or domesticated resources within their own "effective environments."

From Ecology to Ethnoecology

Plant and animal ecologists, phytosociologists, and biogeographers have attempted to classify and understand the variable repartition of plant and animal species within biotic space (see Chapter 1, this volume) ranging from very small areas of several hundred square meters to that of the planet. Patterning has been described at many levels of perception and inclusion, and many terms have been used to refer to similar or related concepts (Whittaker 1962). The most widely accepted conceptual construct now used by ecologists to account for and to communicate about patterning in plant and animal distributions is, and has been for well over 100 years, that of "community" (Daubenmire 1968). While there is still no unanimity of opinion on the scope of the concept, plant communities are generally recognized by appreciating in any assemblage of plants (1) the outward appearance of growth form (physiognomy and/or layering), (2) the fidelity of certain (indicator) species, and (3) the presence of dominant species (Daubenmire 1968: 250ff.).

In theory, a plant community or phytocoenosis is inseparable from the environment (ecotope, biotope, or habitat) it occupies. This environment will always include a zoological component or zoocoenosis (that with the phytocoenosis = biocoenosis), a physical component (edaphotope = soil, moisture, nutrients, etc. in terrestrial systems), and a climatic component (climatotope = sunlight, wind, air temperature, etc.). The combination of components that are ecologically related—that is, linked by "chains of influence" (Daubenmire 1968: 13)—is commonly called an ecosystem. The ecosystem concept is flexible, and it has been applied at many levels of inclusion (Evans 1956). The person choosing to use these concepts must therefore define what set of interconnected phenomena he or she wishes them to be applied to. For heuristic purposes, it is acceptable to separate the floristic and the animal components, or any parts of these components, for any ecosystem at any level of inclusion. Three terms and concepts will be adapted here from ecology to assist in accounting for the manner in which

Alluetais agro-pastoralists conceptualized, structured, and used large parts of the natural and human-modified physical and biotic space known as the valley of Les Allues. These are *ethnoecosystem, folk biotope,* and *folk phytocoenosis.*

Les Allues as Ethnoecosystem

> A natural resource ecosystem is an integrated ecological system, one element of which is a product of direct or indirect use to man ... In all cases, the distinguishing facet of a natural resource ecosystem is that man has a direct involvement in the complex set of ecological interactions.
>
> —Stephen S. Spurr, "The Natural Resource Ecosystem"

> The operation of a factory is somewhat analogous to the utilization of land for agricultural purposes, from intensive agriculture to forest management. Materials and energy flow into the factory. In a variety of steps they pass through the processing divisions of the plant where they are subject to chemical and physical reconstitution; energy is dissipated at each step, and at some points materials may be recycled to an earlier point in the process. Finally, finished products and a variety of wastes emerge from the plant ... Managers of agricultural, semiwild, and wild lands seek ... goals of maximization of output and reduction of costs. Also they have the added responsibility of protecting the producing capital of their lands for use by future generations. Often, management entails not one end product, but several, such as wood, water, wildlife, and recreation. Given the basic abiotic and biotic complexity of land, the phenomena of succession and retrogression, a multiplicity of managerial goals, and a desire for more efficient use of the land, it is obvious that some theoretical framework upon which we can assemble and interrelate these diverse components is a necessity ... The ecosystem concept provides this framework.
>
> —F. Herbert Bormann and Gene E. Likens, "The Watershed-Ecosystem Concept and Studies of Nutrient Cycles"

What Julian Steward called "relevant environmental features" and Netting called "effective environment" (both cited in Netting 1965: 82) were limited in Les Allues to a valley of approximately ninety square kilometers in the northern French Alps. It was overwhelmingly this landscape that Alluetais used for their subsistence-oriented activities much as Bormann and Likens describe, principally by modifying the physical and biotic environments, producing or procuring domesticated and natural resources, transforming them, consuming them, and then rendering waste in the form of recyclable nutrients. While Les Allues as an interacting assemblage of plants, animals, soils, climate, and human beings was not "closed" in any absolute sense, there were readily identifiable barriers—mostly natural, but some social—that regulated the application of knowledge systems, work, and the flow of energy within the valley. For analytical purposes, it is appropriate to view this assemblage as an ecosystem—or rather, because of the focus here on human activities as the integrating feature, as an *ethnoecosystem.*[1]

Not only the Alluetais economic activities but also their classificatory abilities were directed toward this space. Much as the collectivity of plant taxonomists brings order to the world's plant and animal diversity through the process of classification—with the goal of deciphering phylogenetic relations among organisms in an evolutionary context—the Alluetais also brought order to their biotic space, though they were doing so at a local level for more practical reasons. Like members of any traditional society with direct, "non-scientific" ties to biota, the Alluetais were more selective than scientists in the organisms they classified. In fact, the Alluetais carefully attended to only a relatively small part of the biotic diversity surrounding them and, almost exclusively, to that part that had positive (or sometimes negative) economic significance to them. Between 10 and 15 percent of the wild plant species occurring in the valley were specifically named (about 150 of more than 1,000 species; see Meilleur 1986). The Alluetais agropastoralist was thus much more interested in economically relevant biological "information being gathered," organized, and used than in "the process of classification" itself (Raven, Berlin, and Breedlove 1971: 1212).

One way that Alluetais brought order to biotic space was by recognizing and naming some 250 folk botanical taxa (about 150 wild folk botanical taxa and about 100 domesticated folk botanical taxa) and approximately 110 folk zoological taxa (domesticated and wild), with both domains organized at several hierarchical levels of inclusion.[2] Another way was by recognizing some twenty habitat types with which the majority of the 360 folk biological taxa were variously associated. The habitats and associated plant communities recognized and named by Alluetais are called here *folk biotopes,* and the set of folk botanical taxa linked to each folk biotope is called its *folk phytocoenosis.* The modifying term "folk" is added to the ecological head terms to establish analogy and not homology between the scientific and the "native" concepts. However, on the whole, the scientific and the folk concepts of habitat, and of the associated plant community or phytocoenosis, seem to me to be logically similar in several ways.

The Folk Biotope

The folk biotope is proposed here as the basic folk ecological unit employed in Les Allues to cognize and order biotic space at a level more extensive than that of the individual folk botanical and zoological taxa. Some twenty such named categories were recognized by Alluetais, permitting them to discern higher-scale segregates in biotic, mostly botanical, space within the valley of Les Allues (Table 7.1). Elderly Alluetais informants produced patterned responses indicating that these categories were defined by multiple, intersecting features related to vegeta-

Table 7.1 The Alluetais Folk Biotopes *

Dialect Term	Approximate English Equivalent
lèz arkosè	green alder thickets
la brousayè	brushland
lèz éklérsi	forest openings
la foré	coniferous forest
le korti	household garden
le mordjé	cultivated field rockpile
la morin-na	moraine
la moyasir	moderate wetland
le paturadzo, la pravèri	pastureland, prairie
le pra	hayfield
le rèbye	deciduous coppice
le rotirè	marsh
le sètchü	dryland, desert
le tsan	cultivated field
le tsnavyé	hemp plot
le vèrdjé	hayfield-orchard
la vnyé	vineyard

* The defining features of each folk biotope, such as its psychological definition and salience in terms of core versus periphery, were not systematically analyzed; some seem to be "basic level objects" (à la Rosch 1978: 30) while others are derived from these. For example, variants of several folk biotopes were spontaneously created by Alluetais consultants by the addition of terms related to proximity or contiguity (*la bordoré, le bor dè x*/borders, edges of x; *le lon dè l'éva, le lon dè x*/watercourse edge, edges of x; *le tèr dè l'arbé [dè la monda]*/around mountain chalets). In some instances, especially for "around the mountain chalet," the floristic content was especially well-defined. The floristic content of each Alluetais folk biotope, constituting its folk phytocoenosis, can be found in Meilleur 1986, along with their physiognomic and physical properties (when recorded). The language historically spoken in Les Allues, called locally *le patoé dèz aloé* ("le patois des Allues"), was one of scores of dialects traditionally spoken in Savoie and nearby regions and collectively thought to constitute Franco-Provençal, classed by some linguists as a Gallo-Roman language and by others as a highly distinct dialect group of French thought to have begun to separate from standard French in the Middle Ages. The orthography used here has been somewhat modified from that used in Meilleur 1986 by consulting *Le Patoé dèz Aloé*, a dictionary of Alluetais terms and grammar compiled in 2001 by l'Association des Patoisants des Allues. Pronunciation differences were noted among the several clusters of the fourteen Alluetais hamlets, and some of these are reflected in this and the following tables. The list of folk biotopes in Table 7.1 may not be exhaustive. Three terms found in the 2001 dictionary (*la flotsèla* or *la flotèla*/small wood, *le kroké*/unproductive hayfield, and *la tan-na*/underground animal den), not recorded during earlier fieldwork, have the potential to be Alluetais folk biotopes. A vegetation stage (*le tsan en ermo*) with functional importance occurring after a cultivated field was abandoned and before it became a hayfield, was not included in Table 7.1.

tion physiognomy, floristic content, and in several cases salient physical properties (e.g., rocky, wet, or dry soil), as well as by the economic activities performed there. This set of folk biotopes, along with other lexical sets used for conceptualizing and ordering the abiotic environment (such as geographic-topographic, geologic, hydrographic [Tables 7.2 to 7.4], and constructed space), in conjunction with an elaborate set of place names, collectively served as a cognitive and

Table 7.2 Folk Geographic and Topographic Set

Dialect term	Approximate English Equivalent
l'ârétha, l'âréha	*Arête* in French. Mountain ridge, culminating point (falling off rapidly on both sides)
la bourna	Grotto, cave (also refers to entrance of badger (*Meles meles*) and alpine marmot (*Marmota marmota*) dens
le dou	Knoll, butte
le fon dè la komna	*Le fond de la commune* in French. The base or lower reaches of the valley of Les Allues
le golé	Gully
l'indrê	*Endroit* in French. Right bank of the valley of Les Allues, somewhat sunnier than the left bank (uncommonly used term, probably because of the predominantly N-S orientation of the valley)
l'invére	*Envers* in French. Left bank of the valley of Les Allues, somewhat less sunny than the right bank (uncommonly used term)
l'ingolatchu	Naturally-occurring funnel shaped hole (from "*Le Patoé dèz Aloé*, 2001)
le kofat	Naturally occurring hole
le kol	Mountain pass
la komb	*Combe* in French. Large, anticlinal mountainside
la koutha, la kouha	*Côte* in French. Sloped hillside
la krétha, la kréha	*Crête* in French. Mountain crest, more open and expansive than *l'ârétha*
le krwè	*Creux* in French. Depression
la lantse	Steep, vertical mountain corridor
la mintin de la komna	Middle part of the valley
le plan, le platô, la plan-na	Plain, plateau
la rèvna	Ravine
le rouho	Steep hillside (from *Le Patoé dèz Aloé*, 2001)
le sandzon	Summit or head of the valley
le valon	Small valley

Note: This list may not be exhaustive.

Table 7.3 Folk Geologic Set

Dialect Term	Approximate English Equivalent
la bourba	Mud
lo dzi	Clay
la glir	Small island of exposed earth or sand in or at the edge of the Doron (torrent) des Allues
le gravyé	Gravel
la karir	Quarry
---- *dè louzè*	Roofing stone (slate) quarry
---- *dèz épyé*	Rough roofing stone (slate) quarry
---- *dè tsarbon*	Coal quarry
le klapyé	Talus
la nèta	Mud/debris from flooding (? from *Le Patoé dèz Aloé,* 2001)
la périr	Scree
lé rotsé	Large rocks or boulders
la sabla	Sand
la sablir	Sand pit (quarry)
le tō	Tufa
la tsô	Limestone

Note: This list may not be exhaustive

semantic grid that allowed Alluetais to communicate about and interact with diverse elements of their 90 square kilometers of natural and human-modified alpine environment.

The ecological concept and term *biotope* was selected over similar and related terms and concepts for several reasons. Even though most of the twenty-some Alluetais folk biotopes appeared to be defined in large measure by more or less discrete assemblages of plants, and less so of animals, soils, and climatic features (e.g., exposition of the sun) linked through "native" eyes by "chains of influence"—and thus could each legitimately be called an ecosystem, a natural resource ecosystem or a folk ecosystem in its own right—Alluetais nevertheless perceived the folk biotopes as being part of a larger entity, what I call the Alluetais ethnoecosystem. While it is not at all unusual for ecologists to employ the ecosystem concept at many levels of inclusion, from the biosphere as a global (macro) ecosystem to, for example, a vernal pool as an ephemeral (micro) ecosystem (e.g., Odum 1969: 263), it would be awkward to do so at more than one level within such a limited geographical area. Because the Alluetais ecological constructs were viewed emically as integrated components of the larger valley ethnoecosystem, I prefer to

Table 7.4 Folk Hydrographic Set

Dialect Term	Approximate English Equivalent
le bya	Small irrigation channel
la fontan-na, la sorsa	Fountain, source, spring
le glêché	Glacier
la krèvèse	Crevasse
le sèré	*Serac* in French. Glacial ridge
la gouye	"Hole" in the river (preferred fishing spot)
le langô	Puddle
le lê	Lake
le nan	Creek, stream (*le nan dè x*)
le névé	High-altitude patch of hardened snow
le nére	Retting pit
la rè	Rivulet, streamlet
la rigola	Smaller rivulet
la rvire	River (Doron des Allues)
la yaka	Melting, muddy snow (? From *Le Patoé dèz Aloé*, 2001)

Note: this list may not be exhaustive

call them folk biotopes. I find the concept of *biotope* ("[t]he smallest geographical unit of the biosphere or of a habitat that can be delimited by convenient boundaries and is characterized by its biota": Lincoln, Boxshall, and Clark 1982: 34), with the addition of the modifier *folk,* to be most useful for understanding and discussing the combination of cognized and named groupings of plants and animals (= their folk biocoenoses or biological components) and their associated nonliving components (= their folk ecotopes or abiotic content).

In Les Allues, and in other traditional alpine communities (e.g., Meilleur 1985), I employed the folk biotope as a concept and heuristic aide whose existence I hypothesized was predicated upon local, practically motivated interests. By helping economic behavior be directed toward specific (desired) folk biological taxa, which were most often cultural resources (edible plants or animals, medicines, construction materials, etc.), I concluded that Alluetais folk biotopes were, for the Alluetais, logically closer to the wildlife administrator's or the forestry manager's vegetation categories—used mostly to order space for practical management purposes (Van Dyne 1969)—than they were to the abstract vegetation concepts used by many contemporary ecologists for understanding "the sociological interactions of plant species ... within single communities" (Mueller-Dombois and Ellenberger 1974: 9ff.).

Identifying Folk Biotopes

An elicitation framework was devised to gain insights into the manner in which Alluetais structured their effective biological and physical environments for referential purposes above (or at higher levels of inclusion than) that of the folk botanical and zoological taxa. A hypothesis that any higher-order environmental concepts would be related to practical concerns informed such questions as "Where do you find or go for x (= resource)?" or "Where do you practice x (= activity)?" or "Where do people find or go for x (= resource)?" or "Where do people go to practice x (= activity)?" By using the collective subject in queries rather than the second person singular, responses usually moved past place named sites (i.e., the common "individualized" responses to questions framed in the second person singular) to what seemed to be a number of "general-purpose" categories that distinguished functionally important parts of the natural and human-modified biotic and abiotic environments of Les Allues.

Through follow-up questions like "How do you (or does one) recognize x (= the folk category elicited)?" or "How do you (or does one) distinguish x from y (= another of the elicited folk categories)?", "What activity occurs in x (= folk category elicited)?" or "What plants, animals or other resources are found or encountered in x (= elicited folk category)?" I found that Alluetais variously drew from three attributes to describe and define the elicited categories: biological (usually botanical, but sometimes zoological or both), physical, and functional. Sets of categories began to emerge based on the perceived greater or lesser presence or absence of one or more of these attributes. For example, the category *le pra* (hayfield) was seen as possessing well-defined biological, physical, and functional components, while *le kofat* (hole in the ground) was seen as lacking well-defined biological and functional components while possessing a well-defined physical one. The set of categories that possessed well-defined biological components, physical components, and functional components I have called folk biotopes (Table 7.1).

Most activities or "events" (Rosch 1978: 43) involving work with plants or animals, that is, their production, discovery, or acquisition—activities altogether constituting the bulk of economic production in Les Allues—were overwhelmingly directed toward these categories. Each folk biotope was characterized by a variably sized set of folk botanical taxa—its floristic component—which I call its *folk phytocoenosis*.[3] For example, Table 7.5 presents the folk phytocoenosis of *le mordjé* (cultivated field rockpile).

Other environmental categories were found that lacked well-defined biological components. Those that seemed to be employed primarily for general spatial reckoning within the valley, and not for more specific economic purposes, I grouped into a combined folk geographic-topographic set (Table 7.2). Other categories lacking well-defined biological components but possessing important

Table 7.5 Folk Phytocoenosis of *le mordjé*/Cultivated Field Rockpile

lèz anpoè	*Rubus idaeus* L.	Raspberry
lèz anpoè nérè	*Rubus fruticosus* L.	Bramble
lèz épnè	*Rosa canina* group	Dog rose
le frényo	*Fraxinus excelsior* L.	Ash
le savu blan (nére)	*Sambucus nigra* L.	Elder
le savu rodzo	*Sambucus racemosa* L.	Red-berried elder
lé trèpèlète	*Rubus saxatilis* L.	Stone bramble

functional attributes, were grouped into a folk geologic set and into a folk hydrographic set (Tables 7.3 and 7.4). The folk geologic and hydrographic categories came into play in Les Allues when work was directed toward abiotic resources (though fishing was associated by some but not all people with *la rvire*). The clustering of these mostly abiotic categories into folk geographic-topographic, geologic, and hydrographic sets resulted more from an etic assessment on my part than from any concerted effort l undertook to explore whether they were kinds of natural groupings/sets for Alluetais.

In most cases it was not too difficult to separate landscape categories into one or the other of these sets by assessing Alluetais descriptions of their principal activities and by observing their economic behavior. However, in some cases this procedure did not work well. The question then arose of whether Alluetais make qualitative distinctions between the categories in the several groupings that I had made and, if so, whether these distinctions were important to behavior in ways that could justify my introduction of the folk biotope concept. While none of the lexical sets that I presented in Tables 7.1 to 7.4 was labeled linguistically, I believe the answer is a qualified yes to both questions.

The "difficult to class" cases seemed to demonstrate that an emic continuum existed in the perception of presence or absence of biological, physical, and functional attributes of the various landscape categories. A category having a modest biological component, a minor functional component, and a well-defined physical component, like the cultivated field rock piles (*le mordjé*), could not be superficially distinguished from *lé rotsé* (clusters of rocks), which had few if any biological and/or functional attributes but did have a well-defined physical one. Moreover, the Doron des Allues (*la rvire*), while overwhelmingly dominated by the physical aspect of rushing water, nevertheless possessed biological and functional components for some people (mostly a small number of men) when trout were fished from it and a major functional component when water was drawn from it to power mills. In such cases, contextual cues became important in deciding whether to call *la rvire* a folk biotope or *le mordjé* a folk geologic category. In the absence of quantifiable measures to help in making these decisions, there will be cases when the decision to call a category a folk biotope will require subjective judgment. Despite the presence of a biological component in the *la rvire* category, I decided not to call it a folk biotope because the physical component strongly dominated the biological one, because few people fished in *la rvire* (and were thus in contact with the biological component), and because *la rvire* was seen by most Alluetais as the principal component of the hydrographic network, the biological elements of which were seen as mostly inconsequential. In contrast, despite its highly salient physical properties, I accepted *le mordjé* as a folk biotope because multiple folk biological taxa were associated with it (see Table 7.5; however, most of these plants were not economically significant) and because it was closely associated with a biologically and functionally salient folk biotope, *le tsan* (cultivated field).

Despite the potential for such "fuzziness" between two or more of the etic sets I had established, I believe that there is some emic cognitive validity to them, and especially so to the set of folk biotopes. In fact, Alluetais acted similarly toward those categories within a given set and differently toward categories in the other sets: actions involving biotic resources were overwhelmingly directed toward the folk biotopes, actions involving abiotic resources were directed toward those categories in the folk geologic and hydrographic sets, and general territorial reckoning was done by referring to those categories of the folk geographic-topographic set.

Conclusion

Using data from Les Allues, a high-altitude onetime agro-pastoral community in the northern French Alps, this essay has explored some of the processes through which biotic and abiotic resources were procured and/or produced in a small-scale agrarian society. Like other human groups with other economic orientations, as ethnobiologists and other researchers (this volume) have noted, Alluetais developed referential systems that allowed them to relate intellectually and pragmatically to biotic and abiotic space within their effective environment.

I have focused here on the practical aspects of the cognized environment at a level above or more inclusive than the folk botanical and zoological taxa in Les Allues. While the approximately 250 folk botanical taxa and the 110 folk zoological taxa were the targets of most Alluetais economic effort, these taxa were predictably distributed within the valley of Les Allues insofar as they were associated with higher-order folk environmental categories that I call folk biotopes, analogous to the vegetation ecologist's biotope (biotic community + habitat). Once economic action is directed toward a given folk taxon, I believe a cognitive mediation occurs involving the folk biotope to which the taxon is perceptually linked. The value of such a relation in the procurement and/or production of cultural resources is that it permits the resource-seeker, over changing historical, climatic, and/or socioeconomic conditions, to extrapolate "from the uniqueness of (his) past experience to future encounters with reality" (Hunn 1982: 833), thus ensuring a high rate of success when undertaking economic activities. By cognitively linking folk taxa with folk biotopes in a predictable manner as effort is directed toward one or more of the twenty-some folk biotopes, the resource producer or procurer is assured that the spot where work is occurring is appropriate to the production or to the discovery of the desired resource. In Les Allues, once seasonality information was factored in, this also allowed the opportunistic Alluetais to predict which folk taxa were likely to be encountered within any folk biotope at any time. Ralph Bulmer and Michael Tyler (1968: 378) seem to have reached similar conclusions following ethnobiological research among Kalam in New Guinea.

Nevertheless, the reality of human production processes is that they are rarely so clear-cut as the static, linear application of mental and physical effort, from decision-making through mediation by the folk biotope to execution of work directed toward individual folk taxa, as I have summarized it here. At the "mediation by the folk biotope phase" alone, perceptual and cognitive differences among folk biotopes, variants of folk biotopes, and folk biotope-like categories were little explored. Furthermore, folk geographic-topographic categories and place names, both of which can be more or less extensive than folk biotopes in their biotic coverage, were also sometimes used in directing effort toward biotic space in Les Allues, but these were not more specifically analyzed here.

Discoveries pertinent to folk biological classification are perhaps useful in formulating several concluding remarks about what folk biotopes do and do not seem to be. Folk biotopes and folk generics (Berlin, Breedlove, and Raven 1974: 30) are cognitive constructs wherein the human mind has responded to visual and behavioral attributes as perceptual wholes. Both are "not only good to think, they are good to act upon" (Hunn 1982: 833). But the Alluetais folk biotopes do not appear to be understood best as single or even as several contrast sets of some sort of an abstract or intellectually perceived "plant community" or "habitat" domain, but rather as functional categories, more or less adjacent to one another in a spatio-biotic continuum, that are employed in helping guide economic actions toward desired resources. Even though perceptual discontinuities between the Alluetais folk biotopes were sometimes pronounced—and often conspicuously accentuated by human activities—"hybridization" or intergradation between folk biotopes, an exceedingly rare occurrence between Linnean species, was more often the rule than not. This point is underscored by the not infrequent use of the productive, analyzable lexical forms *le bor dè x, le tèr dè x,* or *le lon dè x* ('the edge of x', 'around x', or 'along x'). Such folk biotope-like categories connote proximity, adjacency, and transition, all variants of contiguity. Despite often great effort by the Alluetais to create distinct boundaries, this, and the fact that the overall natural vegetation pattern in the valley was more or less continuous, suggests that appreciations of contiguity (a *is near* b, a *is a part of* b) are more appropriate to understanding classificatory relations among folk biotopes than are appreciations of similarity (a *is like* b, a *is a kind of* b), as is the case in folk biological classification.

Even though many questions remain, the concept of the folk biotope as the basic component of an emic biotic grid of effective environment appears to be a useful tool for understanding the relation between cognitive processes (gnosis) and practical acts (praxis) in traditional societies with close economic relations to land and resources.

Notes

1. Barrau (1975: 16–17) has objected to the use of this term. However, he consistently supported the study of interrelations between human groups and their physical and biological environments using the notion of ecosystem. Burns (1961: 23) portrays the politically, socially, and economically "self-sufficient" high mountain communities of the Queyras region in the southern French Alps as generally occupying "equivalent ecological niches," though each is an "independent niche." Since individual Alluetais families, and collectively the Alluetais community, had direct ecological relations with some twenty folk ecological categories for economic and other purposes, the concept of ethnoecosystem seemed to be appropriate for grasping what was in fact a system in which a well-defined human group occupied multiple niches within a well-defined and integrated but heterogeneous physical and biotic landscape.

2. The focus of this chapter is on the nature and the application of referential categories used by alpine peasants to apprehend biotic concepts above (or more inclusive than and often cross-cutting) the level of individually perceived folk taxa and their classification systems. Time and space do not permit the presentation here of the folk taxa (but see Table 7.5 and Meilleur 1986).

3. While the relation between a given folk biotope and its folk phytocoenosis is important to understanding of how the bulk of economic production occurred in Les Allues, in several cases Alluetais associated ubiquitous taxa not specifically with one but with several folk biotopes. At the other extreme, some highly localized folk taxa—few in number—could not be associated with any of the folk biotopes but were known only from specific places (= place names). The botanical configurations of most folk phytocoenoses must also be seen as cognized means, since interviews were almost always characterized by at least some disagreement among informants about which plants were most closely associated with which folk biotopes.

References

Association des Patoisants des Allues. 2001. *Le Patoé dèz Aloé.*

Barrau, J. 1975. "Ecologie." In *Eléments d'Ethnologie: 2. Six Approches,* ed. R. Cresswell. Paris: A. Colin.

Berlin, B. 1973. "Folk Systematics in Relation to Biological Classification and Nomenclature." *Annual Review of Ecology and Systematics* 4: 259–271.

Berlin, B., D. Breedlove, and P. Raven. 1974. *Principles of Tzeltal Plant Classification.* New York: Academic Press.

Bormann, F., and G. Likens. 1969. "The Watershed-Ecosystem Concept and Studies of Nutrient Cycles." In *The Ecosystem Concept in Natural Resource Management,* ed. G. Van Dyne. New York: Academic Press.

Bulmer, R., and M. Tyler. 1968. "Karam Classification of Frogs." *Journal of the Polynesian Society* 77: 333–385.

Burns, R. 1961. "The Ecological Basis of French Alpine Peasant Communities in the Dauphine." *Anthropological Quarterly* 34(1): 19–34.

Conklin, H. 1954. *The Relation of Hanunoo Culture to the Plant World.* Unpublished PhD diss., Yale University.

Daubenmire, R. 1968. *Plant Communities: A Textbook of Plant Synecology.* New York: Harper and Row.

Drechsel, P. 1985. "Comments to Brown: Mode of Subsistence and Folk Biological Taxonomy." *Current Anthropology* 26(1): 43–64.

Evans, F. 1956. "Ecosystem as the Basic Unit of Ecology." *Science* 123: 1127–1128.

Hays, T. 1982. "Utilitarian/Adaptationist Explanations of Folk Biological Classification: Some Cautionary Notes." *Journal of Ethnobiology* 2(1): 89–94.

Hunn, E. 1982. "The Utilitarian Factor in Folk Biological Classification." *American Anthropologist* 84(4): 830–847.

Lincoln, R., G. Boxshall, and P. Clark. 1982. *A Dictionary of Ecology, Evolution and Systematics.* Cambridge: Cambridge University Press.

Major, J. 1969. "Historical Development of the Ecosystem Concept." In *The Ecosystem Concept in Natural Resource Management,* ed. G. Van Dyne. New York, Academic Press.

Meilleur, B. 1980. "L'ethnobotanique et le Cadastre Sarde." In *Le Cadastre Sarde de 1730 en Savoie, Chambéry, France,* Musée Savoisien.

———. 1985. "Gens de montagne, plantes et saisons: Termignon en Vanoise." *Le Monde Alpin et Rhodanien* 1: 1–79.

———. 1986. "Alluetain Ethnoecology and Traditional Economy: The Procurement and Production of Plant Resources in the Northern French Alps." PhD diss., University of Washington.

———. 2008. *Terres de Vanoise: Agriculture en Montagne Savoyarde.* Collection Le Monde alpin et rhodanien. Grenoble: Musée Dauphinois.

Mueller-Dombois, D., and H. Ellenberger. 1974. *Aims and Methods of Vegetation Ecology.* New York: John Wiley and Sons.

Netting, R. 1965. "A Trial Model of Cultural Ecology." *Anthropological Quarterly* 38: 81–96.

Odum, E. 1969. "The Strategy of Ecosystem Development." *Science* 164: 262–270.

Posey, D. 1984. "Hierarchy and Utility in a Folk Biological Taxonomic System: Patterns in Classification of Arthropods by the Kayapo Indians of Brazil.' *Journal of Ethnobiology* 4(2): 123–139.

Raven, P., B. Berlin, and D. Breedlove. 1971. "The Origin of Taxonomy." *Science* 174: 1210–1213.

Rosch, E. 1978. "Principles of Categorization." In *Cognition and Categorization,* ed. E. Rosch and B. Lloyd. New York: John Wiley and Sons.

Spurr, S. 1969. "The Natural Resource Ecosystem." In *The Ecosystem Concept in Natural Resource Management,* ed. G. Van Dyne. New York, Academic Press.

Van Dyne, G., ed. 1969. *The Ecosystem Concept in Natural Resource Management.* New York: Academic Press.

Whittaker, R. 1962. "Classification of Natural Communities." *Botanical Review* 28: 1–239.

Chapter 8

Life on the Ice

Understanding the Codes of a Changing Environment

Claudio Aporta

"This is true and that is the nature of the moving ice."

—Aipilik Inuksuk, Igloolik, 1988

Introduction

On 27 March 2001 Maurice Arnatsiaq and I were conducting a place-names survey of the island of Igloolik and surrounding area. It was noon, and, having completed the survey of the eastern coast, Maurice guided me onto the sea ice toward Nirlirnaqtuuq, a relatively large island (about twelve kilometers long) a few kilometers north of Igloolik. Known as "Neerlonakto" on topographic maps, Nirlirnaqtuuq, with low topographic features, was barely visible from the distance, forming almost a whole with the sea ice. After taking the island's geographic coordinates, I asked Maurice about a landmark that we could clearly see standing toward the north. It looked like a long, flat hill or a ridge, but I could not see any signs of land on my topographic map, according to which the closest coastline in that direction was thirty-five kilometers away, on the other side of Ikiq (the Fury and Hecla Strait). My companion invited me to board my snowmobile and drive to that place. We drove only four kilometers before arriving at the feature, which, to my surprise, was made entirely of ice. What I had taken for a landmark was, in fact, an icemark. The ice ridge, which was about three meters high and several kilometers long, even had a name: Agiuppiniq.

This name was not on the topographic maps that I carried, neither had it been recorded during a place-names program conducted in Igloolik in the mid 1980s. Yet Agiuppiniq was distinctively recognized in the community as a place name, since several people whom I consulted could readily point to it on the map. I later realized that for some elders it was a very significant place, and that the name Agiuppiniq was frequently mentioned in the database of the Igloolik Oral His-

Figure 8.1 *Agiuppiniq*

tory Project, a collaborative, ongoing project run by the Inullariit Elders Society of Igloolik and the Igloolik Research Centre. Begun in 1986, the Igloolik Oral History Project seeks to document the traditional knowledge and oral history of the Amitturmiut—the Inuit of the Northern Foxe Basin area of Nunavut. Most of the interviews cited in this work (including the ones I conducted) are part of the oral history project and are cited using the name of the interviewee, the year of the interview, and an interview number prefixed by the letters "IE."

In this chapter, Inuktitut place names are used whenever possible. "Iglulik" is used when referring to the traditional hunting camp southeast of the island, while "Igloolik" will be used when talking about the island as a whole, or the town of Igloolik. For practical reasons, place names such as "Baffin Island" and "Melville Peninsula" are also used.

Agiuppiniq is one among hundreds of marks that constitute the topography of the sea ice around Igloolik. Not all of them are alike. Some of them refer to the stable landfast ice; others, to the ever-changing moving ice. Some recur in the same locations every year while others vary in their position from year to year. Only a few of these sea-ice features can be identified as place names in a way that is comparable to the naming of geographical features on land. All of them help the Inuit of Igloolik to remember, live on, travel through, and talk about the familiar topography of the sea ice.

This chapter is concerned with the knowledge of the sea ice as developed and transmitted by the Inuit of Igloolik. This information was obtained from travel, observation, and interviewing that I conducted from October 2000 to May 2001 in Igloolik, as well as from several existing interviews in the Oral History Project database. Inuit knowledge of the sea ice reveals a deep understanding of the complex relationships among ice, currents, the moon, and the winds, as well as a holistic approach to knowledge where classification based on a Western scientific approach becomes difficult, if not counterproductive.

Through detailed knowledge of the ice topography, the sea ice becomes a familiar territory for the Inuit of Igloolik, and through the understanding of the "codes" of the moving ice, its changing nature becomes predictable. This chapter does not pretend to give a full account of a system of knowledge, the understanding and practice of which require a lifetime of apprenticeship. It will only describe some of its elements and offer some insights regarding this complex aspect of Inuit *Qaujimajatuqangit* (Inuit knowledge, also known as IQ).

The Geographic Situation of Igloolik

The community of Igloolik is situated on the island of the same name located in the northern Foxe Basin, between the northern Melville Peninsula and Baffin Island, in Nunavut, the new territory established in 1999 in the eastern Canadian Arctic. Both archaeological evidence and ethnographic accounts indicate that Igloolik has been a center of human habitation for thousands of years.

Dealing with the marine environment is an everyday activity throughout the year in Igloolik. Since little hunting and fishing is carried out within the boundaries of the island or from its shores, most activities require sea travel. The traveling takes place mostly over ice, except for the short periods of open water from July to October. Generally speaking, ice traveling is performed during nine months of the year. Lake fishing and caribou hunting on Melville Peninsula and Baffin Island require the crossing of frozen straits and, therefore, a continuous assessment of ice conditions and an understanding of the behavior of the ice and its interaction with the winds and the tides. Intensive seal and walrus hunting is also practiced on the ice, either at the floe-edge of the neighboring polynyas (sea that remains open throughout the year) or through breathing holes. Not long ago, the ice was also a place to live, where ice camps were established throughout a good part of the spring.

The Ice as Home: A Place to Live

Although the material aspects of ice camps cannot be verified through archaeological remains, both the memory of contemporary elders and the notes of early explorers give account of the yearly events related to life on the ice. The first

Europeans to visit Igloolik, the British explorers Parry and Lyon, remarked in April 1823 how most people had left the large winter settlement of Iglulik to live in smaller camps closer to more-favored hunting areas. Of the three camps mentioned by the explorers—Agiuppiniq, Pingiqqalik, and Uglit—the first one was located on the sea ice. (The explorers did not use the name Agiuppiniq, referring to it only as "the ice-camp.")

Captain Lyon, in his journal entry of 22 March, mentions having heard of "the prosperity of the people on the ice" (Lyon 1824: 250) at about 20 miles (32 kilometers) north of the snow houses of Iglulik (1824: 250) and eight miles to the west of Tern Island (Parry 1824: 423). Lyon himself visited the ice village, where he found twenty-eight people living in five snow houses (Parry 1824: 423). The geographic clues about the position of this camp, provided by both explorers, give clear indication that the camp was located somewhere between Agiuppiniq and an ice lead known in Igloolik as Naggutialuk.

One hundred years later, Mathiassen, a member of the Fifth Thule Expedition, records in April 1923 the existence of a "snow-house village" on the ice north of Qikiqtarjuq ([1928] 1976: 30), at the same general location as the ice village described by Lyon. Mathiassen does not provide a description of the village but notes that the inhabitants hunted "*utoq* seal" (seals that are basking in the sun).

The ice camp at Agiuppiniq is remembered by contemporary Inuit elders. Louis Alianakuluk, a hunter in his mid sixties, remembers his life at the ice camp:

> In my childhood, we used to go to that place, that is Agiuppiniit. From Avvaja we would go to Agiuppiniit to get closer to the floe-edge, the entire family would move to that area … that would become our home for a while. This happened as the sun was getting higher. The main reason was so that we would get to eat different diet than the *igunaq* [fermented walrus meat]. There would be marine animals such as seals and walruses, which were mainly the animals that were hunted during that time. So we used to get closer to the floe-edge by establishing ourselves at Agiuppiniit … at first there would be only a few but the numbers would grow as families moved in, so that would make the place with quite a few people. (2001)

The late Aipilik Inuksuk pointed out that in the spring, people residing at Avvajja (a former settlement located 20 kilometers southwest of Agiuppiniq and about eight kilometers northwest of the present settlement of Igloolik—see Figure 8.2) would move to the landfast ice at Agiuppiniq:

> This location was suitable to the hunters that would hunt on the ice through seal breathing holes at Ikiq [the Fury and Hecla Strait]. The place that I refer to as Agiuppiniq is located in this area … we used to make our home just past that where there is no danger of ice cracking and [of being] carried out to sea. (1986)

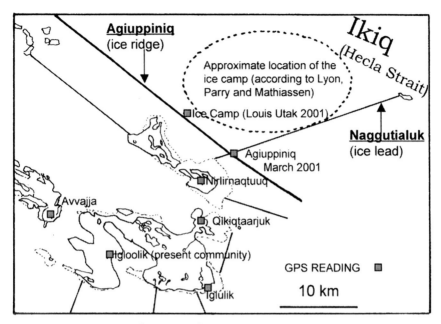

Figure 8.2 *Ice camp at Agiuppiniq*

Michel Kupaaq also remembers that in early spring they would go "to the ice camp at Agiuppiniq where there were numerous people already living there, as we arrived we started to make camp with the building of an igloo" (1987). George Kappianaq (1991a, 1991b) and Noah Piugattuk (1991) also refer to the spring camp at Agiuppiniq where the men would hunt seals at the breathing holes (*agluit*) situated on the cracks of the landfast ice.

All the references cited indicate that the place where the ice camps were established was effectively known as Agiuppiniq. The reasons for choosing that place were related to the safety of the ice, which remains stable throughout the spring, and the abundance of wildlife, especially seals, which were hunted mainly through *mauliq* (hunting seals through snow-covered breathing holes). Referring to the lead known as Naggutialuk (great or large lead) in the proximity of Agiuppiniq, Louis Alianakuluk points out that "from the time of my childhood, I have known it to be hunted" (2001). A great deal of *mauliq* still takes place today at the same lead, especially when the tidal strength of the full and the new moons widens the crack (Ikummaq, 2000a). Another good reason for settling down in Agiuppiniq could have been the large size of the snowdrifts in the area, which provided good material for igloo building (John MacDonald, personal communication).

The evidence shows that the ice camp has been in the same general location since at least the first written record (1823), which indicates that some features of the sea-ice topography have not changed for at least 200 years. Both Agiuppiniq and Naggutialuk seem to have recurred in the same general area year after year throughout this time. Figure 8.2 shows the location of the ice camp as described by Lyon, Parry, Mathiassen, and Louis Alianakuluk. The location of the ice ridge Agiuppiniq is also shown as measured by a global positioning system unit in March 2001. Both lines representing the ice ridge and the ice lead Naggutialuk were drawn by Louis Alianakuluk.

What at first might be regarded as an ever-changing, unpredictable environment appears to be, in fact, fairly stable. The structure of the ice topography is constituted by seasonal icemarks, but many of them are also stable and recurrent. In the memoryscape of the Inuit of Igloolik, the topography of the landfast ice may compare with the topography of the land. In fact, the land environment in the region of Igloolik is also constituted by seasonal landmarks that *disappear* under a thick layer of snow through most of the year. The concept of memoryscape has been defined by Mark Nuttall (1992) not as mere physical territory remembered by a particular individual, but the result of a community's interaction with a place through time.

The ice camp at Agiuppiniq has not been included as an Inuit settlement site in many important anthropological studies (see, for example, Crowe 1969; Rasing 1994) even though Agiuppiniq has been a recurrent spring settlement for at least 200 years of written history. One of the reasons for this absence might be that ice camps leave no archaeological remains. The difficulty in determining Inuit prehistoric use of the sea ice has been discussed by Wenzel (1984). He points out that there are three sources of archaeological data for determining patterns of sea-ice use: house sites, faunal remains, and artifacts. The most relevant (as archaeological evidence) of the three, the house structure, "is as transitory as the sea-ice environment" (1984: 45).

Furthermore, topographic maps do not illustrate the features of the sea ice, and elders have difficulty pointing out such features when commenting on regular topographic maps. However, Agiuppiniq was considered to be not only a good place for hunting but also a pleasant place to live: a place people could call *home*. Remembering the night he spent at Agiuppiniq, Captain Lyon remarks that "I never slept so warmly" (Lyon 1824: 251). Asked to compare dwelling on the sea ice with dwelling on the land, Louis Alianakuluk also says:

> The only thing is that material for bed platform that would cover the bed platform is not available on the sea ice. So they used to go and get gravel for this purpose from Nirlirnaqtuuq. That was the only difference, that is not having any materials that would keep your bedding from the snow bed platform. I believe that was the only thing. It is much warmer than the land when you make your igloos on the ice. This is in the winter. This is something that I have heard and I know that it keeps the igloo much warmer than the land. (2001)

Igloolik hunters still keep their dog teams on the sea ice, not only because it is close to the community, but also because they think the sea ice is warmer than the land (John MacDonald, personal communication). Louis Alianakuluk has fond memories of his childhood at the ice camp, where they would play such games as "*amaruujaq, aammakasauti* (tag), *uqsuutaaq* and other games like *taqqiujaq*" (2001). Life at the ice camp, therefore, was not entirely occupied by hunting. The place offered the conditions for warm shelter and people found the time to entertain themselves. They were at home on the ice.

A Recurrent Topography

There have been several studies giving full or partial attention to Inuit knowledge and use of the ice. Nelson (1969) gives perhaps the most detailed explanation of ice environments yet written, providing an extensive lexicon of sea-ice terms. Nelson also points out that the Eskimos of Wainwright, Alaska, divide the environment into two separate entities: land and sea. In turn, they divide the sea ice into two different conditions: stable and moving (1969: 9). In similar fashion, Collignon, focusing her studies on the Inuinnait of the Victoria Island region, has described the perception of three types of surfaces that are "clearly opposed to each other" (1996: 100). These are *nuna* (the land), *hiku* (the ice cover) and *tariuq* (the sea). Freeman (1987) writes extensively about contemporary use of the sea ice and stresses that both the ice surface topography and ice movement are of "major concern for travelers" (1987: 74). He also illustrates some recurrent ice cracks in the Resolute Bay region (1987: 82).

MacDonald compiled a list of 117 terms related to snow and ice (1989) from three different sources: Schneider's *Ulirnaisigutiit: An Inuktitut-English Dictionary* (1985), a manuscript prepared by Spalding (1979), and Emile Immaruituq (an elder from Igloolik). Müller-Wille published a list of thirty-one Inuit words "which are used in geographical names to identify places according to their particular attributes such as snow and ice" (1986: 56). Of this list, however, only fifteen terms refer to the sea ice. Brody wrote about knowledge of the ice in the Igloolik area, and especially about Igloolik hunters' attitudes toward the moving ice (1976: 164). Riewe (1991) provides convincing data about the extensive use of the sea ice by the Inuit of Nunavut and analyzes Inuit sea-ice technology. He considers that "virtually the entire landfast ice region in Nunavut is currently utilized" (1991: 5). A report on traditional knowledge of Inuit and Cree communities compiled by the Hudson Bay Programme (Arragutainaq et al. 1995) offers a detailed description of Inuit knowledge of the sea-ice structure and dynamics in the Hudson Bay region. The report also provides a list of eighty-three sea-ice terms. Researchers of the Inuit Sea Ice and Occupancy Project are currently preparing several publications that will offer a more comprehensive understanding of Inuit Sea Ice use in Nunavut and Nunavik.

The topography of the ice is created every year when the landfast ice solidifies. Several features of that topography will recur year after year in the same locations owing to the configuration of the coastline, submarine features, and the action of the currents. The Inuit knowledge of this topography is not the result of mere observation and the memorization of ice features. Instead, it is the consequence of a comprehensive understanding of the behavior of Ikiq (the Fury and Hecla Strait) in relation to land formations, winds, currents, and tides.

Theo Ikummaq describes different features of Ikiq as a recognizable "structure of the ice" (2000a). To show the recurrence and stability of ice features, he tells a story about a trip he undertook in 1987 from Igloolik to Greenland. Before departing from the community of Arctic Bay, on the northern coast of Baffin Island, the travel party sought advice from an elder who had made a similar trip twenty-five years earlier. Explaining to the travelers what was the best route to cross from Ellesmere Island to the Greenland coast, he was able to give a detailed account of the topography of the ice. Here is part of the description the elder provided, as recalled by Ikummaq:

> [F]rom Makinson Inlet [on the east coast of Ellesmere Island] you are going to hit a pressure ridge, and then there is going to be less snow once you hit that pressure ridge, because, when the snow accumulates over the year, the older the ice the more snow it has. So there was a pressure ridge, he said, "Don't stop, don't use the first pressure ridge. Use the third one." So we had to cross over this one, cross over the next one, and cross over the next one until we hit solid snow, where it was smooth sailing. And then, you know, that had been twenty-five years prior, and then if you look at that you can pretty well determine it doesn't change much over the years. (2000a)

The travelers found the strait to be exactly as the old man had told them. Cracks, ridges, and snow features were just the same, twenty-five years later.

Inuit have several terms for ice formations and processes, making a simple classification of this lexicon extremely difficult since multiple categories occur depending on the approach of the speaker and, in cases where the terms were given in the context of an interview, on the direction suggested by the question. The terms reflect the dynamics of the sea-ice environment, from its formation in late summer/early fall to its breakup in late spring. Pauli Kunnuk (1990), describing the process of freeze-up at a bay in late summer and early fall, uses the terms *illuvalliajuq* (referring to a portion of the sea where a river flows, which usually freezes before the rest of the sea) and *qainguq* (when shorelines start to freeze). The ice condition *qainguq*, however, when carried away by the wind or the high tide, will be called *quvviqua*. In early fall, when the new ice stabilizes in the bay, *qainguq* will become *sikuaq* (thin ice formation, the first stage of the freeze-up) and in late fall or early winter, *sikuaq* will finally transform into *tuvaq* (landfast ice).

The terms can be organized, then, following different classification criteria. For instance, *iilikulaat, maniilagalaat, maniilait,* and *sikutugait* refer to different types of ice ridges. Leads (*naggutiit*) can be classified in terms of their relative position in relation to the floe-edge (*napakkutit, naggun*) or in terms of their origin (*atuarutit*). Shore cracks are also named separately (*tilliqpaaq, akulliq, salliq*) in relation to their position relative to the shore. Finally, the different structures around the floe-edge are also named. Figure 8.3 illustrates some of the features and processes of the sea-ice environment that have been named.

Classification of the terms is difficult mainly because the Inuit approach to the sea ice is experiential rather than theoretical. In order to make full sense, a term must be identified with the speaker, whose speech, in turn, refers to a specific place at a particular time of the year under particular circumstances. For this reason the exact date, location, and in some cases, the time of day at which some statements were made were recorded in this study. *Tilliqpaaq* ('the one that is higher than the rest'), for example, refers to the closest shore crack (as viewed from the land), and can only be fully comprehended in the context of a conversation.

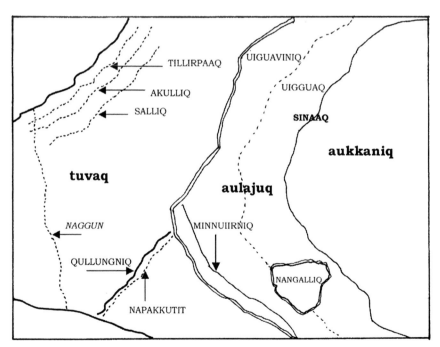

Figure 8.3 *Terminology of the ice: unmarked areas at left are land; double line in center is open lead.*

Place Names on the Ice?

Agiuppiniq, as shown above, is widely recognized as a *named* place. In fact, it is evident that within the community there are several recurrent ice features that are recognized and situated by their names. Theo Ikummaq explained that on 18 December 2000 (the day of the interview) there was a crack that the tidal currents, driven by the full moon, would have opened. It was an important crack, a good place to hunt seals through breathing holes. "If somebody is to ask 'Where is it?'" Ikummaq said, "it's at Agiuppiniq, and that describes that different structure on the sea ice. Where it had been grounding, and then piling and grounding, to the point where is now four or five feet high, and it's got a straight edge. That's what they look for. They go there and that's where they find the crack" (2000b).

Some of the names refer to polynyas (Akulliqpaaq, Aukkarnaarjuk, Aukkarnirjuaq, Kangillirpaaq), others to leads (Naggutialuk), general areas (Sikutuqqijuq), and outstanding ice formations, such as ridges or ice buildups (Agiuppiniq, Ivuniraarjuruluk). Finally, it is common to name portions of landfast ice in relation to a nearby, coastal place name. Igluliup Sikua, for example, refers to the landfast ice off the coast of Iglulik, while the sea ice off the coast of Avvaja is called Avvajap Sikua.

Some *icemarks* are of great importance for multiple reasons. Ivuniraarjuruluk, for instance, is an outstanding ice formation that is used for navigation, to scan the surrounding area and to obtain fresh water:

> If you got carried out to sea on the moving ice, and you were carried out in this general area, the moving ice would hug that place as it moves out ... when the moving ice detaches itself from Iglulik, it will usually move right out to sea, and that also goes for Ikiq. In that area which we refer to as Ivuniraarjuruluk, when the moving ice moves out, it will always rub through that place. This is when the wind is not blowing from **akinnaq** [west, northwest]. (Louis Alianakuluk, 2001)

As noted above, Ivuniraarjuruluk is not only used as a navigational aid when hunters are carried away on the moving ice. Alianakuluk refers to that feature as a good and useful place: "As it is usually high from pressure ridges built up, it is a good place to scan the surrounding area, it is useful as it is pressure ridge built up, a good place to scan with your telescope. This is also a place where you can get unsalted snow, it is a useful place" (2001).

All the place names attributed here to ice formations are descriptive, either of the structures themselves or of their relative positions. It might be argued, then, that some of these names are mere descriptions of ice conditions. However, these names are widely known in the community and, more importantly, people attach to these names concrete and consensual geographic locations. Besides, place names related to land formations among the Inuit of Igloolik are also frequently descriptive. It is important to highlight that neither the names given here nor the ice formations plotted form an exhaustive list. The Inuit Sea Ice and Occupancy Project is presently conducting research on Inuit sea ice use in several communities. For an updated study of Inuit sea ice in Igloolik see Laidler and Ikummaq (2008).

The ice topography is the consequence of recurring factors, and it is the understanding of these factors that makes the topography predictable and, therefore, a place that someone could call *home*. Dealing with the moving ice presents different challenges and also implies an understanding of the complex relationships.

The Moving Ice

If the location of the camp at Agiuppiniq was related to some stable topographic feature of the ice, the location of the other two settlements, Pingiqqalik and Uglit, was connected to an opposite condition: the changing but predictable nature of the moving ice. People who in late winter and throughout the spring moved to those camps were looking to take advantage of the moving and fertile "land of the walrus" (see below), which was brought in and taken away again and again by the tides and the wind.

Parry describes in his journal how, in March 1823, several families would leave Iglulik for Pingiqqalik, "where the walruses were more easily procured" (1824: 418), and where the snow-huts were built "on the ice in immediate contact with the beach and the open water" (1824: 424). When Parry visited Pingiqqalik on 2 April, the open water was about three miles from the camp. He describes how at the end of April some people would move to the shores of the small island of Uglit (1824: 428, 435), where they would be even closer to the moving ice where walruses abounded.

The "land of the walrus" is in fact a large portion of floating ice that moves with the tides and the winds back and forth within the fertile polynya southeast of Igloolik Island. Life at Uglit and Pingiqqalik is frequently spoken of by contemporary Igloolik elders. Noah Piugattuk, for instance, describes in detail how Inuit would favor both Pingiqqalik and Uglit over Iglulik when they wanted to hunt walrus in the spring, pointing out that the floe-edge was closer, especially at Uglit, where "the ice contacts are more frequent in comparison to that of ice edge located more toward the land" (1989). Referring to earlier times, he adds that "with the remains of the sod houses in Pingiqqalik, there is evidence that this location used to have numerous inhabitants same as it is with Igloolik. As well as Uglit, the islands. These locations were central habitats" (1989). The relative positions of the traditional camps of Iglulik, Pingiqqalik and Uglit are directly related to the whereabouts of the moving ice. The Iglulingmiut are widely known for their familiarity regarding the moving ice, in the sense that they have regularly hunted on it and have an understanding of its codes and of the factors involved in its movement.

This intimacy with moving ice involves both knowledge and respect. Aipilik Inuksuk warned that the moving ice is not a good place to live: "It is not a very good place to be in and you can never relax and you have to be ready to escape to more stable ground" (1988). At the same time, however, a good hunter would feel safe while being in such a potentially dangerous place because, if he pays attention to all the factors, "there is absolutely no concern about the dangers regarding the ice" (1988). A competent moving-ice hunter, then, can be defined in this context as someone who understands the factors involved in the movement of the ice and who waits patiently for the right conditions. "It is never safe when people are impatient to use the ice when it is newly formed. As far as anyone can remember people have had accidents on ice due to their impatience and I have known cases of accidents that happened regarding the water and the ice," concluded Inuksuk (1988).

Richard Nelson's notes on the Inupiat of Wainwright, Alaska, permits comparison with another group also oriented to sea-ice hunting. After stating that most hunting and traveling during the fall, winter, and spring is carried out on the solid landfast ice, Nelson points out that "seldom do hunters venture onto the mobile ice pack beyond its margins … too many times in the past men have gone beyond the floe and become trapped on a drifting floe when the wind or current carried the ice away, opening a wide lead which prevented their return to the land" (1969: 33).

In contrast, Igloolik hunters have regularly practiced hunting (mostly walrus hunting) on the moving ice. In the words of Theo Ikummaq, "Igloolik is unique in the Baffin region, where we are the only ones who go out onto moving ice to hunt for walrus. Maybe Repulse does it. Hall Beach does it also, but not to the extent that we do" (2000a). Through generations of careful observation, Igloolik hunters have developed the knowledge to predict the changing behavior of the moving ice. This knowledge is learned from a very early age, beginning with the icing process of lakes and rivers, and finally on the moving ice itself. Aipilik Inuksuk remembered his learning experience:

> [W]hen I was a child I was taught about the ice conditions of the lakes and ponds. I was told that the newly formed ice on the lakes are very hard to crawl out from. This is what I learned when I was a child but it was later on when I was out hunting with the men when I was taught about the moving ice conditions. To this day it seems that I have never really understood the moving ice and its nature. (1988)

Although the factors related to the moving ice are many and their relationship complex, it can be said that, along with the topography of the coastline, there are two main factors hunters consider before venturing on the moving ice: the wind and the tides. The tides, in turn, are connected to the time of the day and the phases of the moon. Not everyone ventures onto the moving ice nowadays in Igloolik, but there are several people (including some young hunters) who have the knowledge and skills to do it. Furthermore, hunting at the floe-edge also requires an understanding of the ice behavior. The following description of the role of winds and tides in the behavior of the moving ice is based on my observations during a trip to the moving ice with Maurice Arnatsiaq in March 2001, as well as interviews and conversations with Louis Alianakuluk, Theo Ikummaq, and George Qulaut, and draws on interviews with Noah Piugattuk and Aipilik Inuksuk in the Igloolik Oral History Database.

The Winds

This chapter will not discuss Inuit knowledge of winds in detail (see Fortescue 1988 and MacDonald 1998). It is sufficient to know that the Inuit of Igloolik designate four primary winds: *Uangnaq* (WNW), *Kanangnaq* (NNE), *Nigiq* (ESE), and *Akinnaq* (SSW) (MacDonald 1998.: 181). MacDonald points out that these winds constitute two pairs of counterbalancing winds, "one on the *Uangnaq-Nigiq* axis, the other on the *Kanangnaq-Akinnaq* axis" (1998:182). He also points out the symbolic value of these opposites, especially in the pairing of Uangnaq and Nigiq, which "are said to retaliate against each other" (1998: 182). As I will show now, this opposition (and the understanding of its occurrence) goes beyond the symbolic to play a leading role in predicting the mood of the moving ice.

The influence of the wind in determining the location of the floe-edge can be observed from the community of Igloolik. When the prevailing wind *Uangnaq* (WNW) blows, the phenomenon known in Inuktitut as *tunguniq* (reflection of the open water in the sky) looks well defined, occupying a good part of the southeast horizon, indicating the proximity of open water. On the other hand, when *Nigiq* blows, the reflection is barely visible on the horizon, which means that the open water is further away.

On 1 April 2001, Louis Alianakuluk described the conditions of the moving ice as follows: "[W]hen the *Nigiq* wind blows, it blows in the moving ice to the edge, now today, it is certain that it is iced over from the moving ice. Then in alignment of that strait the ice tends to part as well. That is the way it is" (2001). People hunting from the floe-edge, then, wait for *Uangnaq* to blow, while hunters wanting to hunt on the moving ice rely on *Nigiq.*

Generally speaking, then, we can say that a northerly wind will carry the moving ice away while a southerly wind will bring it back. For the hunters residing in Uglit and Pingiqqalik, on the other side of the polynya, the situation is reversed. When asking about the influence of the winds on the moving ice, it is therefore important to know *where* the speaker situates himself in his speech. Aipilik Inuksuk described the moving-ice process from Iglulik:

> The north wind [*Uangnaq*] causes the ice to break up and drift away. After the wind, the weather improves and the area that was left by the ice freezes and that is the time when the current will move the ice back and forth to each other. That is the time when

it is the least dangerous to hunt. "Agliurisimajuq." (That is even when there is a south wind [*Nigiq*] the ice still moves back and forth to the main solid ice.) That is the time when you can hunt without too much worries about drifting away. The wind will cause the ice to move together then it will start to drift apart and that is said to be the best time to hunt. This is true and that is the nature of the moving ice. (1988)

Therefore, the northerly winds that take the moving ice away from Iglulik bring it closer to Pingiqqalik and Uglit. Noah Piugattuk remembers that when northern winds blew at Pingiqqalik it was "the prime time to hunt walrus" (1989). That was the time when the moving ice was in a position to make contact with the landfast ice. The hunters at Pingiqqalik would wait until the northwest wind (*Uangnaq*) shifted to northeast (*Kanangnaq*); that was the time to move onto the moving ice. Noah Piugattuk remembers a revealing story from his times at Pingiqqalik, "when the sun was higher" (springtime):

When they wake up, it was a customary practice when one woke up the first thing was to step outdoors and relieve bodily fluids [urinate]. As it was a custom to step outdoors and advise the rest that the winds were blowing from the easterly directions. He will say that the winds have shifted from northerly to easterly directions. The winds are blowing in line with the shoreline from the direction of the northeast. He comes out with this wind direction. The person may be a male or a female. The small lake that is located near the sod houses is shaped so it faces the northerly, it is little long and follows the direction of northerly position. They call the little lake "Qukturaaq" (thigh) when the person goes out and observes the winds. The wind is coming from the direction of the moving ice. This means that the moving ice will land at the fast ice. When he gets back indoors he will announce the thigh has been broken, the person may be a male or a female. If that person was one of the elders he will announce that the thigh has been broken. Then at that instant the men immediately become lively in anticipation, knowing that the moving ice will make contact with the fast ice. This was measured by the lake, the winds were cutting across this little lake named Qukturaaq therefore they say the winds have broken the thigh. (1989)

Noah Piugattuk's story (also cited by MacDonald 1998: 180–181) gives an indication not only of the role of the wind regarding the moving ice but also of a holistic approach that sees connections between the ice on the lake, the wind, and the position and stability of the moving ice. The wind, however, can be deceiving, and Noah Piugattuk warns about the dangers of entrusting one's safety to the wind direction:

The two opposing winds, namely the south and the north. If the wind from either direction had blown for a prolonged period of time, then this will be followed by the shift of wind direction to the opposing direction and it will blow with force. Therefore, after the wind has blown from the direction of the moving ice for a period of time, it is with certainty that the winds will shift to the north, and thus the hunters were advised of dire consequences if they were to remain in the vicinity of the moving ice. The younger people were taught about the sea ice environment. (1989)

In fact, all accounts from experienced hunters make it clear that paying attention to the winds is not enough if one is to safely venture onto the moving ice. As important as the wind is, tidal currents play a central role in determining ice movement.

The Tides

Winds are not always reliable indicators of the future behavior of the moving ice. A sudden change in wind from **Nagiq** to **Uangnaq** would cause the ice to break and hunters would be carried away. The changing behavior of the winds, then, makes it a risky business to go hunting on moving ice. It is for this reason that observation of the tides is essential. Louis Alianakuluk advises that hunters who plan to spend time on the ice must always pay attention to the tidal currents, especially in the winter when temperatures are at their extreme. "The ice tends to break off and detach itself from the landfast ice," he says. "It is dangerous when you do not pay attention; it may even look as if it would not break off" (2001).

How are the tides understood, and how do they influence the course of the moving ice? There are two main factors to take into account: the strength of the tidal currents in relation to the phases of the moon, and tracking the timing of tidal shifts. On 3 April 2001, during a hunting excursion to the moving ice, Maurice Arnatsiaq made a hole in the thin ice with his *savik* (snow-knife) and lay down for several minutes, his eyes fixed on the water as he attempted to determine the tide. He later told me that we would not venture onto the moving ice because the tide was going to take the ice away sometime that evening. The current was already going out.

Aipilik Inuksuk said that "in the morning when it is low tide then the rest of the day the movement of the current will be inwards toward the main ice. That is what happens. Sometimes before the movement of the current is inwards it stays still and there is absolutely no doubt that you can make it to the main ice after hunting on the new ice" (1988). This indicates that when tidal and wind conditions are favorable, hunters can go onto the moving ice even knowing that the tide will temporarily drive the ice away from the landfast ice (note: the technical term for this phenomenon—when low tides stay still—is "slack tide"). Aipilik Inuksuk, situating himself in Igloolik, described a common method of determining the current's direction and how the tidal currents evolve day after day:

You have to pay attention to the current, drop something to find out which direction the current is going, whether it is high or low tide. When it is low tide ... what does it mean? ... It means that you are moving southwards ... That is how the current can tell you ... see what time it is ... Nowadays we have watches ... so [we can] see what time the low tide had stopped. Before they had watches they used to use the daylight ... how dark was it when the low tide started to slow down. For instance it is nearly two so if the tidal current stopped at this hour then the next day the tide would stop at three. The tidal current is always moving ahead almost an hour every day ... You always have to pay attention to the tidal current and mark the time when the current has stopped moving outwards and the time when the current started moving inwards. Always keep an eye on the tidal current. (1988)

Therefore, the moving-ice hunters must know that the tidal shift "moves" ahead about an hour every day. However, knowledge of the tidal shift must be supplemented with the equally important knowledge of the tidal strength, as determined by the phases of the moon (for a more detailed study on the Igluling-miut knowledge of the moon and other astronomical phenomena, see MacDonald 1998). During an interview on 18 December 2000, Theo Ikummaq explained that

if you look at the new moon, at nine o'clock in the morning the tide is going out. At three o'clock in the afternoon the tide is going in. The same thing with the full moon. The full moon and the new moon are pretty much the same, in that at nine o'clock in the morning the tide is starting to go out. I mean, away from the bay. And at three in the afternoon the tide is going in; it's rising. So it helps you to determine at what time you will be going to the floe edge, or it determines how long you can stay on the moving ice, from the phase of the moon. (2000b)

It is important to make clear, again, that the conditions described by Ikummaq are only valid for the particular day on which he made that statement. Aipilik Inuksuk also pointed out that "when there is full moon or no moon at all then those are the times that the current is the strongest" (1988). As the moon moves toward full moon, therefore, the currents become stronger, but as it starts to wane the current slows down. "People sometimes get marooned on ice because of the strong current. When you are at the floe edge you should always watch out for the high or low tide" (1988).

Hunters know with certainty that strong tidal currents will detach the moving ice from the landfast ice. Noah Piugattuk says that "when in the full moon, when the currents pick up strength, it is predictable that it might break up the ice, causing some ice ridges when the ice pans were to gather, which resulted in further breakups" (1989). The predictability of tidal currents gives the hunter assurance that with the right wind conditions, even if he gets carried away on the ice, he will land again at a certain time of the day, when the tide comes in. Of course, it is essential that he observes the evolution of the tidal shift. Furthermore, knowledge of the strength of the tides in relation to the phases of the moon will help him assess the relative stability of the moving ice, by estimating the strength of the tidal currents.

Naming the Moving Ice

The extensive moving-ice vocabulary developed by the Inuit of Igloolik is an indication of the knowledge gained by careful, long-term observation of its behavior. The processes described by these terms are so complex that any attempt to classify them may result in oversimplification. While there are some terms whose meanings are easily translatable into English (e.g., *tuvaq:* landfast ice) there are many others that can only be understood within the framework of the environment as a whole, including people's interaction with that environment (e.g., *tulak:* to arrive). In this section, I comment on a few significant terms. For explanatory purposes, I have separated the terms into four levels.

At the first level there are some general terms that can be easily identified with generalized non-Inuit categories, and that refer to the broad distinction between stable and moving ice. *Tuvaq* means landfast ice, while *Aulajuq* means moving ice. Between these two is *Sinaaq,* the floe-edge. At the same level, there are some words that refer to a subject (a person or an animal) coming from or going to the moving ice. *Tulak* ('to arrive'), a concept applied in the context of marine travel, implies touching a solid object (whether land or landfast ice). Noah Piugattuk says that the word *tulak* is used to describe the movement of marine animals off moving ice and the precise moment when they land on land-fast ice (1989). *Pijuk,* on the other hand, refers to the opposite action: to move onto the moving

Figure 8.4 *Terminology applied to the moving ice*

ice from the land-fast ice at the time when the two—moving ice and land-fast ice—make contact.

A second level of terms refer to several conditions and transformations which take place at the moving ice and the floe-edge. *Uiguaq* is a thin layer of newly formed ice that attaches to the edge of the floe-edge. With the stabilisation of the moving ice (in Igloolik, when there is a southeast wind), *Uiguaq* transforms into *Uiguaviniq*. *Uiguaviniq*, then, takes the place of *Uiguaq*, which "moves" further away from the floe-edge. These two conditions are easily identified. *Uiguaq* is generally a smooth, darker ice, while *Uiguaviniq* is whiter and has a build-up of ice ridges. The latter, of course, is much safer to travel on than the former. *Uiguaviniq* eventually becomes part of the land-fast ice.

A third level of terms refers to ice formations. At the floe-edge, for instance, *Uigutarniq* is constituted by small ice forms on the edge, usually extending a few meters from the floe-edge. *Tuggarniq* refers to ice cracks on the land-fast ice that are produced by the strong southerly winds that have blown the moving ice onto the land-fast ice.

Finally, at a fourth level, a few terms refer to the process of separation (see Figure 8.5). *Iriqqaaq* is the general name for the separation process of the moving ice from the land-fast ice. *Qaattuq* refers to open water along the edge when the

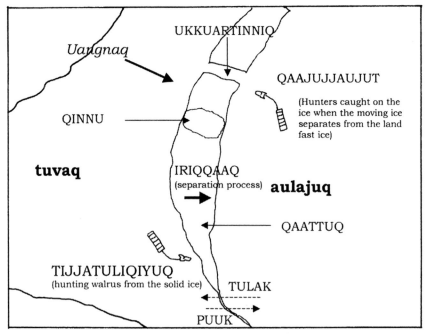

Figure 8.5 *Moving ice features and processes: dark outlined areas at left are land: outlined area in center is open lead.*

moving ice moves from the land-fast ice. *Iriqqaarujjaujut* refers to the process through which soft, compacted snow known as *qinnu* begins to surface as a consequence of the moving ice separating from the land-fast ice (*qinnu* is usually known for presenting dangerous travel conditions). At the same level, a few terms refer to people's relative position at the time of the separation process: *qaajjuj-jaujut* refers to hunters caught on the ice when the moving ice separates from the landfast ice, while *uukkarujjaujut* describes when hunters are caught on ice that has drifted away from the landfast ice after a crack happens on the seaward side of the floe-edge. Both *qaajjujjaujut* and *uukkarujjaujut*, as well as *uukkaqtuq*, refer to hunters being taken away on the moving ice.

Tulak, as described above, means going from the moving ice onto the landfast ice. A way of landing is to gain the solid ice via *ukkuartinniq*, which is a strip of ice that runs from the moving ice to the edge of the landfast ice. A story by Emile Imaruittuq describes this process:

> There is a saying in connection to this word. The boys had a task of blocking the entrance of the igloo when they retired for the night. The word to describe this action is *Ukkuaq*. They were told to do it so that should they ever go adrift on the moving ice, they will be able to reach the landfast ice through an *Ukkuartinniq*. When the hunters are out on the moving ice they are sometimes taken out on it away from the edge. In most of the cases, hunters will usually get on to the landfast ice on these *Ukkuartin-niit*. (1990)

Recurrent Features of the Moving Ice

In the same way that many formations on the landfast ice are recurrent, the moving ice can be spatially identified and located, although only an animated map could illustrate the process in its entirety. In fact, the very existence of Iglulik over a period of at least four thousand years (according to archaeological evidence) is directly related to the recurrence of the southeast polynya.

In his journal, Parry noted "a large extent of sea keeping open to the southeastward through the winter" (1824: 418)—a clear reference to the southeast polynya. Contemporary Inuit such as Joe Iyerak maintain that open water and its associated features always recur in the same general areas (2000). *Sikutuqqijuq* (floating, young ice, where walruses are usually found) is an example of a place that can be identified and even plotted on maps (as Louis Alianakuluk illustrated in Figure 8.3), regardless of the course of its changing but predictable movements. The same can be said about *Sinaaq* (the floe-edge), *Aulajuq* (the moving ice) and even *Uiguaq*.

The changing nature of the moving ice can be understood and, therefore, predicted. This is the way Igloolik Inuit deal with the moving ice on a regular basis, managing and reducing risk to an acceptable level. As pointed out above, this knowledge is partially reflected in the development of an extensive vocabulary, created not only to describe the processes of the moving ice but also people's interaction with it. The hunters' ability to predict the movements of the ice is manifest in Louis Alianakuluk's astonishing recollection of crossing Ikiq (the Fury and Hecla Strait) on the moving ice. The particular moving-ice conditions described in the story were not associated with a polynya but with the early stage of the freeze-up. It was a late-fall crossing in a portion of the strait that usually stabilizes around Christmas. The crossing was about thirty kilometers and required overnighting on the moving ice:

> In fact, I took part once, that is, crossing the moving ice. From the north tip of Nirlirnaqtuuq to Nuvuksirpaaq the currents are stronger as it is almost in line with Aukkarnirjuaq ... They can spend the night on the ice; indeed, they can make an overnight camp on the moving ice. This of course is made possible with dog team. I believe they used to make igloos from thin ice when they needed to spend the night out. It was not crossed every time, only on few occasions, we use to hear certain individuals having crossed the moving ice. This happens when the strait still had not frozen over, but is still moving. (2001)

Louis Alianakuluk remembers that they used to hear people saying "*aulajukkuguuq ikaarqtut*" ('it is said that they have crossed the moving ice'): "There was a common phrase that goes '*ikiruuq nuqarttuq*' [it is said that Ikiq has stopped]. This meant that the moving ice was no longer moving" (2001). Although the crossing was done only on rare occasions, probably associated with food needs, the trip reveals a deep understanding of the currents and, in turn, the hunters' faith in their predictive abilities based on the knowledge of relationships between tides, currents, coastlines, and winds.

A Holistic Understanding

The extensive ice-related terminology, as well as the ice names, indicates a comprehensive understanding of the sea-ice features and behavior. That is, the hunters of Igloolik not only remember recurring icemarks but also know why those icemarks are there. The formation and position of Agiuppiniq, for instance, is understood in relation to the currents coming from the northwest, and that current is in turn understood in relation to the position of the island Similik (Ikummaq 2000a). It is also clear that hunters have decoded the sea-ice behavior through their understanding of lunar phases, tidal currents, and wind.

Inuit hunters would also predict a wind or tidal change by observing the behavior of sea mammals. Noah Piugattuk points out that in the spring, basking seals (*utaq*) were an indication that the water was no longer freezing (1989): the seals would stop going onto the moving ice when the thinning ice was no longer able to support their weight. George Kappianaq says that "game animals are also used to determine what is to come ... The well-experienced moving-ice hunters, who have learned the conditions, used to say that when the wind had just shifted to the *Nigiq* direction, upon reaching the moving ice, they will discover that the walruses are not situated on the edge so they used to go deeper into the moving ice away from the direction of the land" (1997). A simple list of Inuktitut terms, therefore, says little about the Inuit understanding of the ice. Astronomical and biological observations are intrinsically linked to the knowledge that allows the hunters of Igloolik to exploit the moving ice on a regular basis.

The ice vocabulary recorded in my research reveals another particularity: the terms not only refer to ice states and processes but also to human interaction with the sea ice. That is why it is so difficult to merely *talk* (or *write,* for that matter) about this issue. The true nature of the ice can only be understood through a lifetime of apprenticeship. Careful observation, knowledge passed down from generation to generation, and the very experience of hunting, traveling, and living on the ice completes this holistic approach.

The existence of verbs such as **tulak** or **pijuk,** and even others such as **tijjatu-liqiyuq** (hunting walrus from the solid ice), reveals a point of view in which the environment is viewed not as a separate entity but rather in connection with the human experience of it. Compared to some Western classifications of sea ice (see, e.g., World Meteorological Organization 1970), the Inuit approach recognizes the interaction between humans and environment in the process of naming the ice environment.

The knowledge of currents and tides reveals another feature of this approach. It is not only what is seen that matters, but also what is not seen. While talking about navigation, Theo Ikummaq pointed out that "we teach the younger [hunters] what we see but also what is underneath. The dangers sometimes are below and you don't see them" (November 2000, fieldnotes). The knowledge of the currents underneath the ice will help not only in navigation but also in hunting. As Ikummaq stresses: "How you are going to get that seal through a breathing hole is determined by what you know of what the current is, because that seal is going to be facing upwards toward the current. And then when you know that that seal is going to face upwards toward the current, you know where the seal is, so you know where to hit your target. And again, that's one of the reasons why we were taught to keep the current in mind always" (2000a).

The Inuit knowledge of the ice is not only visual. The ice can also be *heard.* Describing a failed attempt to cross the Fury and Hecla Strait at the north tip of the Melville Peninsula, George Qulaut said, "I knew that the ice further out was still going back and forth. We could hear a lot of movement, crashes ... It was almost like if the ground was shaking a bit, the ice was shaking a bit. We could feel it. The ice was breaking up at parts" (2000). George Kappianaq adds that when there are strong winds and the tides are coming in, the moving ice can be pushed toward the landfast ice with such violence that it may cause the moving ice to rift. This violent movement produces streams of open water, making the area extremely dangerous for hunters. Kappianaq says that one can recognize this condition by the sound. The opening of the cracks caused by the tides coming in and strong winds is termed **Qulluaq,** and can be recognized because it is accompanied by "the sound of boiling" (1997).

When the separation process takes place and the *qinnu* (soft, compacted snow) surfaces, the hunter taken by the moving ice can use that snow formation to *tulak* (land on the landfast ice). According to Pauli Kunnuk, this is also a condition that can be identified by its sound: "One can hear it being crushed ... Even if the distance between the moving ice and the landfast ice might cover some distance, as long as one can hear the *qinnu* being crushed, the sound which is usually not a crushing sound but something soft, one can travel on it" (1990). While hunting seals through their breathing holes in thin ice, the successful hunter will also rely more on his hearing than on his sight. Maurice Arnatsiaq says that "the hunter will sense the seal approach" (fieldnotes, 2001).

The nature of the understanding of the sea-ice topography is revealed in the hunters' knowledge of where the recurrent or multi-year ice comes from, as this observation made by Pauli Kunnuk illustrates:

> I do not exactly know where the multi-year ice comes from, that is before the multi-year ice passes through the *Aukkanirjuaq*. The multi-year ice that usually is seen where the walruses usually stay is more yellowish than the multi-year ice that passes through the *Aukkanirjuaq*, and sometimes one can tell that some of the multi-year ice comes from the direction of *Piling* which is rare, but it appears to come from that direction on rare occasion. These multi-year ice [formations] must be the ice that originated from our water, it does not appear to be possible that they would come from different locations like passing from Hudson's Bay, and in addition some of these multi-year ice [formations] appear to be huge floes. (1990)

A hunter, therefore, must learn not only how to identify recurrent features but also where these features come from, and why they recur year after year.

Conclusion

For a good part of their lives, the Inuit of Igloolik move across, talk about, and live on the sea ice. In the past, the sea ice even constituted their home during part of the spring. It is not surprising, then, that they have developed a thorough knowledge of the ice, its nature, and its processes. The ice topography is a changing but recurrent environment. Icemarks are familiar to most people in the community, and some of them have names, as land features do. The ice environment is constituted by both the semipermanent and recurrent landfast ice, and the changing but predictable moving ice. Confidence and competence to travel with minimal risk and inconvenience in that environment involves knowledge of the topographic features and understanding of the ice movement. This knowledge has been developed (and is still developing) and shared through generations.

The Iglulingmiut show a unique attitude toward the moving ice that combines respect with the confidence to venture onto the ice even when the tide takes it away. Risk has been reduced to a minimum through understanding the processes involved in ice movement, and cultivating the patience to wait for the right conditions. This approach reveals a holistic understanding of the environment and the place of humans in it. It is not an environment that is conceived as a separate entity but one in which human interaction is always present.

Acknowledgments

The research on which this chapter is based would not have been possible without the financial and logistical support of the Igloolik Research Center, the Canadian Circumpolar Institute, the Killam Trust, the Wenner-Gren Foundation for Anthropological Research, and the Anthropology Department of the University of Alberta. I would also like to thank Eric Higgs, Milton Freeman, Cliff Hickey, John MacDonald, Alexina Kublu, Lorraine Kiel, and Carol Berger for their helpful comments.

References

Alianakuluk, Louis. 2001. Interview for Igloolik Oral History Project. Igloolik: Archives of the Inullariit Society, Igloolik Research Center (IE-477).

Arragutainaq, Lucassie et al. 1995. *Traditional Ecological Knowledge of Environmental Changes in Hudson and James Bays.* Ottawa: Canadian Arctic Resources Committee; Sanikiluaq, NWT: Environmental Committee, Municipality of Sanikiluaq; Ottawa: Rawson Academy of Aquatic Science.

Brody, Hugh. 1976. "Land Occupancy." In *Inuit Land Use and Occupancy Project, Volume One,* ed. Milton M. R. Freeman. Ottawa: Indian and Northern Affairs.

Collignon, Beatrice. 1996. *Les Inuit: Ce Qu'Ils Savent Du Territoire.* Paris: L'Harmattan.

Crowe, Keith J. A. 1969. *Cultural Geography of Northern Foxe Basin.* Ottawa: Department of Indian Affairs and Northern Development.

Fortescue, Michael. 1988. "Eskimo Orientation Systems." *Meddelelser øm Grønland, Man and Society* 11.

Freeman, Milton M. R. 1987. "Contemporary Inuit Exploitation of the Sea-Ice Environment." In *Sikumiut: People Who Use the Sea-Ice,* ed. Alan Cooke and Edie Can Alstine. Ottawa: Canadian Arctic Resources Committee.

Ikummaq, Theo. 2000a. Interview for Igloolik Oral History Project. Igloolik: Archives of the Inullariit Society, Igloolik Research Center (IE-466).

———. 2000b. Interview for Igloolik Oral History Project. Igloolik: Archives of the Inullariit Society, Igloolik Research Center (IE-478).

Imaruittuq, Emile. 1990. Interview for Igloolik Oral History Project. Igloolik: Archives of the Inullariit Society, Igloolik Research Center (IE-101).

Inuksuk, Aipilik. 1986. Interview for Igloolik Oral History Project. Igloolik: Archives of the Inullariit Society, Igloolik Research Center (IE-004).

———. 1988. Interview for Igloolik Oral History Project. Igloolik: Archives of the Inullariit Society, Igloolik Research Center (IE-035).

Iyerak, Joe. 2000. Interview for Igloolik Oral History Project. Igloolik: Archives of the Inullariit Society, Igloolik Research Center (IE-463).

Kappianaq, George. 1991a. Interview for Igloolik Oral History Project. Igloolik: Archives of the Inullariit Society, Igloolik Research Center (IE-167).

———. 1991b. Interview for Igloolik Oral History Project. Igloolik: Archives of the Inullariit Society, Igloolik Research Center (IE-173).

———. 1997. Interview for Igloolik Oral History Project. Igloolik: Archives of the Inullariit Society, Igloolik Research Center (IE-412).

Kunnuk, Pauli. 1990. Interview for Igloolik Oral History Project. Igloolik: Archives of the Inullariit Society, Igloolik Research Center (IE-093).

Kupaaq, Michel. 1987. Interview for Igloolik Oral History Project. Igloolik: Archives of the Inullariit Society, Igloolik Research Center (IE-117).

Laidler, G. J. and Ikummaq, T. 2008. "Human Geographies of Sea Ice: Freeze/thaw Processes around Igloolik, Nunavut, Canada." *Polar Record* 44 (229): 127–153.

Lyon, Captain G. F. 1970 (1823). *The Private Journal of Captain G.F. Lyon of H.M.S. Hecla During the Recent Voyage of Discovery Under Captain Parry.* Barre, MA: Imprint Society.

MacDonald, John. 1989. "Snow and Ice: Inuit Terminology." In *Touching North,* ed. Andy Goldsworthy. London: Fabian-Carlsson.

———. 1998. *The Arctic Sky: Inuit Astronomy, Star Lore, and Legend.* Toronto: Royal Ontario Museum and Nunavut Research Institute.

Mathiassen, Therkel. [1928] 1976. *Material Culture of the Iglulik Eskimos.* New York: AMS Press.

Müller-Wille, Ludger. 1986. "Snow and Ice in Inuit Place Names in the Eastern Canadian Arctic." *Proceedings of the Forty-Second Annual Eastern Snow Conference.* Montreal: McGill University Press.

Nelson, Richard. 1969. *Hunters of the Northern Ice.* Chicago: University of Chicago Press.

Nuttall, Mark. 1992. *Arctic Homeland: Kinship, Community and Development in Northwest Greenland.* Toronto: University of Toronto Press.

Parry, Sir William Edward. 1969 (1824). *Journal of a Second Voyage for the Discovery of a Northwest Passage From the Atlantic to the Pacific.* London: J. Murray.

Piugattuk, Noah. 1989. Interview for Igloolik Oral History Project. Igloolik: Archives of the Inullariit Society, Igloolik Research Center (IE-054).

———. 1991. Interview for Igloolik Oral History Project. Igloolik: Archives of the Inullariit Society, Igloolik Research Center (IE-181).

Qulaut, George. 2000. Interview for Igloolik Oral History Project. Igloolik: Archives of the Inullariit Society, Igloolik Research Center (IE-476).

Rasing, W. C. E. 1994. *Too Many People: Order and Nonconformity in Iglulingmiut Social Process.* Nijmegen: Rish and Samenleving.

Rasmussen, Knud. 1930. "Intellectual Culture of the Iglulik Eskimos." *Report of the Fifth Thule Expedition 1921–24*, vol. 7. Copenhagen: Gyldendalske Boghandel.

Riewe, Rick. 1991. "Inuit Use of the Sea Ice." *Arctic and Alpine Research* 23: 1.

Schneider, Lucien. 1985. *Ulirnaisigutiit: an Inuktitut-English Dictionary of Northern Quebec, Labrador, and Eastern Arctic Dialects (with an English-Inuktitut Index)*. Quebec: Les Presses De L'Universite Laval.

Spalding Alex. 1979. *Eight Inuit myths: Inuit unipkaaqtuat pingasuniarvinilit*. Ottawa: National Museums of Canada, V. 59

Wenzel, George. 1984. "Archaeological Evidence for Prehistoric Inuit Use of the Sea-Ice Environment." In *Sikumiut: The People Who Use the Sea Ice*, ed. Alan Cooke and Edie Can Alstine. Ottawa: Canadian Arctic Resources Committee.

World Meteorological Organization. 1970. *WMO Sea-Ice Nomenclature*. Geneva: Secretariat of the World Meteorological Organization.

PART 3

Linkages and Meanings of Landscapes and Cultural Landscapes

Chapter 9

Visions of the Land

Kaska Ethnoecology, "Kinds of Place," and "Cultural Landscape"

Leslie Main Johnson

In this chapter I present a synthesis of my understanding of Kaska landscape ethnoecology. I have attempted to discern Kaska "kinds of place" or cultural ecotopes, and I have also sought to gain a broader-scale understanding of the *meaning* of the land, the understanding people have of the land, and how people learn about the land, to enable a fuller understanding of Kaska ethnoecology.

The traditional territory of the Kaska Dena straddles the 60° parallel in western Canada and lies within a mixed region of mountains and broad uplands traversed by major rivers. It is part of the Boreal Forest and also contains areas of alpine tundra in the mountains. The Kaska Dena are Athapaskan speakers and, like other Dene peoples, were and are hunters and fishers. Formerly the Kaska were highly mobile, traveling widely through their territories in a seasonal round but also shifting areas where they hunted and trapped in response to other factors, including changes in environment, animal populations, intergroup and intragroup relationships, and locations of trading posts. Contemporary Kaska who remain in the territory generally reside in larger settlements in the region, though extended hunting trips still occur. Many people still have traplines, camps, and cabins, and take every opportunity to spend time on the land.

The research was carried out in the southern Yukon and adjacent northwestern British Columbia from 1998 to 2004, in a series of visits during the spring to early fall seasons. I attempted to learn about Kaska "kinds of place" or cultural ecotopes through engagement in activities on the land with Kaska elders and other community members and through audio-visual and visual documentation, emphasizing an experiential learning that is culturally appropriate for Dene peoples. I also sought to gain a broader-scale understanding of the meaning of the land, the understanding people have of the land, and how people learn about the land by listening to narratives and participating in activities in the community and on the land.

The focus of my research was "landscape ethnoecology," by which I meant people's cultural understandings of, and practices on, the land, whether these be cognized or embodied knowledge. I began my investigation from the perspective of wishing to learn about *kinds of place* in Kaska understanding, as part of a broader comparative project investigating landscape ethnoecology with several different Canadian First Nations of the general northwestern region of Canada. I initially conceived the need for this investigation when I realized that my notions of *habitat*, shaped both by my participation in Euro–North American culture, broadly speaking, and by my particular education within scientific ecology and natural history, did not seem similar in nature or specificity to those of Gitksan consultants with whom I worked on an ethnobotany and traditional medicine project in northwest British Columbia (Johnson 2000). I realized that I needed to look further into understanding of place kind for the Gitksan and Witsuwit'en in northwest British Columbia (BC) with whom I was then working, but I also felt I needed to look at similar questions in other places and with other peoples to attain a less limited perspective on peoples' understanding of landscape.

The word "ethnoecology" has been used in many ways in recent years. In this work I investigate landscape ethnoecology, which deals with a people's relationship to the land or their environment, and their understanding of the environment and their relationship to it. Landscape ethnoecology as I conceive it thus includes *kinds of place,* cultural ecotopes, or "geographic ontology";[1] *traditional ecological knowledge* (knowledge of the habits and habitats of animals, fish, and birds, the phenology and properties of plants, seasonal knowledge of land, weather, and the habits of animals, and so on); *practices* that flow from and instantiate kinds of place and traditional ecological knowledge; *history* as localized on and situated in the local landscape; *memory* of human life on the land; and a diffuse overarching level of *meaning* often linked to cosmology. The land is also strongly related to identity for people of land-based cultures, and this is an important aspect of and extension of meaning.

The term "cultural landscape" has also been used in a number of ways in recent years. The sense in which I use cultural landscape is similar to the articulations of Iain Davidson-Hunt (Davidson-Hunt and Berkes 2003 and chapter 10 of this volume) and Veronica Strang (1997), and in that sense comprises the larger framework of meaning of land, including cosmology, history, the sacred, and the customary activities and places of activities of the people on the land. It also encompasses environmental values in land and proper behavior on and toward the land. Thus the sum of Kaska understandings of the land and relationship to the land might be conceived of as the Kaska cultural landscape.

The Research Project

My goals in this project were to explore Kaska landscape ethnoecology, that is, "knowledge of the land," to discern Kaska perception of landscape components, habitats, and ecological relationships. This is a large topic, and the present study must be regarded as preliminary. The research proved challenging, as I needed to figure out how to elicit such information without obscuring the (putatively different) ways that Kaska understand their landscape. Little guidance from similar previously published studies was available, as this type of ethnoecological work is relatively rare. As an ethnobotanist, I was aware of the potential of reifying my own categories through the way I pursued my field research. Direct questions beginning with English terms are likely to be counterproductive, whereas figuring out what the significant local categories are generally requires that the researcher have both substantial knowledge of the landscape and a sense of how Kaska interact with it.

I also realized that landscape ethnoecology or knowledge of the land necessarily included dimensions of meaning, and that the questions I initially framed were embedded in larger domains. The meaning and significance of the land, and the context within which subsistence activities and travel occur, comprise other levels and require approaches to learning about them other than eliciting terms for locally significant habitats or kinds of place. For land-based cultures, land is the foundation of virtually all cultural domains, so my investigation has been difficult to bound, and many different kinds of information have bearing on the things I have sought to understand.

My field research has therefore been very open-ended, and has required a reflexive awareness of how I and my background have shaped the research I have undertaken. My previous experience with other Canadian First Nations and with land in northwest BC, alongside my expertise with plants and background in ethnobotany and traditional medicines, has suggested approaches to the investigation. As I am not a fluent Kaska speaker, I converse in English and ask about Kaska-language terms when I conduct my research. The elders with whom I have worked directly have a command of the local Dene-influenced dialect of English in addition to Kaska, allowing me to work mostly without a translator. This does, of course, influence the subtlety and completeness of my linguistic interpretations.

This investigation of landscape ethnoecology of a northern people, then, begins with the facts that I am female and interested in plants, rather than male and interested especially in hunting and animals, and that I am interested in indigenous languages but not a fluent speaker. It has also been shaped by the key introduction to a woman elder who is a local healer and has both a strong interest in medicinal plants and extensive experience working with linguists (especially the colleague who introduced me to her, Patrick Moore) and with "Professor

John" Honigmann, who worked with her family in the early 1940s.[2] Another key elder with whom I have worked is the aunt of a Kaska colleague who has a Master's degree in Canadian studies and held a key position in the local Band administration when I first began work with Kaska. As knowledge and social relations are mediated by social networks and family relationships, I have learned more from some families than others. And because I am a woman, most of my local contacts are also women.

Although, as northern Dene people, Kaska are culturally highly focused on hunting (more a male domain than a female one) and on the knowledge of land required to successfully hunt and fish, there also exists a body of knowledge that pertains to plants and to habitats, which has not been widely documented.[3] Plant knowledge may be particularly associated with women, who are more concerned both with plant food and material gathering, and probably with healing plants. Therefore, my research within the community has often begun with recording knowledge of plants and learning about habitats, ways of understanding, and the significance of land through the activities of plant gathering and documenting plant-related knowledge.

An interest in language preservation and in local efforts to create language-teaching materials for the young provides another way to learn about Kaska concepts of land and its meanings. Through participation in language meetings on the land and in language classes, I have learned terms for kinds of place and acquired knowledge about plants and animals, as well as traditional skills such as hide processing, butchering, and meat distribution. Language-oriented activities also connected me with a number of Kaska who are concerned with preservation of traditional knowledge, culture, and language. A last arena that has yielded significant insights is participation in the Liard Aboriginal Women's Society's an-nual Healing Camps at Frances Lake, where being on the land is a strong aspect of the process of healing.

People are keen observers of the occurrence of plants, and of the ecological relationships of animals to food plants and other aspects of the environment over the pattern of the seasons and over longer periods of changing weather or abun-dance of different species. Some of this is coded linguistically in descriptive terms for kinds of place, and much may be tied to a large inventory of specific named places along habitual routes of travel. Such knowledge accrues over a lifetime and is passed on to younger relatives.

Specific Kinds of Place

In the course of activities on the land, and through listening to natural language to infer place categories, I have gained a sense of some Kaska ecotopes, the kinds of places recognized, and the level of specificity of habitats recognized. Terms that emerge this way include things like tall trees/forest, brush, swamp, lake, "blind lake," "fish lake," high bank, lookout, slough, quicksand, glacier, "slide area," hot spring, and so on. The corpus I have recorded includes sixty terms (Table 9.1). Many of these were elicited on the land, and some were recorded at language meetings (also on the land) with Dr. Patrick Moore of the University of British Columbia. Physiographic and hydrographic terms are prominent, while terms based on vegetation are less developed. A class of ecotopes related to human activities, especially hunting, clearly emerges, consistent with the significance of hunting in the Kaska economy historically and at present. Some terms also describe substrates; these are perhaps somewhat marginal to a discussion of cultural ecotopes and reflect issues of spatial scale (see discussion of this issue in the introduction to this volume). The Kaska dictionary (Moore and Wheelock 1997) contains more land terms, but I have not included them in the present discussion because I have not been able to verify what their referents are or explore their significance. A weakness of my present research is that I have only worked in the spring to early fall period and thus have gained little information about snow and ice terms, which arguably are very significant in a place where snow and ice may be present eight months of the year at low elevations, and longer at high elevations.

As is true in other indigenous North American groups with which I have worked, there is no evident separation of terms that indicate vegetation cover or "plant communities" from other terms for features of the landscape, including land forms and terms for waterways. There is also no obvious appropriate scale of feature, so one can encounter difficulty deciding whether a term names a substrate or an ecotope, as alluded to above.

Although one can record cultural ecotope terms or place kind generics, generic place kinds, as other authors have described (Strang 1997; M. Roberts pers. comm.), may be less salient in small-scale traditional land-based cultures than knowledge of specific places. Toponyms may index a great deal of information about land and resources (Hunn 1996; Kari and Fall 1987; Strang 1997; Johnson 2000; Sillitoe 1996; Fowler this volume). In the section that follows, I will discuss examples of several groups of Kaska ecotopes: wetland terms, vegetation terms, terms relating to landforms, and some key place kinds that are related to hunting.

Wetlands

Wetlands are prominent in the Canadian Boreal Forest, where many low-lying areas constitute some type of wetland in the summer months. "Swamp" is the term generally applied to wet areas by Kaska in English. As I traveled with Kaska teachers and participated in language workshops, I learned that 'swamp' in Kaska is *tūtsel.* I began to learn things about *tūtsel,* including the importance of swamps in the annual cycle of moose. My background in ecology made me aware that

Table 9.1 Kaska Landscape Terminology

English gloss	Kaska terms
water, river	*tū, chū*
lake	*man*
shore line, bank	*tahmā, tūmā*
lake shore	*manmā*
gravel shore, rocky shore, beach	*tsēkāge*
gravel shore, gravel	*tsēza*
rock canyon	*tse dzéh*
rock shoreline	*tsedzeh tamā*
"swamp"	*tūtsel, tūtsel ma*
slough	*tūtsel*
slough	*tu łetese elīn, łetesgwech'edi*
slough	*łíni*
slough (small)	*tsēlē'*
slough	*tū tíli, tū tilį̄*
waterfall	*tū dinídislīn*
confluence	*tūłídli*
water flowing [=creek?]	*chūkéli**
willows growing along creek	*chūkéli gûle neyé*
along the creek	*chū kinēli*
brushy willow area along creek	*chū kinēli gûle dä'a**
where creek comes in [tall trees]	*kudache*
large meander loop 'long way around'	*tū wit'íli, dzínadliín***
island	*etûde, deskwidle*
slide area	*wítl'at*
high rock bank	*tādzai*
high bank	*tl'átagi, bes*
top of the hill	*tlétāgi, gunentsįndle*
lookout	*tlétāgī*
hillside [south side, sun side],	*kéhnesí', kę̂ nazí*
sidehill (open S. facing slope)	
alpine, mountain	*héskage*
alpine, mountain	*nedudze hése*
mountain (esp. rock mountain)	*tsé'*
mountain	*hés chō, hés*

English gloss	Kaska terms
uphill [traveling up hill]	*kŭda digé*
downhill [traveling downhill]	*kúda ats'ắ*
upwards "saddle," 'where mountain joins'	*tātege*
valley, gully, creekbed and along the creek (in the mountains)	*kādejhalé*
valley in the mountains	*tsātl'ah*
clearing, dry hillside, dry clearing	*gunedẹje*
open place	*gunejẹje, wanitsịndle*
burned place [long ago burned]	*chōladé, sa'a gukādek'ān*
small trees, poles on old burn (mountainside?)	*desk'ese tses*
white spruce forest	*gat chō tah, gat tah*
forest, timber, tall trees	*dechen chō, łedu tah, łedu, tahche*
open small lodgepole pine area	*tsezel, detah néhsban*
lodgepole pine stand	*gọdze tah*
"pine brush"	*gọdzets'edlī*
"brush" (thick coniferous growth)	*ts'adli*
"brush" (probably thick coniferous growth)	*naw'a*
willow along creek or river	*gúle' chō tah, gúle' chō****
moss (*Sphagnum* and sites with *Sphagnum*)	*ts'ātl*
trail	*at'ane'*
moose trail	*kedā t'ane'*
rabbit trail	*gah t'ane'*
lick	*alếs*
when it's all sand, sandy area	*witl'ús*
quicksand	*ts'ā zā chāzā*
red ochre	*tsịh*
earth	*nan*
camp	*kọ*

Source: Terms listed were compiled from speakers from several dialect areas and also from the *Kaska Dictionary;* orthography provisional.
* lit. 'along the creek willow down there'
** recorded on site and off site at a later time, referring to the same feature
*** here the focus is on the willows as vegetation, not on their landscape location beside the creek

one could distinguish diverse kinds of wetlands. What kind of wetland, then, were *tūtsel*?

In summer of 2002, immediately after participating in a language class where we explored place terms and habitat words, I focused on trying to document terms for kinds of wetlands. As my elder teacher Mida Donnessey and I traveled collecting medicines and picking berries, I asked whether various wetlands were *tūtsel* and photographed the wetland sites to show their character. Some were open fens with small channels and emergent sedges and rushes (Figure 9.1a). At the site shown in Figure 9.1a, Mida talked about moose and how travel through there is not feasible because the pack dog's pack will get wet; instead one must walk on the ridge beside the swamp. Initially I hypothesized that sedge-dominated fens and sphagnum-dominated muskeg sites might be called by different terms. However, sedge-dominated wet meadows such as the site shown in Figure 9.1b were also *tūtsel*. At a mossy site where we were picking blueberries, she gave the term *tsātl*, 'sphagnum'. Sphagnum was the traditional material for diapering babies. In 2003 we returned to the mossy area and I learned that this site too could be called *tūtsel* (Figure 9.1c).

The Kaska term *tūtsel*, then, appears to encompass both sedge- and moss-dominated wetlands. But the ecological significance of sedge- and moss-dominated stands is not identical. The nutritious water plants moose feed on grow in the sedge- and willow-dominated wetlands, and beavers may be active there.

Figure 9.1 Tūtsel, *'swamp'*
a Fen with sedges and beaver activity

b Sedge meadow seen from the Alaska Highway

c Black spruce/ericoid shrub/sphagnum peatland

Certain medicinal plants may also grow there. These swamps are too wet to walk through. The moss-dominated stands offer other resources: blueberries and cranberries, tamarack (*Larix laricina*) bark for medicine, and, depending on the site, sphagnum for baby diapers or Labrador tea for beverage and tonic (*Ledum groenlandicum,* syn. *Rhododendron groenlandicum*). Many moss areas can be traversed on foot in the open season, though they are never easy walking.

Tūtsel can also designate some sites called 'slough' in local English. Sloughs are cut-off channels of rivers like the Liard. An open, unwooded slough with a sedge-dominated wet meadow fringing the old channel was also referred to by Mida as *tūtsel* (Figure 9.2a). However, Mida gave a different term, '*tsēlē*',[4] when I asked the term for 'slough' at another local slough where we went looking for medicinal plants. This site had wooded banks and lacked the fringing sedge meadow (Figure 9.2b), and the channel was narrower. (When queried why this was *tsēlē*' on a subsequent occasion, she indicated the narrowness of the channel as the salient feature.) A third slough she called *tū tilī*, the term for 'slough' that she had given at a language class the previous year. This shows the complexity of matching Kaska and local English place kind terms, and the need for field reconnaissance and visual recording to document what kind of place is meant by each term.

'Slough' in local English usage has certain entailments: sloughs are good for moose hunting because moose feed there and they may be accessible from the

Figure 9.2 *"Sloughs"*
a An old river channel with a well developed sedge meadow **tūtsel**

b A narrow cut-off channel with wooded banks **tsēlē'**

c A wide scoured more recently abandoned channel **tū tilį̄**

river by boat. They are places where medicines may grow, where beavers are active, and where the river ran in the past. Mida spoke of one particular slough on a number of occasions. She described this site as "*Tütsel* where you get your food," and as we traveled into the area, abundant fresh moose tracks and flightless geese underscored that aspect of the area. Reviewing photographs taken of the trip, Kaska linguist Leda Jules commented that she and her husband had gotten a bull moose at the same place where Mida's son had stopped to look for moose when we traveled with him. Another aspect of this slough that Mida mentioned on a number of occasions was that it was a good place for medicine. In particular, it is an excellent locality for **tl'ōtsan,** 'wild mint', which occurs in abundance only in certain locations. "*Nón nístlǫ,*" 'lots of medicine', she said.

The last kind of wetland discussed here are springs. Kaska distinguish cold springs from hot springs, which Kaska are familiar with owing to geothermal activity in the Northern Rockies and Logan Mountains. Hot springs are a kind of place, and particular hot springs are named places. The Liard Hot Springs along the Liard River and Alaska Highway in northern British Columbia are well known. These springs figure largely in elder Alice Brodhagen's family history. Other hot springs occur on the Coal River in the Yukon and at the site of Tungsten, just across the Northwest Territories border. Cold springs are good sources of drinking water.

Vegetation Terms

The Kaska landscape is largely forested, as it lies within the Canadian Boreal Forest biome. Species diversity is relatively low, and forest stands are dominated by a few species, particularly white spruce (*Picea glauca*), black spruce (*P. mariana*), lodgepole pine (*Pinus contorta*), trembling aspen (*Populus tremuloides*), and balsam poplar (*P. balsamifera*). Birch (*Betula papyrifera*) and balsam fir (*Abies balsamea* ssp. *lasiocarpa*) occur more locally. Alpine areas, grassy south-facing slopes, and areas of seral scrub are other noteworthy local kinds of terrestrial vegetation.

Several kinds of vegetation are distinguished by Kaska, and a general formula for describing the vegetation is to take the name of the dominant type—white spruce (*Picea glauca*), or pine (*Pinus contorta*), for example—and add **tah,** which means 'among'. Alternately, one can say "among big _____". Tall bottomland spruce stands were described as **gatchō tah** or **gat tah.** Pine stands, **gǫdze tah,** another distinctive vegetation type, may be dense and difficult to walk through or open with a lichen carpet, **ajú.** The latter offer feed for caribou traveling in the fall or winter. **Ajú,** 'white moss', is known as caribou food and comprises species of *Cladina* and *Cladonia*. Thick young lodgepole pine and spruce stands are locally known as "brush," **ts'adlī,** which can be specified by reference to a dominant species, for example, **gǫdzets'adlī,** 'pine-brush'. This is good habitat for snaring snowshoe hares.

In local English "brush" focally refers to conifer boughs and does not include deciduous scrub, which is called "willows."[5] Willows (here inclusive of shrubby *Salix, Alnus, Populus,* and *Cornus*) offer a barrier to travel and to visibility, providing both feed and cover to animals such as moose and bear. While walking along the highway near her home I asked elder Alice Brodhagen what she would call the deciduous scrub along Watson Creek. She said, "*Chū kinēli gúle dā'a,*" which means 'along the flowing water willow down there'. Looking at the photo I took, linguist Leda Jules gave the term *gúle chō tah,* 'among the big willows'.

Other Landscape Terms

Some Kaska place terms are relatively unproblematic to translate and document, such as *tamā,* 'shoreline, bank'. Similarly, 'lake shore' can be designated as *manmā.* In the language classes, elders and other fluent speakers had no difficulties listing plants that grow on *manmā* or *hés kage,* 'on the mountain', especially above the timberline, giving a clear sense of their awareness these areas as habitats. *Tsās,* 'bear roots' (*Hedysarum alpinum*) is a plant that is described as growing along the shores of rivers, for example, while wild chives (*Allium schoenprasum*) and coltsfoot (*Petasites sagittatus*) are characteristic plants of lakeshore beach gravels. A plant called "gopher food" was listed as typical of the alpine.

The Alpine

The most important resource of the alpine is caribou. Woodland caribou spend the summer season ranging the alpine meadows, moving to lower ground in the winter. Kaska actively seek caribou in July and August and know what kinds of areas caribou like. On a trip to one mountain area, elder Mida Donnessey commented on the alpine as caribou habitat and called my attention to a small lake in a saddle as a good place for animals, which come to the lake to drink. Certain powerful medicines are also found in the alpine; care and circumspection are required for their safe gathering and use.

On the same massif, Mida pointed out a relatively green slope as good "gopher" habitat and called my attention to the burrows and to the succulent green cushion plant known in Kaska as *tselidié'* 'gopher food' (moss campion, *Silene acaulis*). Gophers and gopher food are both spiritually powerful because of their closeness to the earth, *nan,* which is itself deeply powerful and so must be treated with respect.

Sacred Earth

On quite a different occasion and in a different place, the sacredness of the earth and power of things from the earth was brought home to me. Several years before, as we traveled together on an old truck trail along the Rancheria River, Mida pointed out a deposit of red ochre, a very iron-rich sediment, and we stopped to gather some. Mida admonished me that a tobacco offering must be left when anything is taken from the earth. I learned that the Kaska name for the Rancheria River was Tsįh Tue, 'red ochre river'. Red ochre is widely regarded as a spiritual substance by Athapaskan speakers.

Terms Related to Hunting

Some Kaska kinds of place are especially associated with a hunting way of life. A highly significant kind of place in the Kaska landscape is a "lick," a mineral-rich area where animals congregate to get salt. The term for lick, **alés** or **elés,** is important in toponyms (Moore and Wheelock 1997). Over the course of several years, I photographed three reported licks (Figure 9.3). The first looks like a verdant wet meadow (Figure 9.3a), the second a patterned fen, very wet (Figure 9.3b), and the third an expanse of bare mud or silt pocked by animal tracks (Figure 9.3c). Licks can be spoiled or cease to work. A formerly active lick next to the Campbell Highway was pointed out, and over another willows had grown thickly, obscuring the view. The lick shown in Figure 9.3c is widely known by all the Kaska, and people passing up or down the adjacent road it will be asked if they saw anything,

Figure 9.3 **Alés,** *'lick': three identified mineral lick areas*
a A wet meadow

b A patterned fen

c An area of bare mud with many animal tracks

or if they encountered any animals, at the lick. Licks, then, are important places. They are specific known locations and must be treated with care to remain productive. As is evident from Figure 9.3, they do not share a single visual profile. The status of an area as a lick is recognized more by the *behavior* of the animals that use it than by its appearance or vegetation cover. The inventory of licks will be communicated to younger people as they are trained in skills related to living on the land, as will appropriate human behavior in the vicinity of the lick.

Another important Kaska kind of place related to the hunting way of life is referred to in English as a "lookout."[6] Elder Mida Donnessey took me to several lookouts, places on hilltops or the tops of bluffs that afford visibility of meadow or wetland areas below, where animals may feed. The sites are associated with camps and with old foot trails. Camping too close to the *tūtsel* would drive the animals away, so camping on the hill at the lookout is more appropriate.

These sites emphasize another aspect of Dene ethnoecology: the landscape is a humanized landscape, and the land is seen in relationship to people and their travels over the land, the trails of their lives through space and season. In Athapaskan languages, things must be spoken of in relationship; things do not first exist separately and then perhaps relate. Kaska speech is full of terms called directional words or deictics, which are used both literally and metaphorically (Moore 2000, 2002).[7] Dene ethnoecology tends to be organized in terms of drainage basins, and the concepts of "upstream" and "downstream" are fundamental to the way land is perceived (cf. Kari 1989; Kari and Fall 1987; Johnson fieldnotes re: Gwich'in and Witsuwit'en).

Discussion: Kaska Landscape Ethnoecology

Knowledgeable Kaska integrate their traditional ecological knowledge—understanding of the habits and habitats of animals, fish, and birds, and the phenology of plants and their properties—with an understanding of the cycles of the seasons and the dynamic nature of the land. One must behave in the appropriate way in order to ensure safety and survival, health and success. There are many rules about how one must treat animals, plants, and the earth, and about which things should be avoided. Elder Mida Donnessey calls these rules *á'ī*, which sometimes indicates taboo, and sometimes prescribed behavior. These rules are part of "respect," and form an aspect of social and spiritual relationship to the other entities of the land.

When moving across the Kaska land with Kaska elders, one travels trails of memory and story. Driving by Watson Lake itself (Łuwe Cho, 'big fish [lake]'), one learns both that it was named for the productive and abundant whitefish (*Coregonus* sp.) run, now spoiled by pollution and environmental change, and that it is the place where a brave and resourceful youth saved his people from a killer elephant[8] by tricking the elephant into going out on thin ice, whereupon it fell through and drowned. It was also the site of an ancient conflict between groups from the Stikine River and other areas that led to a massacre. More recently, it was the place where returning Klondike miner Frank Watson settled down with his Kaska wife and, in the 1930s, the place where the first airstrip in the region was constructed by the Watson family and other local people.

Kaska relationship to land is also constitutive of identity, which plays out in sociopolitical arenas and the contemporary politics of Canada. It is striking that the annual General Assemblies of the Kaska Dena take place on the land, hosted every year in different parts of Kaska Territory by successive local groups. A variety of issues of governance and land management are debated at the General Assemblies. Elders, political leaders, and others make impassioned speeches about themselves as Kaska, and about the significance of the land to the health and well-being of Kaska. At the 2003 General Assembly traditional knowledge policy as well as economic development and co-management agreements were presented and discussed.

Conclusions

Kaska landscape ethnoecology consists of detailed knowledge of place and season nested within an overarching "cultural landscape" (cf. Davidson-Hunt and Berkes 2003 and this volume; Strang 1997). Kaska knowledge of the land is often embodied rather than cognized and verbally encoded. Knowledge of land and competence on the land is fundamental to Kaska identity and to the sense of security, particularly of older people, whose knowledge of the land is a resource they can rely on in the face of uncertainty. The Kaska understanding of land integrates people and their specific knowledge of place with movement over the land and response to changing opportunities with place and season and through time. Fundamental to the Kaska relationship with land is respect for other beings on the land. The land embodies history and power, and living the right way on the land is an important part of *Dene K'é,* the Dene Way.

Acknowledgments

I would like to acknowledge first the people and elders of the Liard First Nation and other Kaska with whom I have worked and shared time on the land, especially elders Mida Donnessey, Clara Donnessey, Alice Brodhagen, Eileen Van Bibber, May Broadhagen, and linguist and elder Leda Jules. The students and elders involved with the Kaska 100 course taught by Josephine Aklak and Dr. Patrick Moore for the University of British Columbia on-site at Frances Lake in 2002 also taught me a great deal. I would like to acknowledge Linda McDonald for her insights and continued support, and for reviewing this manuscript. I would also like to acknowledge the support of the Liard First Nation Lands and Resources Department and Education Department, and linguist Dr. Patrick Moore. I would like to acknowledge financial support from the Social Science and Humanities Research Council of Canada, the Canadian Circumpolar Institute, and the Athabasca Research Fund.

Notes

1. *Geographic ontology* is the term used in geographic information science (GIS) for the set of basic place kinds that are used as categories in GIS analysis (Mark and Turk n.d., Mark and Turk 2003).
2. Honigmann produced a number of pioneering ethnographic studies of the Kaska based on field research conducted in the 1940s. Key works include Honigmann 1947, 1949, 1954, and 1981. A description of his fieldwork experience may be found in Honigmann 1970.
3. The underrepresentation of female knowledge in ethnobotany has been problematized and addressed in a preliminary way in Howard's 2003 volume *Women and Plants: Gender Relations in Biodiversity Management and Conservation,* though with the exception of Turner's chapter (2003) the studies included in the book do not emphasize traditional nonfarming northern societies.
4. Transcription and exact meaning of this term is uncertain; linguist Patrick Moore did not recognize the term. It may be a variant of "creek," which is *ts'elį* in the Fort Ware dialect (Kaska Elders 1997).
5. Kaska do distinguish the various genera of deciduous shrubs by individual names and have specific uses for various kinds. However, they all appear to be grouped as an ecological type called "willows," which I interpret as an intermediate in Kaska ethnobotanical taxonomy.
6. The term I learned for lookout in Kaska, *tlétāgī,* appears to be related to the word for 'top of the hill' or 'high bank'. In Witsuwet'en, an Athapaskan language spoken in west central British Columbia, the word used for such places is *co'ënk'it,* which refers to looking out (Johnson and Hargus 2007).
7. See Moore references for a complete description of the Kaska system and for their metaphoric functions in narrative. Hargus (2007) describes the deictic system for Witsuwit'en, and similar systems are described for Gwich'in and other Athapaskan languages.
8. This has been interpreted by some scholars as likely referring to a mammoth. Notwithstanding the absence of proboscideans from the Yukon during the Holocene, "elephant" is the English term consistently used by Kaska to refer to this monster.

References

Davidson-Hunt, Iain, and Fikret Berkes. 2003. "Learning as You Journey: Anishinaabe Perception of Social-Ecological Environments and Adaptive Learning." *Conservation Ecology* 8(1): 5. URL www.consecol.org/vol8 and revised version, this volume.

Hargus, Sharon. 2007. *Witsuwit'en Grammar: Phonetics, Phonology, Morphology.* Vancouver: UBC Press.

Honigmann, John J. 1947. "Witch Fear in Post-contact Kaska Society." *American Anthropologist* 49: 222–243.

————. 1949. *Culture and Ethos of Kaska Society.* Yale University Publications in Anthropology No. 40. New Haven, CT: Yale University Press.

————. 1954. *The Kaska Indians: An Ethnographic Reconstruction.* Yale Publications in Anthropology 51. New Haven, CT: Yale University Press.

————. 1970. "Fieldwork in Two Northern Canadian Communities." In *Marginal Natives: Anthropologists at Work,* ed. Morris Feilich. New York: Harper and Row.

————. 1981. "Kaska." In *Subarctic: Handbook of North American Indians,* vol. 6, ed. June Helm. Washington, D.C.: Smithsonian Institution Press.

Howard, Patricia, ed. 2003. *Women and Plants: Gender Relations in Biodiversity and Conservation.* London and New York: Zed Books.

Hunn, Eugene S. 1996. "Columbia Plateau Indian Place Names: What Can They Teach Us?" *Journal of Linguistic Anthropology* 6(1): 3–26.

Johnson, Leslie Main. 2000. "'A place that's good,' Gitksan landscape perception and ethnoecology." *Human Ecology* 28(2): 301–325.

Johnson, Leslie Main, and Sharon Hargus. 2007. "Witsuwit'en Words for the Land- a Preliminary Examination of Witsuwit'en Ethnogeography." In *ANLC Working Papers in Athabaskan Linguistics, Volume 6,* ed. Siri Tuttle, Leslie Saxon, Suzanne Gessner, and Andrea Berez. Fairbanks: Alaska Native Language Center, University of Alaska.

Kari, James. 1989. "Some Principles of Alaskan Athabaskan Toponymic Knowledge." In *General and Amerindian Ethnolinguistics, in Remembrance of Stanley Newman,* ed. Mary Ritchie Key and Henry M. Hoenigswald. Berlin and New York: Mouton de Gruyter.

Kari, James, and James Fall. 1987. *Shem Pete's Alaska: The Territory of the Upper Cook Inlet Dena'ina.* Fairbanks: Alaska Native Language Center, University of Alaska.

Mark, David, and Andrew G. Turk. n.d. "Ethnophysiography." Pre-conference paper (e-document) for Workshop on Spatial and Geographic Ontologies, 23 September 2003 (prior to COSIT03).

————. 2003. "Landscape Categories in Yindjibarndi: Ontology, Environment, and Language." In *Spatial Information Theory: Foundations of Geographic Information Science,* W. Kuhn, M. Worboys, and S. Timpf, eds. Berlin: Springer-Verlag, Lecture Notes in Computer Science No. 2825, pp. 31–49.

Moore, Patrick J. 2000. "Kaska Directionals." Paper presented at the American Anthropological Association Meetings in San Francisco, November 2000.

————. 2002. "Point of View in Kaska Historical Narratives." PhD diss., Indiana University.

Moore, Patrick J., and Angela Wheelock, eds. 1997. *Guzāgi Kúge;' Our Language Book.* Watson Lake: Kaska Tribal Council.

Sillitoe, Paul. 1996. *A Place Against Time: Land and Environment in the Papua New Guinea Highlands.* Amsterdam: Harwood Academic Publishers.

Strang, Veronica. 1997. *Uncommon Ground: Cultural Landscapes and Environmental Values.* Oxford and New York: Berg.

Turner, Nancy. 2003. "'Passing on the News': Women's Work, Traditional Knowledge and Plant Resource Management in Indigenous Societies of North-western North America." In *Women and Plants: Gender Relations in Biodiversity and Conservation,* ed. Patricia Howard. London and New York: Zed Books.

Chapter 10

Journeying and Remembering

Anishinaabe Landscape Ethnoecology
from Northwestern Ontario

Iain Davidson-Hunt and Fikret Berkes

Introduction

Ethnobiology has expanded our knowledge of how people classify individual organisms into taxa and how taxa are grouped hierarchically. But only recently has ethnoecology turned its attention to the question of whether societies systematically classify landscapes (Johnson this volume; Hunn and Meilleur this volume). The way we approach this question in this chapter is to begin by asking how a particular society, the Anishinaabe, pattern space and time so that such patterns provide a means to convey knowledge of their landscape amongst individuals and across generations.

In the ethnoecological literature the knowledge of plant harvesters is often tied to a specific location. Plant harvesters are often asked to map out where they harvest a specific plant. That approach, however, generates a problem. What if a plant harvest location is destroyed? How does an individual harvester then find another location where that plant is growing? We hypothesized that one way would be through the recognition of mature landscape patterns like forest types, cliffs, wetlands, and other types of places, or what Johnson, and Hunn and Meilleur (this volume) term ecotope, and the relationship between those ecotopes and a specific plant.[1] We also realized that we could not just consider how space might be patterned by such ecotopes but needed also to consider changes over time, in particular the successional process by which a mature ecotope is disturbed and then reverts to a mature type over time. An Anishinaabe ecology of landscape would consider how both individual biological organisms and nonbiological elements are patterned in space and time and the processes by which those patterns change.

This chapter is organized into four sections allowing us to develop our perspective on an Anishinaabe ethnoecology of landscape. We begin by considering an Anishinaabe way of knowing landscapes, followed by discussions of spatial and temporal perception of landscapes, and concluding with a focus on remembering landscapes. Our approach to knowing landscapes begins with what others have called an Anishinaabe land ethic, with a particular focus on the ontology and epistemology of landscapes (Overholt and Callicott 1982). Landscape perception necessarily begins with an examination of spatial and temporal perception of landscape pattern, but we also include cultural perception of these dimensions, leading to a holistic ecology of landscape. This is similar to what Sauer (1956) proposed when he reintroduced the concept of cultural landscapes. Finally, we consider the remembering of landscapes through journeying and how that process brings together perception of places with knowing landscapes gained through journeys. We use the term journey not in the sense of passing through, but in the sense of repeatedly traveling in an area in such a way that an intimate relationship with the land is developed. The term travel refers to the physical movement of a person from one place to another. This leads to the final section, in which we discuss an Anishinaabe ethnoecology of landscape. Prior to turning to this four-part discussion of Anishinaabe landscape ethnoecology, we introduce the place and people discussed in this research and the methodology that was followed.

Iskatewizaagegan Anishinaabek and Iskatewizaagegan

Research was undertaken with the Anishinaabe (also known as Ojibway, Ojibwa, Saulteaux, Chippewa) people of Iskatewizaagegan No. 39 Independent First Nation (IIFN), located near the border of Ontario and Manitoba, Canada. IIFN is one of two First Nations with permanent communities on Shoal Lake, with a combined population of 530 on-reserve band members and some 300 members living off-reserve. In this research we worked with a number of people from the community. Brennan Wapioke was the community researcher and translator for the project. Ella Dawn Green, the late Robin Greene, Walter Redsky, Jimmy Redsky, and the late Dan Green participated as elders.

Anishinaabe is an Algonquian language. Algonquian is one of the largest indigenous language groups in North America. In the written historical record, the presence of Anishinaabe people in the region dates back to the early 1600s (Lund 1984). They were important participants in the fur trade in the sixteenth to nineteenth centuries, and they signed a treaty with the government of Canada in 1873. They journeyed through the land and moved with the seasons according to their economy based on hunting, trapping, gathering, and horticulture. Over time these repetitive movements built a cultural landscape and defined their territory. A land-based economy continues to be an important part of IIFN livelihood and identity, although their territory is now shared with settler populations of Canada.

Iskatewizaagegan (Shoal Lake) is described as part of the Boreal Forest and the Lake of the Woods ecoregion (Perera, Euler, and Thompson 2000). The natural history of Shoal Lake is notable, as it brings together three great biomes: prairie, the Great Lakes–St. Lawrence Forest, and the Boreal Forest. This is partly a result of the geology of the region. The thin, acidic soils of the Precambrian shield give way to the deeper and more basic soils of the Prairie Biome as one moves in a southwesterly direction. The region has a mean summer temperature of 15° C and a mean winter temperature of −13° C. Precipitation is evenly distributed throughout the year and totals about 600 mm.

This unique set of biophysical features allows for a biologically diverse mixture of vegetation. Tree species typical of the Great Lakes–St. Lawrence Forest coexist with those of the Prairie Biome and the Boreal Forest. A rich mix of shrub and herbaceous species, extensive wetlands, treed bogs, and fens add to the diversity of the flora. The fauna is likewise diverse and includes large mammals such as ungulate species, small game species, fur mammals, waterfowl, and upland game birds. The Shoal Lake region is noted for its fish. This diversity of plants and animals has provided the people of IIFN with a wealth of resources for their livelihoods in the fur trading era and the contemporary period.

Typical of many Canadian First Nations, the people of Shoal Lake have a mixed economy consisting of wage employment, transfer payments, and living off the land (Berkes et al. 1994). Although trapping has diminished due to the collapse of the fur economy in the mid 1980s, most households on the reserve exhibit the lifestyle described in this chapter. However, only a few individuals are considered to be experts in bush knowledge. A concern about diminished activity on the land, and subsequent loss of the terminology and associated knowledge, was a primary reason cited by elders when they were asked why they agreed to participate in this study.

Research Methodology

The research reported in this chapter was undertaken during two field seasons, May to October 2000 and July to October 2001. Verification workshops to check the findings with the community were held in January 2001, and again between January and April 2002. The research was undertaken in cooperation with a community researcher, the Shoal Lake Resource Institute, and elders and local government of IIFN (see Davidson-Hunt and O'Flaherty 2007).

The methodology was a combination of site visits to a place where elder community members knew the plants, and transects determined by known travel routes. During these site visits and transect walks, research themes were used to generate conversations. Discussions of the major themes of plant nomenclature and plant uses occurred during site visits, while habitat descriptions, biophysical landscape nomenclature, and place names were discussed during both site visits and transect walks.

Plant vouchers were taken for species identifications. Photos, videos, GPS readings, and informal interviews were utilized to obtain specific information about plants. Photos and/or videos, along with GPS readings, were taken of different types of habitats, biophysical landscapes, plants, and named locations. As new habitats, biophysical landscapes, plants, and named locations were encountered, informal conversations were initiated about those topics. Map interviews were held with elders about place names. These discussions often led to stories about places and the activities that used to occur in specific places. On a few occasions, such interviews led to new site visits. Essentially, the methodology could be summarized as one in which the researcher approaches a knowledgeable person and explains what he would like to learn. This knowledgeable person then structures the learning experience regarding the steps to be taken and the places that must be visited, so that learning can occur.

Knowing the Land

Before discussing Iskatewizaagegan perception of space and time, it is important to consider the ontological and epistemological land ethic as we have come to understand it. Our Anishinaabe colleagues suggest that knowledge resides within the landscape itself. Institutions of knowledge, which we have characterized as knowledge as revelation, place-based knowledge, holistic knowledge, and embedded subjects in Table 10.1, provide the ontological rules on the "nature" of true knowledge. Truth cannot be discovered but is revealed as part of a person's development within a web of relationships of a place (Ingold 2000). There is no separation between society and individual, culture and nature, or society and environment. These categories individually, not as oppositional pairs, are ontologically significant in an Anishinaabe worldview. Individuals move in and out of networks, operative for specific places, as they undertake their own projects (Davidson-Hunt 2005, 2006). They become more competent as individuals as knowledge is revealed to them through their participation in life experiences and subject-to-subject relationships. This is quite different from the perspective on "truth" that is adhered to by many scientists. In a Cartesian perspective, truth is to be discovered in mathematically described, universal, simplified, and subject-to-object relationships (Ingold 2000). However, the view of many contemporary scientists differs from the idealized Cartesian model. It has been suggested that the Anishinaabe view may have parallels to the dialectical thinking of such scientists as Levin (1999).

How truth statements may evolve in Iskatewizaagegan society is based upon their epistemology: knowledge is progressively revealed to individuals through

Table 10.1 A conceptual framework of adaptive learning based upon Anishinaabe philosophy and institutions of knowledge

Philosophical principle	Institution of knowledge	Teaching from Shoal Lake elders (represented by core statements)
Knowledge Resides in the Land	*Revelation*	People are gifted for different things. Beings reveal knowledge to people through visions and dreams (Ella Dawn Green).
	Place-based	When traveling, the person who knows the land best always leads the way (Walter Redsky).
	Holistic	When I am healing a person, a plant will reveal itself in a dream and then I know I should use it for this person (Ella Dawn Green).
	Embedded Subjects	A powerful person has learned to show respect to other beings and has developed a finely tuned awareness of his land. Other beings begin to reveal themselves so he becomes a better hunter, fisherman, or healer (Robin Greene).
Knowledge Is Progressively Revealed through Experience on the Land	*Direct Coupling*	You should know where everything is on the land with which you are familiar. My father, in the middle of winter, he could go to the exact place; thrust his hand through the snow, and dig up the root he needed for healing (Walter Redsky).
	Empirical Observation	When you are on the land with someone, you should always be watching where you are and what the other person is doing (Walter Redsky).
	Personal and Collective Ceremonies	You can do ceremonies by yourself or as part of a group. Ceremonies are necessary to show respect to others for what you are about to undertake (Robin Greene).
	Social Gatherings	We used to always get together as a group in the summer to harvest fish, blueberries, and wild rice (Walter Redsky).
	Self-awareness	After harvesting a lot of something, like birch bark, you should go back and leave cloth and tobacco. Let the birch trees know you respect them by giving something for what you were given (Ella Dawn Green).
	Mentoring	The right way to be taught is not from a book. It is your aunties, uncles, mom, dad, and other people who know the land (Jimmy Redsky).
	Language	Our language is very descriptive. It tells us things like how one thing might be related to another, or the way that things look on the land (Dan Green).
	Narrative	Our people never wrote anything down. We know the land and our histories from the stories we tell (Jimmy Redsky).

their guided experience on the land. An individual is expected to learn through participation in experiences on the land under the tutelage of a knowledgeable person while also engaging in collective experiences. Again, institutions of knowledge such as direct coupling, experience on the land, self-awareness, mentoring, language, and narrative provide a set of rules by which perception of the landscape matures. These institutions also allow for individual perception of changes in the land to lead to changes in individual behaviors and practices. Other members of Anishinaabe society will consider change suggested by a person who has followed these institutions while undertaking practices on the land over an extended period of time. Since truth is linked to knowledge of the land, changes in truth statements are linked to changes in the land. Anishinaabe philosophy builds a direct linkage between landscapes, practice, and knowledge. Knowledge can be gained through ceremony, embodied experience, songs and narratives, and in many other forms, as examples in Table 10.1 show. Some people might consider the Iskatewizaagegan philosophy regarding landscape, practice, and knowledge to be a phenomenological philosophy of landscape.

Spatial Perception of Landscape

Ecological perception is usually considered to lie along two axes, spatial and temporal. Spatial perception relates to landscape patterning. Ecological theory requires categories that can be used to describe spatial distribution, interrelationships, and properties at a specified level of a spatial scale (Cash et al. 2006; Levin 1999). Likewise, units can be created that describe distribution, interrelationships, and properties for a temporal scale. As discussed previously, knowledge of where a plant is growing allows that resource to be harvested as long as that plant community does not change. But what do people do if a fire destroys that particular locality? As previously mentioned, we hypothesized that recognizing abstract categories such as a forest type, or a wetland, would facilitate finding a new resource harvest area. A standard ethnobiological approach to this type of question is to look for terminology ("linguistic signifiers") denoting cognitive categories related to spatial and temporal patterns of the landscape (Berlin 1992).

This section presents the results of our exploration of these questions with our Anishinaabe colleagues. As we discuss, our Anishinaabe colleagues concurred that their language described the spatial and temporal patterning of the landscape. However, they insisted that knowledge of where a resource might be found also required an understanding of the social and cultural patterning of particular landscapes.

Biophysical Landscape Patterning

Figure 10.1 demonstrates the complexity of spatial patterning of the biophysical landscape perceived by our Anishinaabe colleagues. Obvious landscape features are perceived and named, for example, lakes (*zaagaigan*), rivers (*ziibi*), and hills (*pikwapikaa*). More interesting is the evidence of relational concepts such as river inlet (*saagichi-ogiima*) and river mouth (*saagidawang*). Important functional features such as a spring (*mookojiwanibiing*) and complex relational features such as a rocky slope on the banks of a water body (*niisapkaang*) are also part of the perception of landscape patterns. The language also provides evidence that relationships between physical features and biological structures are recognized. A point that contained a specific type of tree was described by combining the name of the physical feature with the vegetation that was growing. For example, a bur oak point was referred to as *giineyaamitigomizhiikang,* which would be directly translated as 'there point bur oak place'.

Landscape patterns that are made up of biological structures are also demonstrated in the language. Forest patches are denoted by the term *okwokizowag.* For instance, a grove of birch trees is called *okwokizowaag wiigwaasatigoog.* Another example is the use of the linguistic unit -*kwaa,* and variants, to describe patches of vegetation. A blueberry patch, for example, is denoted by the word *miinikaa,* which combines the word for blueberry with that of the descriptor for patch, thus 'blueberries growing in a patch'.

The Anishinaabe language suggests that the recognition of physical features, biological structures, and their interrelationships allows for the recognition of spatial patterning of the landscape in categories and subsequent functions.

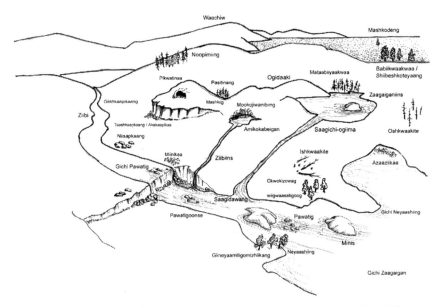

Figure 10.1 *An idealized schematic of Anishinaabe biogeophysical knowledge. A sample of terms.*

Our thinking led us to suggest that for research on birch we could look for *ok-wokizowag wiigwaasatigoog*, 'birch groves'. This would be facilitated by our knowledge that *okwokizowag wiigwaasatigoog* was equivalent to vegetation-type V4 of the Ontario Ecological Land Classification (OELC) (Ontario Ministry of Natural Resources 1996; Sims et al. 1989). By looking for the ecosites (mapping scale of the OELC) with V4s in the Ontario Forest Research Inventory, we would locate our research sites. Our Anishinaabe colleague Walter Redsky did not concur. Walter perceived birch groves when encountered on the landscape; however, he did not organize his knowledge of the spatial distribution of birch patches as a map. He utilized a different system to construct a pattern of the distribution of birch within the landscape.

Cultural Landscape Patterning

Jimmy Redsky, another Anishinaabe colleague, listened respectfully as we presented our results regarding Anishinaabe terms for ecological units and OELC equivalents. After a time he suggested that we had not paid enough attention to the history of the land. While he agreed with the linguistic description of the land we presented, it seemed incomplete. He suggested that we should consider how Anishinaabe people drew upon their history in order to begin to understand how they perceived the spatial patterns of the land.

One way to link the history of a people with their occupancy of a landscape is through place names. For people who do not utilize maps, place names provide a mental map of the land—how particular places within the landscape look, how they are related to other places, what occurred at those places, and/or what might be found at them (see Hallowell 1955: 186–189). For instance, **Gichi neyaashi-ing** describes a big point, **Aagimakobawatig** a place where black ash (*Fraxinus nigra*) grows beside a rapid, **Azaatiozaagaigan** a lake where trembling aspen (*Populus tremuloides*) grows, **Ogishkibwaakaaning** where wild potatoes (*Helianthus tuberosus*) grow, **Gitiganiminis** an island where gardening has occurred, **Animoshiminis** where the howling of dogs is said to have been heard in the past. Place names are the fixed nodes, reference points, upon which the creation of spatial patterning of the landscape is built (Hallowell 1955; Johnson 2009).

Place names do not just mark places but bring places together in relation to each other, linked by paths of travel. The Anishinaabe traveled using waterways, lakes, portages, and winter trails, pathways along which either a canoe or a dog team could pass. Trails link different nodes together. Places and trails provide the references for remembering the path of a journey from one place to another. The landscape, in this perspective, becomes a network of nodes and trails that orient a person in physical, social, and cultural space. Thus, spatial patterning does not exist independently of the journeyer, nor does it exist until the journey occurs. As both the journeyer and journey are physical, social, and cultural in nature, the knowledge of the spatial pattern of a landscape requires access to these three axes. Thus Walter and Jimmy thought that maps and habitat categories were a crude and limiting way to find where birch groves were located. Once they knew what we were looking for, they could get us to such places situated in the appropriate physical, social, and cultural space for our task at hand.

Temporal Perception of the Land

Landscape patterning becomes increasingly complex as one adds temporal dynamics to the spatial distribution of plant communities and resources. We began our discussion with Anishinaabe colleagues through a process similar to that described above. We searched for linguistic markers of the temporal dynamics of the landscape. We thought the best way to generate such discussion would be to visit sites a given number of years following a disturbance. A primary post-disturbance activity, blueberry-picking following fire and logging disturbances, was chosen to focus the discussion. We then asked our Anishinaabe colleagues for a word that they would use to describe such sites.

Disturbance Dynamics

During verification workshops held in the fall and winter of 2001–2002, Ella Dawn Green and Walter Redsky were asked to construct a temporal forest cycle. The different names that were gathered during field research were put together into a diagram, Figure 10.4, and the processes of forest changes were discussed. The cycle they constructed contained a number of different categories of forest from a disturbance event (*ishkote*) through to a mature forest (*noopimiing*) as shown in Figure 10.2.

Our Anishinaabe colleagues also noted different pathways that could follow a disturbance event. Fire could be initiated by lightning or by a person. In Anishinaabe philosophy both events require agency, but each event occurs due to the agency located in different beings. From this perspective the category "wildfire" does not exist. In either case, disturbance could follow a pathway back to mature forest or could lead to an area cleared for a garden. The process of burning and planting progressively leads to an area that is free of roots and easily planted. Eventually, such sites would revert back to mature forest when abandoned. Another pathway that was described is that which follows logging disturbance. Robin Greene thought such sites would revert to mature forest. However, they are still waiting to assess the long-term outcome of this cycle, as logging is a relatively recent activity. Different categories and pathways produce sites for different activities and intensities of use. Some were useful for gardening and some for blueberry harvesting, while others were good for hunting moose.

The construction of such temporal cycles was a rewarding process for our participation in the research. However, our Anishinaabe colleagues again challenged

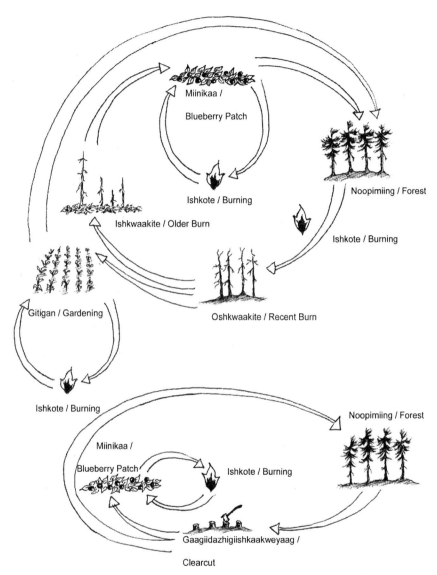

Figure 10.2 *A schematic of an Anishinaabe perspective on forest disturbance cycles. Arrows represent relative temporal scale of a cycle but not a temporal metric.*

our description of how they perceived temporal dynamics. Linguistic evidence that we collected had demonstrated that there is a perception of the temporal dynamics of landscapes. However, the cycles that our Anishinaabe colleagues felt were more important were marked by words that denoted an interconnected shift in the landscape that brought together the biological, social, and cultural dimensions to form a dynamic cultural landscape.

Cultural Landscape Dynamics

Harvesting birch bark is an activity that demonstrates how Anishinaabe perception of the dynamics of the cultural landscape operates. Walter Redsky and Ella Dawn Green mentioned that birch bark must be harvested when it is ready to be peeled. The timing of the birch bark harvest does not correspond to the calendar, but to the development of the birch tree in a given year. There is a time in the early summer when the bark loosens from the tree and can be peeled. Before or after this time, the bark does not peel cleanly and rips when harvested. This time corresponds to the time when raspberries ripen. Perception of temporal dynamics is linked to the awareness of changes that occur in the landscape and the appropriate adjustment of social and cultural activities (see Lantz and Turner 2003). Such awareness comes through acute awareness of the land and experience on the land, and is marked in different ways in the language.

One way in which temporal dynamics is marked in the language is through the concept of the six seasons, which reflect changes in the biogeophysical environment. Figure 10.3 outlines the cycle of the seasons, which are denoted according to changes in the landscape. *Tagwaagin,* for instance, begins when the leaves turn color and fall from the trees. *Tagwaagin* turns into **oshkibiboon** when all the leaves have fallen from the trees and the first snows are falling. **Biboon** turns to **ziigwan** when the ice on the lakes begins to melt and break up. Rather than sharp edges delineating seasons, there are periods of transition from one season to another. A season changes more quickly in a year when there are quick changes in the ambient environment. Periods of seasonal change are keenly observed and warrant frequent comment. Differences in how seasons are changing are measured against other years and provide a baseline for noting seasonal changes that are considered to be anomalies.

The yearly passage of time is also linked to changes in appropriate activities. Feasts that are held during these periods of seasonal change mark cultural activities. Such feasts teach that certain activities and behaviors are appropriate for different times of the yearly cycle. A person will hold a feast to respect the beings who made life possible during the season past and those who will make it so in the season to come. For instance, as **ziigwan** ('breakup') turns into **miinokamin** ('berry season'), a feast to celebrate the first fruits is held. This feast shows respect to the plant beings that share with the Anishinaabe during berry-picking season. Changes are perceived not only in the biological environment but also through the social and cultural activities that orient a person to the temporal dynamics of a cultural landscape (Hallowell 1955: 226–230; Malinowski 1927).

Another system utilized by the Anishinaabe to mark the temporal dynamics of seasonal change is the moon cycle. This provides a more standardized way to note the passage of time during the year. *Migizi giizis,* 'bald eagle moon', marks the time when the bald eagles return. *Maangwag giizis* denotes the time when loons return and begin to nest. Other moons, such as **Miinikaa giizis,**

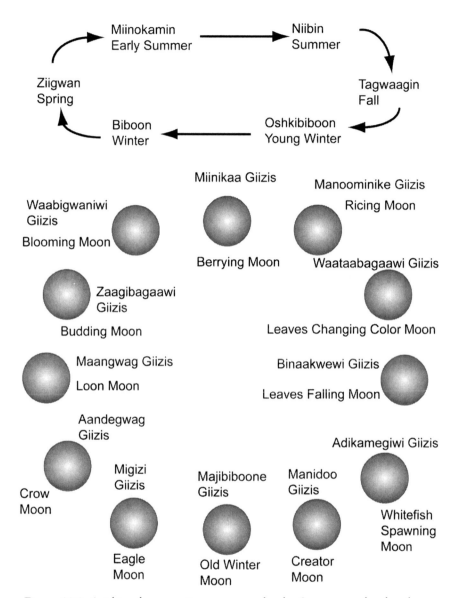

Figure 10.3 *Anishinaabe perspectives on temporal cycles. Seasons are related to changes in the biological environment whereas lunar cycles provide a fixed metric.*

'blueberrying moon', and ***Manoominike giizis,*** 'ricing moon', mark the time period when these different livelihood activities are undertaken. Other names for the moons refer to the time when certain ceremonies are undertaken, such as ***Manitoo giizis,*** 'Creator's moon'. Similar to the six seasons, the moon cycle links the biological, social, and cultural to mark the temporal dynamics of the cultural landscape.

When we began to search for linguistic markers of temporal dynamics, we thought that the existence of such categories would help us understand Anishinaabe perception of temporal dynamics. Recognition of temporal dynamics was considered to be important to a theory of adaptive learning. People would need to be able to distinguish between the temporal patterns of cyclic change and change that required more lasting changes in social and cultural behavior. However, our Anishinaabe colleagues took a different approach to this question. Consider, for example, the linguistic marking of different plant communities following disturbance. The perception of temporal dynamics also emerges out of experience on the land that is simultaneously biological, social, and cultural. Perception does not exist independently of the person, nor can it occur prior to a person's participation in an event. It is embedded in the temporal patterns that emerge from the biological, social, and cultural interconnections of a people and landscape.

Remembering Landscapes

Remembering, for the Anishinaabe people with whom we worked, begins with a person situated in a spatial and temporal context. This is different from a Cartesian perspective, in which one assumes that people are situated outside of a spatial and temporal context. Spatially, the Cartesian perspective is like the view from an airplane. Temporally, it can be considered as the perspective gained from a time series[2] sequence of photos where time can be compressed into the present. This view allows a person to distinguish replicating patterns of biogeophysical structures (landforms) established by biogeophysical processes. These replicating patterns can then be turned into categories of habitat types, so that some categories can be grouped, while others are differentiated. Temporal pathways can then be established for each category of habitat type. Habitats can then be related and mapped in Cartesian space, described in terms of biological, geological, and physical structures and processes, and described by change over time. These spatial and temporal categories become containers for holding information, which can be mapped in Cartesian space and time. Memory becomes detached from a location in an environment to a category located outside the landscape.

Anishinaabe ways of remembering are akin to the experience of journeying within the land, on a journey that is situated in both space and time. The practices, moons, seasons, and ceremonies that mark the passing of diurnal, yearly, and life stages structure the journey temporally. Spatially, the paths of travel link places that can be revealed and described as they are encountered. Places change in yearly and longer-term cycles. These changes are observed and remembered through frequent journeys. Places are known in relation to the paths of journey and reference points along the way that orientate a person within the land. Memory is embedded in the land and the people. The spatial and temporal locations of things can never be forgotten as long as the journeys continue.

This perspective on memory is reflected in the way by which Anishinaabe elders teach about plants and where to find them on the land. They do not prioritize categories of habitat so that the location of plants might be remembered. Rather, they describe the location where they remembered having encountered a plant in the past. Such locations are described by recalling the journey along paths of travel and places that were encountered along that path. Stories about places turn locations into places of remembering and points of reference within the land. Anishinaabe ways of remembering bring to mind events that occurred in the past, stories, place names, physical features, and biological features, not just landforms.

Figure 10.4 presents an idealized schematic on the way in which this perspective renders the spatial and temporal dynamics of the land. Campgrounds, trails, portages, cabins, planting areas, cultivated fields, ricing lakes, hills, mountains, habitat patches, rapids, stony slopes, springs, "thunderbird nests," homes of the "little people," and many other biological, physical, and cultural features come together to form different places on the landscape. It is the spatial and temporal dynamics that create places, places that are remembered as they are encountered along the paths of remembering. The land, the people, and their histories become

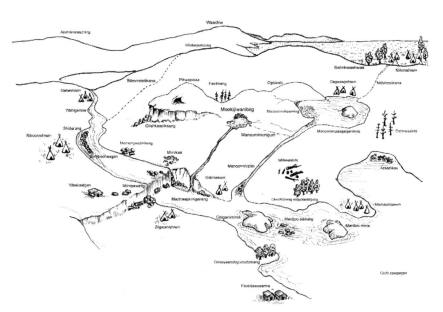

Figure 10.4 *An idealized schematic of an Anishinaabe cultural landscape. Cultural landscapes are a mixture of biogeophysical, artifactual, and known (named or unnamed) features. A sample of terms.*

interwoven into a landscape in which Anishinaabe people undertake daily practices, learn, and create new memories.

Fashioning Landscapes

A landscape in an Anishinaabe perspective is spatially and temporally patterned by physical, biological, social, and cultural structures and processes as described previously. The landscape becomes as it is shaped through a society's practices and remembered through its stories. We draw upon one such story to illustrate the nuances in this way of perceiving space and time and the way by which space-time places become. The fur trade that emerged in northwestern Ontario in the seventeenth century led to a number of novel opportunities for Canada's First Peoples (Davidson-Hunt 2003). Often these new opportunities were based upon resources and practices that had existed prior to the fur trade and now required an intensification of production. One such product was sugar made from the sap of maples (*Acer saccharum* and *Acer negundo*) and birch (*Betula papyrifera*). The sugar was used to sweeten the pemmican made from dried blueberries (*Vaccinium* spp.) and dried buffalo (*Bison bison*) meat. As a "sugaring bush" did not exist on Shoal Lake, ancestors of current Iskatewizaagegan families transplanted *Acer negundo* to an island on the lake. Before this time they would travel to what is now southwest Manitoba to produce sugar for their own use. By transplanting the trees they were able to create a sugar bush on Shoal Lake near to where they practiced other livelihood activities. This also allowed the families of Iskatewizaagegan to control and intensify production of sugar for trade. The movement of the sugar bush to a place where it did not previously occur required knowledge of the biophysical conditions under which *Acer negundo* would produce in a new locality. This sugar bush, while not currently utilized, still exists on an island of Shoal Lake.

Another important plant that has been traded since before the written historical record up to the present is ***manomin*** (*Zizania aquatica*), also called wild rice. This plant was propagated and traded extensively by Anishinaabe people throughout northwestern Ontario (Vennum 1988). However, at the turn of the century the wild rice in many lakes and rivers was destroyed by the construction of dams. This occurred on Shoal Lake and Lake of the Woods when dams were constructed in the early 1900s at the outlet of Lake of the Woods. Dams raised water levels, but more destructive for wild rice was the reversal of the natural water cycle. Prior to damming, the water level would rise in the spring and drop gradually throughout the summer. After damming, the water level would begin to drop in mid summer but then would be allowed to rise to provide storage for winter power needs. Wild rice is a determinate plant. It stops growing when day length begins to shorten and the "floating leaf" emerges. The floating leaf is critical, as it continues photosynthesis during seed fill. When the water level rises in July and August, the floating leaf is submerged, the plant is drowned, and there will be no rice harvest.

In order to transplant wild rice it is necessary to find a lake that has the right type of soil on the bottom and the right depth of water in spring and fall (Vennum 1988). Wild rice will grow only in lakes with a "muck" bottom and a specified range of water depth and clarity. The seed requires light of a specific range of wavelength to begin germination. Depth and clarity of water affect the wavelength range that will reach the seed. Shoal Lake families sought out lakes in Manitoba that were not connected to the flooded water systems and that had the right characteristics. These lakes produced wild rice throughout the first half of the 1900s. However, in the mid 1900s a provincial park was established, and Shoal Lake people were no longer permitted to harvest wild rice in the park.

Shoal Lake people decided at this time to look for a lake near the newly established reserve. Crowduck Lake, just north of Shoal Lake, was underlain by the right type of soil and had a "muck" bottom. However, it did not have the right water depth. Walter Redsky, who was the chief at the time, convinced the Ontario Ministry of Natural Resources to blast a channel through the rock at the outlet of the lake. With the drop in water level, this lake became a productive *manomin* lake for Shoal Lake people. During the 1960s, 1970s, and 1980s this lake was their main source of commercial rice harvest.

The Iskatewizaagegan Anishinaabe perceive landscape pattern through the interactions between biophysical, social, and cultural structures and processes. The patterning of the landscape emerges out of these interactions so that an Iskatewizaagegan ecology of landscape incorporates humans into a complex network of dynamic relationships. Changes in these patterns also emerge out of these interactions and may require the agency of a number of actors within the landscape.

Landscape Ecology from an Anishinaabe Perspective

Anishinaabe elder Walter Redsky's father grew up "on the land." In the middle of winter, Walter related, his father could travel to a place, thrust his hand through the snow and pull up the exact root he needed for a medicine. At any time of year, his father knew exactly where animals, plants, and fish would be located. Things changed during the year and from year to year, but Walter believed that his father knew where animals and plants were, as he was always traveling on the land, undertaking ceremonies, and ensuring the survival of his family. This was Walter's way of saying that he did not need to know that sage (*Artemisia frigida*), for example, grew on a certain type of rock face. To be sure, he could describe the rock face and other landscape categories; however, his father taught Walter that the land is not a stranger made up of abstract categories. Land is known intimately through experience and journeying. Each place is unique, as an aspect of creation, in biophysical characteristics, names, and histories. These teachings by Walter and other Anishinaabe elders tell us something about their ecology of landscape.

When we began this research, we hypothesized that an Anishinaabe landscape ethnoecology would be rooted in a perception of biophysical elements that would be structured along spatial and temporal scales. However, as our research progressed, we came to the understanding that for Anishinaabe elders, landscape patterns do not preexist human action but are fashioned and known through it. Landscapes are as much social and cultural as they are biophysical. The Iskatewiz-aagegan notion of *Akii,* often translated as land, is the closest parallel concept to what some definitions of landscape emphasize.

Akii is a concept that emphasizes an interacting set of organisms and can be utilized across a range of spatial scales. It can be used to refer to a specific hand-ful of earth that animates life, it can refer to a bounded territory that animates the life of a society and provides all that society needs, or it can refer to all of the life that is found on what we call planet Earth. It is impossible to begin under-standing an Anishinaabe ecology of landscape without understanding that it be-gins from the perception of landscapes as wholes with constituent elements and can be thought about at a range of spatial scales. Each whole explicitly includes people, their culture, and history, which effectively amounts to an ecosystem ap-proach (*sensu* Berkes et al. 1998).

It is this philosophical premise of the whole that gives meaning to the vocabu-lary used to describe the spatial and temporal dynamics of Anishinaabe cultural landscapes. The terminology is part of an Anishinaabe system of ecology that helps guide the Anishinaabe as they journey through the landscape and act to secure their survival within that landscape. A person knows a landscape through individual action situated in a landscape. A person does not learn a classification of ecotopes in the abstract, but learns about habitats through experiences on the land. A place on the land is not just described as a category of habitat, but as a place with attributes of physical and biotic characteristics and history. These places become known as a person journeys over the landscape, which is thus more like the medium through which one moves than the view that one paints. Or as the Cree people of eastern James Bay put it, "the land gets to know a person" (Berkes, unpublished field notes). It is through journeying on the land that one begins to perceive its spatial and temporal dimensions rendering the places of remembering visible and fashioning the cultural landscape over time.

Acknowledgments

We dedicate this paper to the late Robin Greene who was a faithful guide in our work with the Iskatewizaagegan Anishinabek. We could not have completed this research without the assistance of the Shoal Lake Resource Institute of Iskatewizaagegan #39 Independent First Nation (IIFN), IIFN elders, successive chiefs and councils of IIFN and a team of IIFN Ojibway language teachers who reviewed the terminology used in this chapter. Support for the project was also provided by individuals from the Ontario Ministry of Natural Resources, Canadian Forest Service, and United States Forest Service. Davidson-Hunt and Berkes received financial support from the Sustainable Forest Management Network (SFMN) and the Social Sciences and Humanities Research Council (SSHRC) for this research.

Notes

1. In this chapter we discuss three levels of spatial scale. The smallest level is what we term a landscape element. A landscape element, in this chapter, can be an individual physical or biotic structure that may or may not be modified by people but is given meaning by humans. For example, it could be a rock or a cabin, or a plant or an animal. The next level in our spatial scale is the assemblages of landscape elements that Johnson and Hunn refer to as ecotopes. In this chapter we refer to this level as habitat, site, and place; in landscape ecology this level would also be that of the patch. In this chapter our intent is to understand the spatial level that has been termed landscape. We start from the assumption that a landscape is the patterning of landscape elements and ecotopes at a larger spatial level.

2. A time series sequence is a series of photos taken at different points of time but then viewed at a point in the present. Thus the series of photos compresses time into the present and makes those different points in time analytically comparative, and thereby takes the subject out of the environment to look at time outside of experience. This is in contrast to how time is experienced as an embedded subject, where time can only be experienced in the present; thus a person never steps into the same river twice.

References

Berkes, F., P. J. George, R. J. Preston, J. Turner, and B. D. Cummins. 1994. "Wildlife Harvesting and Sustainable Regional Native Economy in the Hudson and James Bay Lowland, Ontario." *Arctic* 47: 350–360.

Berkes, F., M. Kislalioglu, C. Folke, and M. Gadgil. 1998. "Exploring the Basic Ecological Unit: Ecosystem-like Concepts in Traditional Societies." *Ecosystems* 1: 409–415.

Berlin, B. 1992. *Ethnobiological Classification: Principles of Categorization of Plants and Animals in Traditional Societies.* Princeton, NJ: Princeton University Press.

Cash, D. W., W. Adger, F. Berkes, P. Garden, L. Lebel, P. Olsson, L. Pritchard, and O. Young. 2006. "Scale and Cross-scale Dynamics: Governance and Information in a Multilevel World." *Ecology and Society* 11(2): 8. [online] URL: http://www.ecologyandsociety.org/vol11/iss2/art8/

Davidson-Hunt, I. J. 2003. "Indigenous Lands Management, Cultural Landscapes and Anishinaabe People of Shoal Lake, Northwestern Ontario, Canada." *Environments* 31(1): 21–42.

———. 2005. "A Contribution to Anishinaabe (Ojibway) Ethnobotany of Northwestern Ontario, Canada: Toward a Holistic Representation of Iskatewizaagegan (Shoal Lake) Plant Knowledge." *Journal of Ethnobiology* 25: 189–227.

———. 2006. "Adaptive Learning Networks: Developing Resource Management Knowledge through Social Learning Forums." *Human Ecology* 34(4): 593–614.

Davidson-Hunt, I. J. and R. M. O'Flaherty. 2007. "Researchers, Indigenous Peoples and Place-Based Learning Communities." *Society and Natural Resources* 20(4): 1–15.

Hallowell, A. I. 1967 (1955). *Culture and Experience.* New York: Schocken Books.

Ingold, T. 2000. *The Perception of the Environment: Essays on Livelihood, Dwelling and Skill.* London: Routledge.

Johnson, L. 2009. "Trail of Story: Gitksan Understanding of Land and Place." In *Trail of Story, Travellers' Path: Reflections on Ethnoecology and Landscape.* Edmonton: Athabasca University Press.

Lantz, T., and N. Turner. 2003. "Traditional Phenological Knowledge of Aboriginal Peoples in British Columbia." *Journal of Ethnobiology* 23(2): 263–286.

Levin, S. 1999. *Fragile Dominion: Complexity and the Commons.* Reading, MA: Perseus.

Lund, D. R. 1984. *Lake of the Woods II: Earliest Accounts.* Staples, MN: Adventure Publications.

Malinowski, B. 1927. "Lunar and Seasonal Calendar in the Trobriands." *The Journal of the Royal Anthropological Institute of Great Britain and Ireland* 57: 203–215.

Ontario Ministry of Natural Resources. 1996. Terrestrial and Wetland Ecosites of Northwestern Ontario. NWST Field Guide FG-02. Thunder Bay: Ontario Ministry of Natural Resources.

Overholt, T. W., and J. B. Callicott. 1982. *Clothed-in-Fur and Other Tales: An Introduction to Ojibway World View.* Washington, D.C.: University Press of America.

Perera, A. H., D. L. Euler, and I. D. Thompson. 2000. *Ecology of a Managed Terrestrial Landscape: Patterns and Processes of Forest Landscapes in Ontario.* Vancouver: UBC Press.

Sauer, C. O. 1956. "The Agency of Man on the Earth." In *Man's Role in Changing the Face of the Earth,* vol. 1, ed. L. Thomas, Jr. Chicago: University of Chicago Press.

Sims, R. A., W. D. Towill, K. A. Baldwin, and G. M. Wickware. 1989. *Field Guide to the Forest Ecosystem Classification for Northwestern Ontario.* Thunder Bay: Ministry of Natural Resources.

Vennum, T., Jr. 1988. *Wild Rice and the Ojibway People.* St. Paul: Minnesota Historical Society.

Chapter 11

What's in a Name?

Southern Paiute Place Names as
Keys to Landscape Perception

Catherine S. Fowler

Through the years, ethnographers and linguists working in the Great Basin have recorded place names from several groups speaking Numic languages, including Owens Valley and Northern Paiute peoples (Kelly 1932; Fowler 1992; Steward 1933), Western, Northern, and Wind River Shoshone groups (Miller 1972; Shimkin 1947; Steward 1938), Southern Paiute (Kelly 1964; Sapir 1930–31), Chemehuevi (Laird 1976), and Ute speakers (Goss 1972; Givon 1979). However, most have gathered these data as adjuncts to general ethnographic or linguistic work, and thus the names, although they are occasionally numerous, were rarely the focus of specific data gathering efforts or analyses. Only Isabel Kelly undertook a project that was specifically ethnogeographical, lasting from 1932 to 1934 among some fifteen remnant Southern Paiute groups of Utah, Nevada, Arizona, and California. During the course of eighteen months in the field, Kelly collected various kinds of data on land use, including roughly 1,500 place names or toponyms. Unfortunately, she was unable to synthesize and publish her data before her death in 1982, and thus the names remain largely as lists scattered in her field notes or as notations on maps (Kelly 1932–34).

Kelly also did more general ethnographic work among all of these groups, and she gathered extensive collections of their material culture for museums. For the past few years, I have been compiling and editing for publication the records in the Kelly Southern Paiute archive.[1] While the publication will focus on a basic presentation of her work with some editorial comment and annotations based on my own intermittent field studies since 1961, possibilities for deeper analyses continually suggest themselves. The data that follow offer a view of some aspects of Southern Paiute landscape perceptions as seen through their toponymy, as well as through other aesthetic features such as myth and song that serve to provide context for the names and other observations. Following Tilly (1994: 23), landscape is approached here as having both a natural or physical component and a human component, with culture providing the filter for interpretation. Or, as seen by Ashmore and Knapp (1999: 1), "landscape is an entity that exists by virtue of its parts being perceived, experienced and contextualized by people."

For the purposes of this analysis, the focus will be on the semantics of toponyms for the Southern Paiute people in general, but a comparison will also be drawn of those for groups in two different landscape settings, the Mojave Desert and the Colorado Plateau. Although all Southern Paiute subgroups share some common principles in naming, as well as a focus on what is named, there appear to be a few differences in the semantics of the names that may be related to their different environmental settings. These differences are also detectable in the broader data from myth and song, thus suggesting an important role for these cultural forms in helping to contextualize the names and to define a Southern Paiute sense of place. As Basso (1996) notes, through story even unaltered parts of the natural environment can be filled with deep cultural meanings, which in turn anchor memory, serving to maintain relationships and preserve cultural values.

Place Names as a Field of Study

Although there have been many important studies of toponymy through the years for North American indigenous peoples, in many ways their study as keys to environmental perception was pioneered by Edward Sapir among the Southern Paiute in an important comparative paper titled "Language and Environment" (Sapir 1912). After work in 1910 with young Carlisle Indian School student Tony Tillohash from the Kaibab subgroup of the Southern Paiute, Sapir took specific note of what seemed to be the unique specificity of environmental terminology in place names in the Southern Paiute language as follows:

> In the vocabulary of this tribe we find adequate provision made for many topographical features that would in some cases seem almost too precise to be of practical value. Some of the topographical terms of this language that have been collected are: divide, ledge, sand flat, semicircular valley, circular valley or hollow, spot of level ground in mountains surrounded by ridges, plain valley surrounded by mountains, plain, desert, knoll, plateau, canyon without water, canyon with creek, wash or gutter, gulch, slope of mountain or canyon wall receiving sunlight, shaded slope of mountain or canyon wall, rolling country intersected by several small ridges, and many others. (Sapir 1912: 228)

All of this, Sapir remarked, reflects the interest of the people in such environmental features—"accurate reference to topography being a necessary thing to dwellers in an inhospitable semi-arid region; so purely practical a need as definitely locating a spring might well require reference to several features of topographic detail" (Sapir 1912: 228).

Part of Sapir's interest in this case was in the interactive role of language in a group's environmental adaptation, and thus the utility (and necessity) of its study as a reflection of a group's culture. In addition, though, he was also pointing to the role that the grammar of a language might play in setting at least some of the parameters for naming. Franz Boas (1934) would take up the latter theme in his classic study of place names among the Kwakwaka'wakwa (Kwakiutl) of the Northwest Coast, and in his comparison of cultural and linguistic features of these with Yupic (Eskimo), Tewa (after Harrington 1916) and other American Indian languages. In this work Boas illustrated the different "feel" but also "look" of place names in these different languages based on the peoples' interests in what is named, but also on differences in the grammars of their languages; e.g., the nominal vs. verbal character of names, ease of compounding and/or nominaliza-tion, development of locative devices, deixis, etc. Thus, Boas concluded, place-naming becomes a complex interactive process involving language, culture, and environmental setting.

In the years since these early works, interest in the general study of place names in North American Indian languages has at times waxed and waned. However, in recent years it has increased in importance as the languages have faced extinc-tion and the cultures have become increasingly detached from earlier modes of environmental interaction and awareness. Significant and extensive catalogs and analyses of place names and place-naming such as those by Kari and Fall (1987, 2003) on Alaskan Dena'ina, Hunn on Sahaptin (1990, 1991) and Basso (1996) on Western Apache have carried forward the ideas of Sapir and Boas, as well as added important new insights that can be derived from these data. Basso (1996), in particular, has illustrated very eloquently what it is like to be part of an active place-naming system through which people maintain a deep attachment to place as part of maintaining their sense of self. He shows how these names, and the stories surrounding them, are still used to teach important moral and social les-sons to those willing to listen and learn. Hunn (1996) also has called for deeper cross-cultural study of place-naming systems toward the identification of possible universal principals in their semantics. In all, this area of research is still viable and should have something to contribute, perhaps especially to the people whose systems are in need of reclamation as part of their processes of reattachment to the land and language and culture revitalization (Fowler 2004: 120; Hunn 1996: 4).

The Southern Paiute Data

In addition to the data originally collected by Sapir (1930–31) from Tony Tillo-hash, other field workers who have recorded Southern Paiute (including Cheme-huevi) place names include John Wesley Powell (Fowler and Fowler 1971), J. P. Harrington (n.d.), C. C. Presnall (1936), Catherine Fowler (n.d.), Carobeth Laird (1976), and Pamula Bunte and Robert Franklin (1987; Franklin and Bunte

1991). Although these sources span more than one hundred years, and all make important contributions to the overall data base of Southern Paiute place names, even combined they are not as extensive nor as geographically widespread as the materials collected by Isabel Kelly (1932–34).

During her roughly eighteen months in the field, Kelly traveled the length of Southern Paiute territory, accumulating some twenty-one field notebooks of ethnographic material, including the place names. She also made several large and detailed maps, with names and other reference information attached. Unfortunately, she rarely indicated what base map she was using, although given that her work was conducted in the early 1930s, few choices were likely available.[2] Also unfortunately, the names are often without larger ethnographic contexts into which to place them. It appears from her notes that she was using the extensive place-name data to develop a feeling for subsistence and settlement strategies, as well as travel and trail networks.[3] Her procedures for gathering the data are also not stated, although it appears that she traveled to at least some of the areas mapped with consultants, who even at that time were few in number for the tasks she had in mind. It is also likely that she and her more house-bound consultants looked at maps and even drew maps, in some instances, to build the data base. She occasionally noted in her field correspondence the remarkable memories of some of her consultants regarding place names, even though some were not at the time actually living in their home territories (Kelly 1934b).

Franklin and Bunte (1991) have identified some of the common devices used in the Southern Paiute language for constructing place names (Table 11.1). The resulting structures are nominalizations of various types, some based originally on verb stems, others on nouns or various noun-verb complexes. Many are binomials in the sense used by Kari and Fall (2003: 37) and Hunn (1996: 12), involving some type of generic stem (such as 'to sit' [mountain], 'water comes out' [spring], or 'canyon') plus an attributive or descriptive stem ('coyote', 'cattail', 'end of the cliff').[4] These are combined with various other grammatical elements in complexes that serve as the overall nominalized forms. The specificity of some of the generics and attributives is what attracted Sapir's attention, and we will return to some of these after a brief characterization of the semantics of the overall system. Etymologies for the place names provide at least some data on what Southern Paiute people chose to name in their environments as well as their principles of nomenclature—part of landscape perceptions.

A general overview of Kelly's 1,500-plus place names, and especially of those for which there are reasonably complete etymologies (roughly two-thirds), indicates that people preferred to name mountain peaks, ranges, saddles, and margins; knolls, hills, and plateaus; islands and parts of islands; streams, rivers, washes, and hot and cold water springs and tanks; salt lakes and pans; areas with rocky pediments, unique rock formations or rock types; camps where water was sufficient for planting; some valleys, especially if they have a characteristic cover plant or

Table 11.1 Common Place Name Stems (after Franklin and Bunte 1991)

Verb nominalizations

-kadi- , *-yuk*^w*i-*	'to sit' (sg., pl.), "hill," "rock," "mesa," "mountain"
-abi-	'lie down' (sg.), "long or wide mesa," "ridge," "plateau"
-wini-, *-waŋ*^w*i-*	'to stand' (sg., pl.), "pinnacles"

Other verb stems

-aboaga-	'to be semicircular', "box canyon"
-paa-tsi-	'water' (dim.), "spring"
-nuk^w*ani-*	'jut out', "promontory," "headland," "point"
-n-a-xa-	'have/be', "place where there is/are ___"
-patoni-	'round', "sugarloaf"
-(paa)-nuk^w*i-*	'water-flow', "stream," "creek," "river"
-k^w*iok*^w*i-*	'to be hollow', "hollow place"

Noun bases

-paa-(tsi-)	'water' (dim.), "small spring," "well"
-kaiba-	'mountain', "mountain"
-pikabo	'hard-round', "water pocket," "tank," "pothole"
-wi-pi, *-wiŋ*^w*a*	'wash', "wash," "canyon"
-kani	'house', "house," "home"
-poo	'line', "trail"

sg. = singular; pl. = plural; dim. = diminutive

other distinctive feature; trails and parts of trails; and a few caves. Of these, names for springs and mountains are most numerous (nearly one-third are springs and one-third mountains; the remainder are other features). Names derive from various sources: animals, plants, human body parts, spirits, and descriptive features of various kinds. Rarely if ever are personal names used in place naming, although Kelly did obtain detailed information on the ownership of springs, thus suggesting that opportunities to use owners' names did present themselves.

Kelly's etymologies do not bear out quite the level of specificity in naming noted by Sapir, but some are only partial, and additional analysis may produce better and more thorough translations. A sample of etymologies for "spring" names (built primarily on the stems *-paa*, 'water,' *-paatsi*, 'water + diminutive', or *-paatsipitsi*, 'water + diminutive + comes out,' includes the following:

From animals or animal referents: "Dog Spring," "Coyote Spring," "Coyote Nose Spring," "Badger Spring," "Rabbit Trail Spring," "Rattle Hiss Spring," "Big Horn Drinking Spring"

From plant referents: "Brown Willow Spring," "*Chia* Spring," "Has Willows Spring," "*Apocynum* Spring," "Mesquite Spring," "Cactus Drinking Place," "Underbrush Spring," "Has Indian Ricegrass Spring," "Has Serviceberry Spring," "Rabbitbrush Spring," "Brushy Spring," "Cane Spring," "Cottonwoods Surround It Spring," "Arrowweed Spring," "Cattail Spring"

From bird referents: "Golden Eagle Spring," "Quail Spring," "Waterbird Water"

From human or supernatural referents: "Boy Spring," "Mountain Spirit Spring," "Navel Spring" (a rock tank), "Water Baby Cries Spring," "Crying Spring," "Shit Spring"

From miscellaneous geographic or other referents: "Joining Spring," "Separating Spring," "Sand Boils Spring," "Mountain Spring," "White Mountain Base Spring," "Dampness on Grass Spring," "Lava Rock Spring," "Special Rocks Spring," "End of the Wash Spring," "Other Side Spring," "Summit Spring," "On the End of Lava Spring," "Round Hole in the Ground Spring," "Red Ochre Spring," "Water Cave Spring," "Dark Saddle Spring"

A small sample of etymologies for other place names includes: "Sandstone on the End" (a mountain base), "Spotted Lizard's Back" (a rock formation), "Black Serrated" (a mountain crest), "Frost Sits on the Ground" (a mountain valley), "Doctor's House" (a cave), "Gypsum Sitting" (a mountain), and "Willows Standing in a Line" (a stream bank).

Beyond identifications and etymological analysis, which give at least some minimal feel for landscape perceptions, there are some additional data in Kelly's notes that suggest that the people had more complicated views of their territory. These are data concerning hunting songs and ancient mythic journeys, many of which contain place names and provide a broader context. Carobeth Laird (1976), in her rich volume on the Chemehuevis based on George Laird's narratives, speaks of the importance to them of certain hereditary songs in chartering a man's (and occasionally woman's) rights to hunt certain animals in certain territories. Although George Laird could recall only small snippets of such songs, he was aware that they contained many references to specific places, and that they actually charted journeys (as well as shortcuts) to these places for those entitled to the songs. He recalled that there were at least a Deer Song, Mountain Sheep Song, Salt Song, Quail Song, and Day Owl Song. The idea for such songs as well as some of their content is something shared with the Mojave people from whom they may derive. Nonetheless, they are well integrated into the cultural patterns of the Chemehuevi and other Southern Paiute groups, especially in the Mojave Desert region (Las Vegas–Pahrump, Moapa, Shivwits people).

Although equally fragmentary in 1933 when Kelly was in the field, she was able to record the following from a Las Vegas consultant:

In the deer song, the deer travels around Charleston range looking for food. The snow is deep and it goes from place to place. It starts way up on top of Charleston peak; then it comes through the snow, finally out of the snow and down the valley. Comes down through *tsoariuwav* (Joshua tree valley), between Charleston Range and Tule springs. They sing all this in the song; name every place he stops, everything that he eats. (Kelly 1934c: 98)

Kelly then gives two samples of parts of the song, both of which name three places where the deer stops. Kelly reports that an abbreviated version was sung for dances and funerals, but that hunters who owned it sang the full version upon request by other hunters who were about to go out for game. Kelly adds that the Las Vegas mountain sheep[5] song starts from Coachella Mountain near Los Angeles, travels to San Bernardino Mountain and then to two other mountains for which she gives Southern Paiute equivalents, and ends at Charleston Mountain immediately west of Las Vegas. And she adds: "They arrive here in the early morning; are maybe 200 different verses all told; lasts all night, until sunrise" (Kelly 1934c: 100). All of this indicates that the proper singing must have been an exceedingly rich and informative experience, both in terms of places as well as foods for the mountain sheep: a virtual environmental inventory. Others have recorded similar accounts (see, for example, Hunn 1996: 70–71 for Jim Yoke's narrative).

Briefly, two other song cycles not related to hunting territories but also rich in place names and communicating a sense of cultural landscape are the Salt Song and the Talk Song, also known to other Southern Paiute subgroups. Kelly sketched out both with her Las Vegas consultants.

[The Salt Song] concerned the travels of 2 sisters *Yarik*, (wild goose) and *Avinan-kawatsi* (a small unidentified water bird). Lived at the mountain called *Agai* [New-berry Mountain], between Searchlight and Ft. Mohave. They sang en route as traveling along, naming everything they saw—mountains, water, everything. Traveled to Ft. Mohave on the other side of the river. Crossed to the other shore at Ft. Mohave and came up the river on the east side, at a place called *Mowavit*. Crossed the Colorado at the junction of the Virgin; went up to the salt cave there and named it; from there came to Charleston Peak, then to Ash Meadows; then to the salt lake below the town of Shoshone called *Panigi*; went to Blythe, crossing the river once more. Came up the east side, arriving just before daylight at *Kwinava*. Went into these mountains in the morning; there is a large cave there, 2 in fact. They entered one of the caves, thereupon the tale ends. (Kelly 1934c:102)

And a small piece of the Talk Song:

This comes from the ocean, this song. In the mornings the ocean is covered with mist or steam rising. In the beginning white birds, large ones called *parosabi* are in the fog. The man stands in his dream and watches the birds. They come out on dry ground, flapping their wings. As the birds fly out they name a mountain (***Ikanavanti***) in Ca-huilla country. As they fly over the mountain, the longest feather swept the top of the mountain. As the bird passed over the mountain he said, 'I am passing through a land of jimson weed.' But this is not so; there was only one plant there. The bird passed over ***Osapigamanti*** and right on the plain where there are no rocks he sees a ***pita***, eagle feathers tied together to make warriors headgear. He sees this and picks it up. The bird is traveling east. He flies over a wash west of ***Nantapiagant***; he looks at his shadow below and sees that the shadow of his wings just reaches from one end of the wash to the other. (Kelly 1934c: 104)

Unfortunately, Kelly did not have a tape recorder and was only able to transcribe fragments of this long tale and song. But what she did get goes on to take the bird many more places, ultimately ending in Hopi country and naming all the while. This song is at least some indication of the wider traveling habits of the Las Vegas people and their knowledge of areas well outside their traditional territories, as well as their landscape perceptions. All of the songs serve to illustrate at least some of the contexts in which place names were used in this region.

Although hunting territory and salt songs appear strongest among the groups in and near the Mojave Desert, other types of songs, including round dance songs, which are interspersed in myths and referred to as "recitatives" by Sapir (1910), and songs that are apparently composed and sung on a variety of other occasions also help to provide context for place names and additional data on landscape perceptions. Like the hunting songs, in translation they provide poetic images of the landscape and observations of it that are culturally sensitive. Powell (Fowler and Fowler 1971: 121–128) recorded several examples of these in the 1870s, as did Sapir in 1910 (Bunte and Franklin 1994). Some examples include:

(1) The crest of the mountain, forever remains, forever remains; though the rocks continually fall (Fowler and Fowler 1971: 123) [erosional forces];

(2) The edge of the sky; is the home of the river (Fowler and Fowler 1971: 122);

(3) The reeds grow in the mountain glades; and the poplars stand on the borders; they [deer] eat the reeds and get shade in the aspens (Fowler and Fowler 1971: 123);

(4) The cherty limestone yonder; on the Colorado [River] is very steep (Fowler and Fowler 1971: 123);

(5) Approaching the ***Paunsagunt*** [plateau], I met a fierce wind [downdraft] (Fowler and Fowler 1971: 124);

(6) It rains on the mountains; it rains on the mountains; a white crown en-circles the mountain [referring to the clouds] (Fowler and Fowler 1971: 125);

(7) On Elk Mountain, on Elk Mountain, moving through the red pines, the wind passes quickly, the wind passes quickly, carrying snow on its head,

carrying snow on its head [a medicine song] (Bunte and Franklin 1994: 638);

(8) Moving through the Mountain Plateau, moving through the Mountain Plateau, the crowned one, he peeps out now and then as he goes [Ghost Dance Song] (Bunte and Franklin 1994: 655).

Combining the data from place name etymologies with the data from hunting songs and other songs and myths gives a broader picture of landscape perception for the Southern Paiute groups than do the etymologies alone. The songs and stories likely were a primary means by which people learned to connect themselves to these territories, and also learned what was available in terms of water and other resources. Through the use of the names, especially in songs and stories, they were learning practical lessons in how to survive in these difficult environments, but also the spiritual lessons that helped them identify with the land and with themselves in the land.

Mojave Desert/Colorado Plateau Comparisons

The fifteen Southern Paiute subgroups that Isabel Kelly visited and worked among live in an expanse of country that is characterized by different topography, elevation, climate, and to some degree, resources (Figure 11.1). Groups in the far West, in what is now California and southern Nevada, are within the Mojave Desert, one of the driest areas in North America, while those in the middle territories (parts of Nevada, southwestern Utah) live within the Great Basin Desert, often characterized more as desert steppe or cold desert. Both of these locations are within the Basin and Range physiographic region, and are characterized by fault block mountains separated by long and broad interior draining basins. The remaining groups to the east, in south central Utah and northern Arizona, are in country that is part of the Colorado Plateau physiographic region, characterized by horizontal sedimentary strata cut by many steep-walled canyons.

Thus, the Southern Paiute groups were seemingly ideal for Kelly's studies in ethnogeography, as she could compare the cultural adaptations of the different subgroups within varied environmental situations (Fowler and Kemper 2008). These aspects should also lend themselves to comparisons of place-name terminology and other aspects of cultural aesthetics for potential differences in landscape perceptions. However, we should note at the outset that in spite of the relative richness of the data base for the groups, the materials are not quite comparable,[6] and what can be said must be qualified. On the whole, probably because of shared language, culture, and principles and concepts of naming, more similarities than differences are seen when data are compared for Mojave Desert dwellers (Las Vegas–Pahrump, Chemehuevi) and Colorado Plateau dwellers (Shivwits, Kaibab).

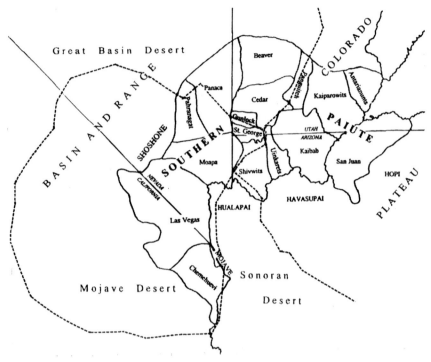

Figure 11.1 *Southern Paiute Territory (after Kelly 1934a) with approximate locations of the Mojave, Great Basin, and Sonoran deserts and the Basin and Range and Colorado Plateau physiographic regions.*

Given that place names often refer to plant and animal resources present in the environment, one obvious difference might be in the plant and animal resources named. Although there are plants that are common in the territories of these groups and that show up in the names—willow (*Salix* spp), cottonwood (*Populus* spp.), rabbit brush (*Chrysothamnus nauseosus*), cane, Indian hemp (*Apocynum cannabinum*), cattail (*Typha* spp.), agave (*Agave* spp.), etc.—there are also a number that are distinctly different in their geographic ranges. For example, in the Mojave Desert, names for plants such as chia (*Salvia columbariae*), mesquite (*Prosopis glandulosa*), screwbean (*Prosopis pubescens*), arrowweed (*Pluchea sericea*), paloverde (*Cercidium floridum*), ironwood (*Olneya tesota*), cholla (*Opuntia* spp.), etc. are found as attributes of spring names and names of formations, while in Colorado Plateau areas there are names commonly incorporating plants like Engelmann spruce (*Picea engelmannii*), ponderosa pine (*Pinus ponderosa*), cliffrose (*Cowania stansburiana*), oak (*Quercus gambellii*), lemonade berry (*Rhus trilobata*), serviceberry (*Amelanchier alnifolia*), wild rose (*Rosa woodsii*), etc. As the process of locating as many of these springs and other formations continues, it will be interesting to see if these species still remain in these locations, or if they have been extirpated by grazing or other environmental changes.

Fewer differences are found in names with animal referents, although for the most part they seem to name the most common animals (coyote, rattlesnake, jackrabbit, quail, owl, mosquito, ant, etc.). A few names from the Kaibab and Shivwits areas contain references to beaver, elk, antelope, and bear, which would not generally be found in the Mojave Desert except in select locations.

Based on preliminary assessments, a few physiographic references seem to differ in these two regions as well. Although both use 'mountain' (*kaibɨ*) in place names, there are far more of these terms in the Las Vegas–Pahrump and Chemehuevi areas, where mountains are a more obvious feature. On the Colorado Plateau (Shivwits, Kaibab), where peaks are rarer, such names are fewer. Groups in the Mojave Desert also use an alternative term for 'mountain' based on the verb 'to sit' (usually sg.), *-kati,* to yield 'hill' or 'knoll'. In the Colorado Plateau it is more common to find names based on the verb *-avi-,* 'to lie down' (sg.). In this region, large plateaus and mesas literally "lie" rather than "sit" on the surrounding landscape. In addition, although the term is used in both regions, names based on the verb *-wini-,* 'to stand' (sg.) are more common here, applied to the several different upright rock pinnacles characteristic of the region. Likewise, there are many more terms for 'canyons' (*-wi-p(i)*) and streams or washes (*-nukwi-ntɨ*) among the Colorado Plateau groups than among those in the Mojave Desert.

One rather unique form that they both share is a separate noun stem meaning 'water pocket, pothole, or tank' (*-pikabo*). This term, more in keeping with the uniqueness ideas put forward by Sapir, serves to label critically important water resources in both of these desert regions—i.e., catchment basins in rock formations. The Chemehuevi, in particular, name a number of these in their habitat.

If we reexamine some of the songs given above for some unique environmental perspectives, a few likewise become possible. For example, in (1) "The crest of the mountain forever remains, forever remains, although the rocks continually fall," there is possibly a very beautiful and accurate characterization of geological processes on the Colorado Plateau. Rather than eroding to form hills or slopes, the edges of the parent sandstone break off in vertical slabs that fall to the talus, leaving steep cliff faces in their wake. Other poetic allusions occur in songs and stories that bear out the deep knowledge and attachment to the land in both regions. And names like "Adobe Hanging Like Tears," "Closed Itself with Mesquite," "Fire Valley," "Night Hawk's Throat," "Where They Clubbed Each Other," "High-on-the-Side Tank," "Deer Carrying Net Spring," etc., are poetic allusions that carry the sense of landscape perception and attachment even further. Some of these may once have had story or song connections; others may be purely aesthetic responses to the beauty of the landscape.

Conclusion

Thus the task: can these data be used as at least partial keys to landscape perception for the Southern Paiute people? And further, can a deeper sense of place be reconstructed from these data? I think that with a lot of work (and ground time) both can be achieved, at least in part. At a minimum, these data give a view of a past time that is available to fewer and fewer Southern Paiute and Chemehuevi people today due to profound language loss and trends away from intimate landscape experiences. It is sometimes said in anthropological circles that "there is nothing as dead as old field notes," because once divorced from the context in which they were gathered, they become meaningless. But in some instances at least part of that context can be retrieved and perhaps even revitalized. The work involves partnering with Southern Paiute and Chemehuevi people in using modern maps to locate as many as possible of the places Kelly cataloged, and then visiting them to see what may be retrieved and what may have stimulated the names. A second is to find any remnants of the songs and stories preserved in living memory, as it obviously is the songs and related tales that are rich sources for tying the place names to real activities and a fuller sense of geography (Basso 1996). One example is the Salt Song (Salt Song Project 2005), which is presently being documented and revitalized.

Perhaps these approaches, plus additional work on etymologies, will replace something of the soul of these materials, giving all of us, but particularly the grandchildren and great-grandchildren of the original place-makers, an enhanced view of their country. A number of young Southern Paiute people today are trying very hard to develop their own sense of place, reclaim their history, and continue their journeys in place-making. These data from the distant past could help in part to serve that purpose.

Notes

1. The Isabel T. Kelly Archive is held in the DeGolyer Library, Southern Methodist University, Dallas, TX, under the control of Dr. Robert Van Kemper, Department of Anthropology, SMU. Permission by Van Kemper to cite is gratefully acknowledged, as is support from the Wenner-Gren Foundation for Anthropological Research.
2. The maps available to Kelly in 1932 included some of the USGS quad sheets, along with miscellaneous highway and spring maps, depending on location. Some are preserved in the Kelly Archive. Other maps seem to be based on Kelly's sketch maps made in the field.
3. Kelly was a student of Carl Ortwin Sauer and A. L. Kroeber at the University of California, Berkeley, having completed her dissertation in 1932. Although she does not define what she means by "ethnogeography," she was taking her approach and methods from those suggested in joint seminars held by these men at UC-Berkeley (Fowler and Kemper 2008).
4. Throughout the use of single quotes denotes etymology and double quotes denotes translation.

5. Mountain sheep are desert bighorn sheep.
6. For a number of groups, Kelly was able to find only a single person to interview for these types of data; for others, she had three to four people. The amount of field time and travel also differed for each group.

References

Ashmore, Wendy, and A. Bernard Knapp. 1999. "Archaeological Landscapes: Constructed, Conceptualized, Ideational." In *Archaeologies of Landscape: Contemporary Perspectives,* ed. Wendy Ashmore and A. Bernard Knapp. Malden, MA: Blackwell Publishers.

Basso, Keith H. 1996. *Wisdom Sits in Places.* Albuquerque: University of New Mexico Press.

Boas, Franz. 1934. *Geographical Names of the Kwakiutl Indians.* Columbia University Contributions to Anthropology No. 20. New York.

Bunte, Pamela A., and Robert J. Franklin. 1987. *From the Sands to the Mountain: Change and Persistence in a Southern Paiute Community.* Lincoln: University of Nebraska Press.

———, eds. 1994. "Southern Paiute Song Texts." In *The Collected Works of Edward Sapir IV: Ethnology,* ed. Regna Darnell and Judith Irvine. Berlin and New York: Mouton de Gruyter.

Fowler, Catherine S. n.d. Southern Paiute Field Notes (1961–2000). Unpublished MS in author's possession, Reno, Nevada.

———. 1992. *In the Shadow of Fox Peak: An Ethnography of the Cattail-Eater Northern Paiute People of Stillwater Marsh.* Cultural Resources Series No. 5. Portland: U.S. Fish and Wildlife Service.

———. 2004. "Material Culture and the Marking of Southern Paiute Ethnic Identity." In *Identity, Feasting, and the Archaeology of the Greater Southwest,* ed. Barbara Mills. Boulder: University Press of Colorado.

Fowler, Catherine S., and Robert Van Kemper. 2008. "A Life in the Field: Isabel T. Kelly, 1906–1982." In *Out on Their Own Frontier: Women Intellectuals and the Re-visioning of the West, 1900–1960,* ed. Shirley Leckie and Nancy J. Parezo. Lincoln: University of Nebraska Press.

Fowler, Don D., and Catherine S. Fowler, eds. 1971. *Anthropology of the Numa: John Wesley Powell's Manuscripts on the Numic peoples of Western North America, 1868–1880.* Smithsonian Contributions to Anthropology No. 14. Washington, D.C.

Franklin, Rob, and Pam Bunte. 1991. "Place Names Data for 'San Juan Southern Paiute Toponymy.'" Friends of Uto-Aztecan and SSILA Joint Meeting, handout.

Givon, Talmy, comp. 1979. *Ute Dictionary: Preliminary Edition.* Ignacio, CO: Ute Press of the Southern Ute Tribe.

Goss, James A. 1972. *Ute Lexical and Phonological Patterns.* Unpublished PhD diss., University of Chicago.

Harrington, J. P. n.d. "Chemehuevi Place Names." MS, National Anthropological Archives. Reproduced in *The Papers of John Peabody Harrington in the Smithsonian Institution 1907–1957,* ed. Elaine L. Mills and Ann J. Brickfield, vol. 3, White Plains, NY: Kraus International Publications [1985].

———. 1916. "The Ethnogeography of the Tewa Indians." *Bureau of American Ethnology Annual Report (1907–1908)* 29: 29–618.

Hunn, Eugene. 1990. *Nch'i-Wána "The Big River": Mid Columbia Indians and Their Land.* Seattle: University of Washington Press.

———. 1991. "Native Place Names on the Columbia Plateau." In *A Time of Gathering: Native Heritage in Washington State,* ed. Robin K. Wright. Seattle: Burke Museum and University of Washington Press.

———. 1996. "Columbia Plateau Indian Place Names: What Can They Teach Us?" *Journal of Linguistic Anthropology* 6(1): 3–26.

Kari, James, and James A. Fall. 1987. *Shem Pete's Alaska: The Territory of the Upper Cook Inlet Dena'ina.* Fairbanks: Alaska Native Language Center, University of Alaska and The CIRI Foundation.

———. 2003. *Shem Pete's Alaska: The Territory of the Upper Cook Inlet Dena'ina.* Fairbanks: University of Alaska Press.

Kelly, Isabel T. 1932. "Ethnography of the Surprise Valley Paiute." *University of California Publications in American Archaeology and Ethnology* 31(3): 67–210.

———. 1932–34. Southern Paiute Field Notes. Unpublished manuscripts, Southern Methodist University, Dallas. Copies in C. S. Fowler's possession.

———. 1934a. "Southern Paiute Bands." *American Anthropologist* 36(4): 548–560.

———. 1934b. Letter to Jesse Nusbaum, 29 January 1934. Laboratory of Anthropology Archives, Museum of New Mexico, Santa Fe.

———. 1934c. Southern Paiute Field Notes: Las Vegas II. Unpublished MS in DeGolyer Library, Southern Methodist University, Dallas.

———. 1964. "Southern Paiute Ethnography." *University of Utah Anthropological Papers* 69. Salt Lake City, UT.

Laird, Carobeth. 1976. *The Chemehuevis.* Banning, CA: Malki Museum Press.

Miller, Wick R. 1972. "Newe Natekwinappeh: Shoshoni Stories and Dictionary." *University of Utah Anthropological Papers* 94. Salt Lake City, UT.

Presnall, C. C. 1936. "Paiute Names for Zion Canyon." *Zion and Bryce Nature Notes,* 8(1).

Salt Song Project. 2005. *The Salt Song Trail: Bringing Creation Back Together.* San Francisco, CA: The Cultural Conservancy and the Salt Song Project.

Sapir, Edward. 1910. "Song Recitative in Paiute Mythology." *Journal of American Folk-Lore,* 23(90): 455–472.

———. 1912. "Language and Environment." *American Anthropologist* n.s.14: 226–242.

———. 1930–31. "The Southern Paiute Language." *Proceedings of the American Academy of Arts and Sciences* 65(1–3).

Shimkin, Demetri. 1947. "Wind River Shoshone Ethnogeography." *University of California Anthropological Records* 5(4).

Steward, Julian H. 1933. "Ethnography of the Owens Valley Paiute." *University of California Publications in American Archaeology and Ethnology* 33(3): 233–350.

———. 1938. "Basin-Plateau Aboriginal Sociopolitical Groups." *Bureau of American Ethnology Bulletin* 120.

Tilly, Christopher. 1994. *A Phenomenology of Landscape: Places, Paths and Monuments.* Oxford: Berg.

Chapter 12

Managing Maya Landscapes
Quintana Roo, Mexico

E. N. Anderson

Introduction

Landscape has been a key concept in geography for decades, and recently has become important in anthropology as well. The theory of landscape in geography was developed by Carl Sauer (1925, 1963). His students, notably Yi-fu Tuan (e.g., 1977, 1979, 1990), have built on his foundations. Landscape, in the Sauerian sense, comprised the landforms, waters, living things of the land, and the people, including their manipulated environments and their understandings of the land. Sauer saw landscape as a result of human management of nature—planned use and unplanned consequences. Nature was a player; human managers had to consider climate, landforms, soils, waters, and vegetation.

Sauer recognized that people and the nonhuman environment interact to produce a particular landscape, and that people's cultural ideas lie behind the strategies they use to deal with the environment. He was probably the first major scholar to realize how much the environment was modified by small-scale indigenous societies; among other things, he was among the first to understand the enormous importance of carefully timed and managed burning in shaping North America's forests and prairies (Sauer 1971; also see Doolittle 2000). Sauer was consciously uniting biology, geology, and agricultural science with German idealist philosophy deriving from the work of Immanuel Kant (Kant 1970, 1978 [1798]; cf. Merleau-Ponty 2003). Daniel Janzen (1998) memorably called the world a "garden"; Sauer, earlier, saw a planet of "landscapes."

In *Loving Nature* (2002), Kay Milton took a phenomenological position (following Kant and Merleau-Ponty; see also Abram 1996). She built on the perception that people relate emotionally as well as cognitively to landscapes, just as they do to other things; in other words, thought requires integration of emotion and cognition (Damasio 1994; Zajonc 1980). She argued that love for nature is required for environmental activism, but that definitions of "nature" then become extremely important, with differing definitions leading to conflict. I shall apply her insights below, in developing Sauer's concept of landscape as lived and experienced.

Some of the most ambitious recent work in landscape studies concerns the Mesoamerican region, including the Maya peoples. A major recent project in the tradition of Sauer is a three-volume series on cultivated landscapes in the Americas (Denevan 2001; Doolittle 2000; Whitmore and Turner 2001).

The Yucatec Region

Some 1,000,000 Yucatec Maya dwell in the Yucatán Peninsula, most of them in the Mexican states of Yucatan, Campeche, and Quintana Roo. Among the most traditional and solidly Maya-speaking areas is the "Zona Maya" of central Quintana Roo, a broad band across the state. As recently as 1980, one-third of the population of the area of this research reported on the census that they had "no income," i.e., that they were strictly subsistence farmers (INEGI 1986: 445).

This agricultural lifestyle has endured for at least five thousand years in the area. It is based on swidden cultivation of maize. This is locally known as *milpa* agriculture, from the Nahuatl word for "maize field." Swidden, also known as slash-and-burn, is only one technique for using the landscape. Other important components of the Maya system include permanently cultivated gardens and orchards, mixed cropping in the more fertile parts of milpa fields, managing regrowth of the milpa so that it has more useful species, and harvesting the forest intensively, often with constant, unobtrusive but vitally important deliberate modification of the forest vegetation. This produces a landscape that Scott Fedick calls "the managed mosaic" (Fedick 1996).

Slash-and-burn agriculture creates irregular light gaps in the forest. These are clustered around settlements, which in turn have to be near permanent water, a scarce commodity in most of the Yucatan Peninsula. At first, people were confined to the vicinity of natural lakes and sinkholes (Spanish *cenotes*, from Maya *ts'oonot*). The Classic Maya learned to deepen natural lowlands and transient lakes (*aguadas* in Spanish), containing them in stone walls and to sealing them with clay to prevent water leaking out. Eventually the Maya constructed completely artificial reservoirs.

An example of what happens without the managed mosaic occurred when Hurricane Gilbert blew down the forest of northeast Quintana Roo in 1988. The next dry season, fire swept the area, burning thousands of hectares in one great conflagration. Under Maya management, the mosaic of often-bare fields and lush, wet regrowth does not carry wildfire. Fires soon die out. Without this fire-breaking mosaic, much of the northern Yucatan Peninsula would burn to the point of regressing to savannah.

The ancient Maya have received much more attention in this regard than the modern ones (though see Faust 1998; Faust and Bilsborrow 2000). *Cultivated Landscapes of Middle America on the Eve of Conquest* (Whitmore and Turner 2001) includes much on the Maya—alleging that the Maya did themselves in by overusing the environment, a notion strongly contradicted by later research that reveals major drought as at least one cause (Gill 2000). Other recent literature sees drought as a precipitating factor but environmental overuse, as well as warfare, lying behind the collapse that drought caused (Demarest, Rice, and Rice 2004; Diamond 2005; Folan et al. 2000; Webster 2002). Most spectacular are the sacred mountains the Classic Maya created in this mountainless land. The huge Classic pyramids rival Yucatan's natural hills in size, and most of the pyramids stand on flatlands where they are visible for miles, rising over the forest canopy. From the top of a pyramid, a priest could look over even the tallest trees and see the next pyramid, ten or twenty miles away.

People did not disappear with the fall of the Classic cities. All through the area, especially in regions of low hills and wide flat valleys, there are small post-Classic sites—typically, a single farmstead on a hill overlooking a wide fertile valley. Small villages occur around water. These unobtrusive but ever-present farmsteads indicate a fairly high population, possibly rebounding after 1200, when the drought slackened. On the other hand, the massive landscape modification is gone. One assumes that the farmers caught rainwater in jars for storage; sites are dominated by thick utilitarian sherds (as I have observed dozens of times in the Chunhuhub area).

Contemporary Maya Knowledge

Walking forest trails with a Maya farmer is a fascinating experience. One's guide is constantly clearing the trail, pointing out fruit trees (many originally planted in a now-vanished field), weeding around desirable trees (such as young mahoganies), noting game trails for future reference, assessing the abundance of birds and of wild fruit, checking the soil, collecting herbal medicines for future use, assessing possibilities for milpas in the future, and otherwise manipulating the forest physically and cognitively. Such a farmer knows perhaps 500 animal names and almost a thousand plant names, and knows uses for perhaps half of the plants. Many plants have multiple uses. Moreover, different farmers have different areas of expertise, and many are recognized by their communities as specialists in particular areas, such as apiculture or herbal medicine.

The total knowledge pool in a Maya community is staggering—fully comparable to the knowledge pool of a university professor, though the realms of expertise are different. All Maya knowledge (like all politics) is local. Knowledge of the landforms, soils, and ecological formations provides a grid on which more detailed knowledge of plants, animals, and farming can be mentally projected. Since colonial times, outside observers have been struck by the precise and careful classification of soils and landscapes encoded in Maya.

Chunhuhub

My knowledge of this realm was learned largely in Chunhuhub ('trunk of the wild plum tree'), a town of about 6,000 people. I have spent a total of a year and a half there, distributed over three major field sessions and many short visits. Almost all the residents are Maya, predominantly engaged in agriculture for subsistence and sale. Most people raise at least some of their own maize, beans, squash, fruit, vegetables, and domestic animals, and get cash for clothing and other needs by selling agricultural produce, primarily fruit. Small-scale specialization and complementarity exists; people are constantly selling or trading small lots of produce or animals.

Chunhuhub is one of a long chain of towns that lie at the foot of the hill range that separates the flat coastal lowlands from the rolling interior of the peninsula. These towns take advantage of the rich alluvial soil that has washed off the hills and collected in low-lying, damp areas. The settlement pattern is an old one; from Chunhuhub southward, one can trace a line of Classic Maya sites—including several large and splendid cities—along this narrow band, with its rich alluvial valleys. (For full data on Chunhuhub and its natural environment, see Anderson 2003, 2005; Anderson and Medina Tzuc 2005.)

The Yucatan Peninsula is a rather homogeneous environment. It is a limestone shelf with shallow red soils. Broadly, these soils are "rendzinas"—soils developed on limestone and affected by cultivation (Fedick 1996). Surface water is scarce, because rain disappears into the porous limestone. Chunhuhub has no permanent surface water of any kind. Moreover, the peninsula is solidly covered by forest dominated by leguminous trees. This forest grades from dry and scrubby growth in the northwest to tall rainforest in the far south. Chunhuhub lies in an area of medium-height, seasonally dry forest. The canopy of old-growth forest is about twenty to twenty-five meters high, and the trees tend to shed their leaves during the dry, hot period from March to mid May.

The Spanish developed a major center at Chunhuhub that rose from a hacienda to a substantial town, growing sugar cane and distilling rum (Anderson 2005). The sugar industry and the extent of livestock noted in colonial records indicate a high population, a rather rich town, and a thoroughly developed countryside. The sugar must have been raised in the flat, fertile plain five miles south, now irrigated for orchards and row crops. The area was depopulated in 1846 in the War of the Castes and not repopulated till the 1940s, when settlers trickled in from old "rebel Maya" towns farther east. The new landscape was initially one of a small subsistence hamlet in a vast thick forest.

After the Mexican Revolution, local indigenous and mestizo communities were given title to lands. Most often, the titleholding entities were collective landholding units known as *ejidos*. Many *ejidos* were constituted in the heady days after the Revolution, in the 1920s, but the real wave of *ejido* formation came in the late 1930s and the 1940s, during and just after the presidency of Lázaro Cárdenas. *Ejidos* were the logical heirs of the old communal landholding communities that began in pre-Columbian times and were recognized under Spanish law in the colonial period. During Mexico's days of independence, most of these collectives were abolished. The poor, the indigenous, and the rural lost their lands and their access to lands. Alienation of land reached a peak in the Porfirio Díaz era, and it was a major cause of the Revolution (1910–1921).

Among the many beneficiaries of *ejido* creation were the Maya of central Quintana Roo. These were heirs of the Caste War (1846–1848), the spectacular independence movement among the isolated Maya of south and east Yucatan state and of the land that was to become the state of Quintana Roo. While what is now Yucatan state was soon reconquered, the Maya held out for almost a hundred years in the wilds of what is now Quintana Roo.

One of the results was vast donations of land in the 1940s. Huge *ejidos* were created from the trackless and unexplored forests of the area. *Ejido* formation continued, sporadically, for many years. Chunhuhub today is a town of 6,000 people, with most urban amenities. Many of its people and most of its leaders are highly aware that a good modern education is the main way to succeed in today's Mexico, so Chunhuhub has become a major "exporter" of skilled person-power. Teachers, mechanics, and computer workers from Chunhuhub are found all over Mexico.

Landscape Classification

Landform classification in Chunhuhub is fairly simple. Hills are **wits**. There seems to be no special Maya word for flatlands. (Spanish terms like *tierra baja* tend to be used.) More important than the hill features are the wetlands. The most dramatic of these are limestone sinkholes (*cenotes*)—places where the overlying rock has collapsed. Where these are deep, they reach to groundwater, and provide a source of fresh clean water for surrounding communities.

Far more important in central and southern Quintana Roo are the larger wetlands known in Maya as **ak'alche'** and in Spanish as *aguadas*. These occur when fine clay washes into a depression, usually an oval area that is the direct or indirect result of faulting. The clay seals the bottom, thus retaining water. **Ak'alche'** can have actual lakes (**k'anab**) or temporary ponds, marshes, or moist woodlands. **Ak'alche'** vary from fertile alluvial soils to barren marls to highly acid black clays.

Another landscape term is **ya'ax joom,** "green hollow" (**joom** means a hollow, ditch, nest). This is a low area that remains permanently moist, has rich soil, and grows lush vegetation, staying green even in the dry season. Usually, the soil is made up of rotted plant material, which is fibrous and holds water well. **Ya'ax joom** grade into **ak'alche'.**

Maya soil classification is primarily based on color, but each soil color term implies data about texture and quality as well. Variations occur from place to place and even from speaker to speaker. In Chunhuhub, the classification is more or less as follows:

> **Chak lu'um** "red soil" is more or less the basic or unmarked category. It is the ordinary, shallow, gritty, poor-looking but surprisingly fertile rendzina that blankets most of the landscape.
>
> **K'an lu'um** "yellow soil" is a yellower, less fertile, often more clayey soil, very similar to the preceding, and grading into it via
>
> **K'an chak lu'um** "yellow-red soil," a relatively fertile soil compared to **k'an lu'um.**
>
> **Ek' lu'um** "black soil" is black alluvial soil.
>
> **Box lu'um** "black soil" is also black alluvial soil.

The Maya words **ek'** and **box,** however, though both meaning "black," are not quite synonyms. In this case, **ek' lu'um** is often used for the infertile, acidic, aluminum-rich soil of the boggy and infertile type of **ak'alche'.** Such soil is dominated by a tree which, appropriately, yields a black dye, and thus is known as **ek'** (it is the dyewood *Haematoxylum campechianum*). **Box lu'um** is more apt to be used for the rich, fertile type of **ak'alche'** black alluvium. However, the term **ek' lu'um** is widely reported in the peninsula for the more fertile type of black soil.

> **Chak ek' lu'um** inevitably occurs as a term, and refers a very fertile soil (in spite of the use of **ek'** here)—a rich soil developed on good sites in old-growth forest, full of leaf mold and loamy in consistency.
>
> **Tsek'el** is very rocky soil, almost bare rock. It occurs on steep slopes.
>
> **Sajkab** is rotted limestone—not yet weathered to red, but still whitish or yellowish.
>
> **Sajkab** ranges from nearly solid rock to completely weathered white soil. (This word is pronounced *saskab* in other areas, and the word has been borrowed into Yucatecan Spanish as *sascab;* in this sense it is used in Chunhuhub for rotted limestone excavated for use as road-grading material. Thus the two words, formerly synonyms, coexist with different meanings in Chunhuhub Maya. A road-material quarry is a *sascabera.* This Maya word with a Spanish ending is

a typical example of the linguistic mix known as *mestiza Maya*, "mixed Maya" or "the Maya of mixed-origin speakers.")

K'ankab is similar but more weathered: a hard rocklike or hard-clay soil of a reddish-yellow color.

Tunchi' is virtually bare rock, with some soil in pockets, from *tun* (stone) and *tunich* (a rock), usually understood to be limestone (the only other locally occurring rock is flint or chert, *tok'*).

This terminology may be compared to the terminology recorded in the nearby and very similar town of Xocen. Silvia Terán and Christian Rasmussen, in their superb study *La milpa de los Mayas*, list the following (1994: 140; my translations):

Chak kankab (i.e. *k'ankab*), red, poor soil

K'amas lu'um (lit. "termite soil") or *chak lu'um,* red, poor, deeper than above

K'an lu'um, yellow, fertile, deep

Box lu'um or *ek' lu'um,* black, fertile, shallow, rocky, dryish

Ek' lu'um or *box lu'um,* black, fertile, shallow, moister and less rocky than above

Kakab lu'um (also called *ek' lu'um* or *box lu'um*), black, fertile, deep

Sahkab lu'um or *ek' lu'um,* black, deep

Ch'ich lu'um or *box lu'um,* black, fertile, with small stones, deep

Of these, *kakab* or *kajkab* is a term I have heard, but rarely; it is widely used in the peninsula, for various relatively fertile soils. "Termite soil," however, seems to be a Xocenism, unless it is a misunderstanding of a local term for soils enriched by debris of termites or ants—such soils being popular for use, and sometimes deliberately collected to fertilize gardens.

In Xuilub, another nearby town (closely related historically to Xocen), Natividad Herrera found the major types of soil recognized were *chak lu'um, box lu'um,* and *saaskab,* all used as in Chunhuhub.

Terms for vegetation types next concern us:

Kool, 'field', 'milpa'

Solar, 'garden' (there is no special native Maya word for 'garden', *kool* being used for both temporary and permanent fields and garden areas)

Sak'a' 'milpa in very early regrowth stages'; Spanish *cañada*

Huubche', 'regrowing milpa' (standardly used for all stages of regrowth—in contrast to the general usage in Yucatan state, where stages of regrowth are distinguished); Spanish (from Nahuatl) *huamil* is often used

K'aax 'forest', 'woodland', 'thicket'; Spanish *monte* (for short trees or brush), *selva mediana* (medium-height forest; this latter term is not used by the Maya)

Ka'anal k'aax, 'high forest' (fully grown forest; tall old-growth forest); Spanish *selva alta*

Plant communities are recognized, but not usually by Maya names. They have to be described. If there is one dominant species, the community is apt to be called by a Spanish name. Spanish has a useful habit of naming plant associations by adding *-al* to the name of the dominant plant. Californians are aware, for instance, of "chaparral," from Spanish *chaparro,* 'scrub oak' (or by extension anything short and scrubby). Thus, in Yucatan, the solid thickets of *ek'* trees that form in acidic **ak'alche'** are called *tintales,* from Spanish *tinte* ("dye").

Finally, an opposition of importance in describing the biota is **akbij** or **alakbij** "tame, domestic" vs **ba'alche'** or **k'aaxij,** "things of the trees, i.e. wild." (The second of these three is not to be confused with **baalche'** "honey mead.")

The above words provide the terminological grid for describing the landscape. A regrowing field may be described as a **sak'a'** originally cut from **k'aax** made up mostly of **ja'abin** (*Piscidia piscipula*) and **tsalam** (*Lysiloma bahamense*) on **tsek'el** soil on the side of a **wits.**

One surprising aspect of landscape language is the absence of place names. Communities (ranches, villages, towns, camps) are named. Most of them are near *aguadas* or *cenotes,* and the *aguada* is named from the community or vice versa. Otherwise, there seem to be very few place names. Hills and valleys have no names. Neither do small hollows and temporary water sources. Archaeological sites are named from the nearest settlement (or, in other parts of Quintana Roo, sometimes with a descriptive label). Otherwise, places are located with reference to communities, roads, and individual milpas—"the hill where Don Jas cut his milpa two years ago" or "the high forest at the southeast corner of Chunhuhub's community land." I suspect this lack of place names is due to two things. First, Chunhuhub was deserted for almost a century, from around 1847 to 1940. The whole area was abandoned, and any minor local names were forgotten. Second, the nature of Quintana Roo Maya shifting cultivation makes fixed landmarks rather meaningless. It is more sensible to describe the land in terms of what is fixed (communities and borders) and, within that grid, in terms of the current milpa situation, which is anything but fixed. Realistically, it is easier to find something when it is located in terms of Don Jas' old milpa or Don Yu's new orchard. An outsider can always ask around and get guidance.

A milpa is cropped for two years, then allowed to regrow. It is not, however, abandoned. Some crops, notably beans, manioc, and sweet potatoes, live on. Fruit trees keep yielding. Moreover, the regrowth is managed. Valuable trees, such as wild fruit trees and thatch palms, are left standing. Less valuable but still useful trees are coppiced. Weeds that have some use are allowed to grow. These include medicinal plants, foods (such as wild jicamas) that can be eaten if the crops fail, and even rat poison—the common local cycad, *Zamia lodigesii,* is encouraged because its root provides an effective rodenticide.

Trees killed and hollowed out by fire receive regular visits because bees nest in them and provide honey. Old wood from the initial burning and clearing is gathered for firewood. Also important is the nature of regrowth. Cultivation exhausts soil fertility, and second growth has trees that either live on very limited resources (including *Cecropia* spp., *Bursera simarouba,* and *Metopium browni*) or, much more often, on trees that provide their own fertility by fixing nitrogen from the air: leguminous trees, especially *Piscidia piscipula, Lysiloma bahamense,* and various species of *Acacia, Lonchocarpus,* and several other genera. They form a fast-growing forest. Their leaves, rich in nitrogen and in minerals extracted from deep layers of rock, form a very fertile but thin and fragile soil. Eventually they are replaced by shade-tolerant, slow-growing, older trees often carefully preserved from the last burning, such as chicozapote, *Manilkara zapota,* which makes up 15 percent of the trees in old-growth forest that has been cropped for thousands of years. The milpa cycle takes anywhere from eight to over a hundred years, but ideally about thirty.

Obviously, thousands of years of this lead to a forest consisting almost entirely of three types of plants: those that are human-encouraged, those that are nitrogen-fixers, and those that are tolerant of poor soil in open locations. Small shade-tolerant but fertility-loving plants survive in pockets of long-uncut forest. Many of these are valuable for medicine, so are encouraged too.

The land is a palimpsest; one digs through, or looks over, layer after layer of different occupations. Yet, through it all, slash-and-burn maize cultivation, supplemented with beans, chiles, cotton, squash, and the other famed New World crops, has remained by far the major influence on the land. This is true today: only 5 percent of Chunhuhub's 14,330 hectares of *ejido* land is under modern techno-agriculture, and none of the neighboring *ejidos* has any significant area of land under cultivation by nontraditional methods. Indeed, on all the *ejidos* of central Quintana Roo, the vast majority of the land is still under forest. Chunhuhub, the most densely populated, is still 75 percent forested. The forest is all second growth, ranging from newly regrowing to very old; some of it has not been cut since before the area was abandoned in 1846. The Maya are always out in it, hunting, gathering medicinal herbs, recutting old trails and boundaries, looking for good soil for future milpas, or just wandering to see what might be there.

Other Ways of Looking at Landscape: Aesthetics, Values, and Emotions

A quite separate way of thinking about the landscape is the aesthetic, emotional, and evaluative mode. This ties back to Kay Milton's point in *Loving Nature*

(2002). Most of the older and more rurally oriented Maya love the forest—its flowers, trees, and animals. They also love to see flourishing fields and orchards, neatly cleared trails, and other marks of successful human interaction with the landscape. Strikingly absent is the opposition of man and nature so familiar to English speakers, let alone the stronger and more salient distinction of *hombre* and *naturaleza* typical of Spanish speakers. Animals, even those that would appear to be pests, are viewed and treated with reverence. Some less traditional farmers burn out leafcutter ant nests, but most refuse to follow this lead, preferring to spare the ants in spite of their devastating effect on orchard trees.

Do the Maya of Chunhuhub love nature? The question is problematic. The Maya language has a word for loving something (*yakuntik*) but no word for "nature," since the Maya manage all aspects of their landscape (Fedick 1996; Gomez-Pompa et al. 2003). Maya recognize that domestic as well as wild beings have an integrity, perhaps even a personhood, quite separate from and outside of human lives. Rarely would a Maya say *in yakuntik le ba'alche'oob o* ("I love the things of the trees"), except, perhaps, when prompted by a conservationist. However, my collaborator Don Felix Medina Tzuc said, contemplating the magnificent old-growth forest of Nueva Loria: *"Me encanta la selva, me encantan los arboles"*—"the forest enchants me, the trees enchant me." He swore he would never cut such a forest down. The Maya recognize that they are tightly bound up with those wild lives. They feel a real and lively affection for them. They might say they love birds and flowers. But they do not think of wild things as a single category that is somehow lovable as a whole. Such a view of "nature" is probably held only by those who are rather cut off from the wilder world—people in industrial and postindustrial civilizations.

On the other hand, many Maya clearly do love the forest—its trees, flowers, animals—in much the same way that preindustrial English loved the "greenwood." They also share, in a somewhat attenuated way, the religious construction of flowers and vegetation described by Jane Hill for Uto-Aztekan societies (Hill 2001; Taube 2004). Love thus is felt as, and shows itself as, a caring, caretaking relationship combined with intense emotional and aesthetic responsiveness to the wild. A word that *is* used all the time is the Spanish verb *cuidar,* 'care for'. It implies a warm, caring, emotional bond, like that of parent to child. Unlike words for love, this word is constantly used in discussing human-nonhuman relationships. The commonest Maya not-quite-equivalent, for tame animals and plants, is *alak-* "raise" (as in *alakbij* "raised" or *alaktik* "to raise something"). It too implies a relationship like that of a parent raising a child.

The amount of care varies (see Anderson 2005). My friend and coworker Felix Medina Tzuc and my landlord Pastor Valdez certainly feel both deeply loving and intensely responsible toward the forest, as I know from hundreds of field hours with them. Most older Maya agree, but a few Maya share some or much of the traditional Hispanic-Mexican value system, within which the wild is something to be transformed into civilized terrain. Ironically, whereas Hispanic Mexicans are rapidly becoming environmentalist, indigenous people, who are becoming more Mexicanized, may absorb the traditional Hispanic-Mexican antipathy to nature.

A Maya household is usually surrounded by flowers, and the house yard is almost sure to include pets of various species. Weeding is done selectively: weeds are not usually killed but simply cut back, and useful wild plants are allowed to flourish in the field. Traditional Maya of Chunhuhub are extremely kind and gentle with all these lives, whether tame or wild. I have seen Maya discover a rabbit's nest while weeding, carefully cover it up, and weed around it—even though rabbits are a terrible pest of maize. Maya who pick up ants to show me are careful to put down the ant, uninjured, heading the same way it was going when picked up. Maya are aware that everything can be useful, but there is more to it than that; the Maya are aware that we all depend on each other, and the world goes best if we all take care of each other.

The Maya lack a "Western"-style concept of "nature" (however defined, however essentialized). This does not preclude them from recognizing that there are differences between the wilder and the more cultivated sides of the world, or from loving the wilder side. Quite the reverse: since the wild is integrated with the sown, people care for them all, as part of one system. Neither is regarded as a mere resource, and neither is regarded as something alien to be aesthetically enjoyed but not used for living. The world is home (significantly, Maya, like Spanish and English but unlike French and many other languages, differentiates "house," **naj,** from "home," **otoch**).

Spiritual Connections

This appreciation of the landscape is connected, in some sense, with the religious representation of it. The traditional folk religion of Yucatan has been, for centuries, a syncretism of ancient Maya religion and Spanish Catholicism. Catholic ritual and language, as well as Catholic beliefs, were grafted onto the Maya religious stock. The traditional elements that remained were a tight complex related to farming, land, and landscape. The religious practitioner is known as a **jmeen,** which means simply a 'doer.' This officiant has been called shaman, priest, and witch doctor, but he is none of the above. He is simply a ritual officiant—an expert in organizing and carrying out formulaic rites. These rites ask the spirits of the land for help with rain, fertility, and increase, and thank them for favors rendered (Redfield and Villa Rojas 1934).

Rites are of three broad types:

Ch'a chaak. This ceremony is a major prayer for rain, originally lasting two to several days, now completed in a day (though the preparations still take several days). It involves offering food and drink, including honey mead, on an altar decorated with fresh *ja'abin* foliage. The spirits addressed are the **chaak,** the rain gods. Traditional sacred breads are cooked in the *piib* (earth oven), and a stew of poultry, colored with achiote, is made.

Janlikool. This ritual asks for help—notably protection and fertility—from the spirit or spirits of the milpa. They are fed, and so are any and all relatives and neighbors of the person whose milpa is involved. A large altar is prepared, and a great deal of food is cooked in the underground oven.

Loj. This rite is a general request for help, protection, removal of bad luck, fertility, or other desired supernatural commodities. It is held for orchards, domestic animals, and other major rural concerns. It is basically the same as the preceding, but somewhat smaller in scale; in fact, a *janlikool* might be considered a special *loj* for a milpa. The spirits involved in a *loj* are various spirits of the land. Among these the most powerful are the **yuntsiloob** or **yum il k'aax** ("lords of the forest," in Spanish *dueños del monte*). Various minor spirits such as the **alux** (elves) may also be addressed.

Within the last generation or two, this syncretic synthesis has been subjected to several pressures. First and most important has been the coming of Protestantism. Though most Chunhuhubians are nominal Catholics, most of the actual churchgoers are Protestant. Second has been the Catholic response, which includes a dramatic shift (not confined to Mexico) toward a more puritanical, radically anti-syncretic form of the faith. Catholic priests and other religious arbiters are far less tolerant of syncretism and folk belief. Third has been the general trend of modernization—the "disenchantment" that Max Weber described (Gerth and Mills 1958: 155).

In some neighboring villages, tradition continues strong. Terán and Rasmussen (1994; see also Dzib et al. 1992) report that Xocen remains highly traditional. Its Protestants moved out after a land dispute and settled a new town near Chunhuhub. In Xpichil, a community with a strong heritage of independence and of local religion, the Catholic circuit-riding priest was told not to come back after he tried to "purify" the local faith in the name of modern Catholicism. The folk of Xpichil simply took over the church. They now use it for **ch'a chaak** and other traditional rituals.

Chunhuhub has tended to drop the old rituals, or to reduce them to brief forms. However, syncretism and reverence for the land and its products are still very much alive. The *loj* and *janlikool* rituals continue, often under the (traditional) Spanish term *actas de gracia.* Grace being more a Protestant concern than a Catholic one, it is perhaps not surprising to find these more commonly among the Protestants than among the Catholics. They have been reinterpreted as celebrations of thanks for God's grace, and supplications for its continuance. At one I attended, a local Protestant minister replaced the *jmeen* as speaker, and a talented Gospel music group performed with electronic amplification—rather a change from the soft chants of the *jmeen.* The food, order of activity, and general *mise-en-scene,* however, were fully traditional.

Religious rituals of this kind bring the community together, and represent it in both direct and symbolic ways. They also stimulate, integrate, and harmonize particular emotions; in particular, they stimulate good fellowship and other pro-social feelings. As such, they are important to community identity and survival. This is generally and explicitly recognized by the Maya, who state that these rituals are valuable for bringing people together and maintaining ties of friendship and mutual support.

What is more important for present purposes is that the same rituals integrate people and the land. Even in the Protestant version, in which supplication is to a remote high god rather than to the local spirits, people are brought into a close, direct, intimate relationship with the land, and also with their economic activities therein. Not only social activity, but also farming and ranching, are represented in ceremony and symbol. Not only human society, but also the whole ecosystem and the realm of interaction between humans and nonhumans, is directly and intensely involved. This is not a use of abstract "nature" as a source of symbols of the social group; this is actual representation of the interactive realm of people, plants, and animals. The Maya regard milpa agriculture as life itself, rather than a mere economic activity (Redfield and Villa Rojas 1934; Terán and Rasmussen 1994). For traditional Maya farmers, the milpa/forest management complex has a meaning in the sense of Viktor Frankl (1988), a deep and basic meaning that involves them as total human persons.

Conservation and Implications for Sustainability

One might expect that the Maya of Quintana Roo would be conservation-oriented. They have lived in the forests of the region for a very long time. They are secure in their control of those forests. They understand the resources there and know how to manage them. They live in small, scattered communities that have a great deal of independence. Indeed, a traditional conservation ethic, backed by powerful supernatural sanctions, is universal among traditional cultivators of Quintana Roo and neighboring areas (see, e.g., Dzib et al. 1992; Llanes Pasos 1993; Terán and Rasmussen 1994). The ethic is simple: take only what you and your immediate family need, now; never take very many items at one time; leave some for others; take things in their proper season and condition; never waste. People are on their own recognizance; there are no wardens, though public opinion is strong.

However, there are also supernatural game wardens and forest guards. These include such powerful deities as the "Lords of the Forest." Among these are masters of game, such as the Spirit Turkey or Leaf-Litter Turkey, a giant turkey who takes revenge on everyone who has been hunting too many turkeys (Llanes Pasos 1993: 83; Llanes Pasos is a highly traditional Quintana Roo Maya). When a hunter has taken several animals, he must hold a ceremony and wait for a while before hunting more. Similar ceremonies fine-tune management of trees, fields, and beehives. My friend Don Jacinto Cauich Canul once succumbed to temptation and shot five pacas (basically, giant guinea pigs, *Cuniculus paca*) at one time. Suddenly he saw that a huge ring of animals of all kinds had formed, silently, around him. They were perfectly silent and totally menacing. After a while he heard a reassuring whistle above him. The Lord of the Forest, who communicates by whistling, somehow told Don Jacinto that he would be let off this time–but he had to be careful in the future. Don Jacinto never shot many animals at a time again, and I believe he never again hunted pacas at all.

In the past, fear of the forest spirits stopped most people from overhunting and also led them to stop others who were less afraid of the forest spirits. Social sanctions were thus adequate to save the animals. The majority, afraid of general failure of game stocks and general punishment by supernatural beings, could be trusted to enforce communal will on cheaters. However, with higher population and more guns and chainsaws, Quintana Roo today needs much more scientific management—stringent limits, seasons, careful records of what is taken, and formal enforcement. Moreover, the old conservation ideology is breaking down. The younger generation no longer takes the Lords of the Forest very seriously. Many of them do not even know the details of their traditional religion. If they are religious at all, they follow one or another "modern" sect that, unfortunately, may have forgotten the biblical teachings about stewardship of the land. This urban religiosity is more exclusively focused on God and Jesus (and, for Catholics, Mary and the saints) as transcendent, divine beings. Not only are the old gods banished, but even the Christian figures become remote in Heaven, no longer involved with the immediate lives and fortunes of the community. Today, individuals submit to Jesus rather than working with the community to bring the saints right out into the field, or to give a feast that humans and forest spirits can enjoy together.

In 1993, Mexico amended its Constitution. Article 27 allowed for, and encouraged, privatization of *ejido* lands. By that time, population explosion and devastatingly bad land use had rendered many *ejidos* unviable. They were simply too small, overpopulated, and eroded to be of significant value. Moreover, in the 1990s, Mexico was in the grips of "neoliberalism." This economic philosophy was neither new nor liberal; it was simply a return to the policies of the Porfirio Díaz regime of the early twentieth century. Foreign investment was encouraged. Resource extraction got a new lease on life. The communal lands of the rural poor were privatized; rural residents lost out. In 2004, Chunhuhub voted to divide its lands into private (or at least long-leasehold) parcels. Part of the reason was flight from the land by many of the young people, leaving many old settler families without labor power to work the milpas. Remaining families want large parcels for commercial enterprises.

Three drastically different ideologies now impact the Maya. First is their own traditional view, with its different spheres of management and different amounts of effort invested in management (least in deep forest, most in gardens), but all the landscape managed, and all part of the human realm. Maya households, even today, are made up of small buildings set among huge fruit trees and thick vines and shrubbery.

The traditional Spanish view, transplanted to Mexico from Spain (and originally from the Roman Empire), contrasts strongly. Slash-and-burn farming is condemned as "primitive" and "backward." Though 500 years of effort have shown that intensive cultivation always fails in the Yucatan and that the Maya milpa always succeeds unless overdriven beyond all bounds, urban Mexican development agents still promote intensive "modern" agriculture.

The third approach is broadly based on Northern European norms and values but has taken on a life of its own. This is the approach of the newest generation of developers and change agents, both Mexican and foreign. They see "conservation" and "sustainable development" (*desarollo sostenible* in Spanish) as the goal. They value wilderness and wild animals for aesthetic and spiritual reasons. They often idealize Maya agriculture, but they also tend to see it as something whose day has passed. This is partly because of the overuse problems noted above, but partly because they move in a conceptually different realm. They do not believe in the spirits of the forest and game. They usually do not understand the depth and thoroughness of Maya forest management. They learn only slowly that people who love and value animals can also overhunt them. The reality of starving children, and the ideology of *using* resources *carefully* rather than simply preserving everything, are similarly alien to the experience of these recent change agents.

In most places, Maya agriculture goes on, impacted increasingly by population pressure. In some places where traditional urban Mexican values dominate, the land is turned into cattle pasture or mechanically cultivated row crops. In some places, vast reserves have been set aside. Often the Maya who lived there are thrown out, or may find themselves barred from most hunting and collecting. When this happens they may lose stake in conserving the reserves and severely overharvest there, as happened on one *ejido* we studied (Anderson and Medina Tzuc 2005) and as is happening now in some larger reserves (Haenn 2005). Ironically, these reserves exist because the Maya were protecting those same resources before.

Signs of Hope: The Plan Forestal and Cooperative Approaches to Conservation and Sustainability

In a few places, enlightened dialogue by biologists and Maya community members has worked out a system whereby traditional Maya management can go on. This necessarily involves some modern activities too: logging, cash cropping, limited livestock raising. It requires informed help from biologists in setting logging levels, hunting limits, and so on. Given the limited supply of funds and of competent biologists in the area, this scenario remains rare.

Yet positive collaborations could greatly increase (see, e.g., Faust 1998; Faust, Anderson, and Frazier 2004; Primack et al.1998). Hugo Galletti, one of the editors of the Primack volume, was one of the "enlightened biologists" noted in the preceding paragraph, and several of the authors in Faust et al., including Faust and coeditor John Frazier, have long, exemplary records of development work in the Yucatan Peninsula. Many other names could be mentioned. The federal and state governments have cooperated on a forest plan, the Plan Forestal, that stresses conservation and sustainable cropping. (On this plan see Flachsenberg and Galletti 1998; Galletti 1998; Kiernan and Freese 1997; and Murphy 1990. It was called the "Plan Piloto Forestal" when many of these authorities wrote.) The Plan Forestal, and several smaller plans like it, work with local *ejidos* and private property–owners to manage logging and hunting. Government workers assess the amount and quality of timber, and the amount of precious and valuable woods (mahogany, Spanish cedar, etc.), that can be cut in a given year. In a few *ejidos,* they have done the same for game animals.

The Plan Forestal works primarily with *ejidos.* This plan has a large, well-staffed field office in Chunhuhub. Chunhuhub has not seen fit to cooperate with the plan, because Chunhuhub has too little valuable timber left and because the plan charges a small fee for services. Most of the cooperating *ejidos* and ranches are small and isolated communities that have tiny populations and huge expanses of land in relatively mature forest. However, after a record of errors and diversions (Anderson 2005), Chunhuhub has also developed sound management.

The outcome of these debates differed in different *ejidos,* according to the realities of the situation. Conservation ethics won out in proportion to how much land there was per capita, how much help came from truly dedicated government experts, and how united the *ejido* was. Conservation lost where relations with the government were adversarial instead of cooperative. There was sometimes a sharp contrast between Plan Forestal workers, who care about countryfolk, and certain other bureaucrats who regard the Maya as stupid Indians who must be pushed around for their own good. Betty Faust (1998) has reported at length on the way bureaucrats badger the Maya of Pich, Campeche, and Nora Haenn has reported similar problems (and resistance) from elsewhere in that state (Haenn 2005). Similar problems occur in Guatemala (Atran et al. 1999; Schwartz 1990). Chunhuhub's people often reject governmental initiatives, and politeness is not always a high priority in the rejection process. I have seen one high official brought to spluttering and impotent fury by merciless cross-questioning in an open *ejido* assembly. He was reduced to addressing the people as "amiguitos" (little friends). This word is quite insulting (because it is belittling and patronizing) in Quintana Roo Spanish, but it is not insulting enough to be "fightin' words." Nobody missed the message: the official had completely lost his cool, but was too scared to get tough. The policy he came to defend was not adopted. Thus expert advice is vitally important, but so is listening to the people on the ground (see also Anderson 1996). Indigenous and local knowledge remains important (on these issues see Sillitoe 2004, 2007).

The Maya conceive of everything as part of the landscape—part of the great web of being. The sense is that the animals are all part of the forest (*selva* in Spanish or *k'aax* in Maya) and one cannot manage landscape without conserving its parts. Thus conservation, or failure to conserve, is done with a full sense of the ecosystem involved and can be considered part of the relationship to the landscape. Traditional cultures sometimes conserve resources but sometimes do not. Michael Alvard (1995), among others, has criticized traditional ecological management for neglecting to concern itself with matters of season, sex ratio, and animal breeding condition. Traditional moral codes may not easily handle such technicalities, which may require tools such as statistics and (often) mathematical models, as well as a professionally trained enforcement staff, as in the Plan Forestal. Perhaps the ancient Maya fell because they tried their best to save their ecosystem but could not muster the expertise at the local enforcement level.

The more the pressure on the resource, and the more control the group has over it, the more management one can expect (Low 1993). Elinor Ostrom (1990) and Sharon Burton (1992) have elaborated similar theories. Burton finds that a group will work to manage resources in so far as the group is a continually interacting, face-to-face group, in so far as they see the resource is in trouble, and in so far as the resource is identified, understood, and controllable. If such decisions are to be rational, fair, and just, they must be made by equals—that is, people with equal power in the negotiating process. The decisions must be made in groups of reasonably manageable size (Burton 1992). One pathology of a resource management system is top-down control and relatively low connection to specific landscape and community. Actual cases of "tragedies of the commons" are often cases in which a higher authority has taken over control of the resource and has then acted irresponsibly—not cases in which local control has broken down due to local causes (Anderson 1987; cf. Anderson and Anderson 1978; Ostrom 1990).

Reflections

Landscape theory provides a good theoretical way of integrating humanistic, historical, and ideological research with hard science. It also allows integration of agent-based and systems approaches. It has the immense benefit of deliberately transcending the "people vs. nature" or even the "people in nature" attitudes, which have cost us so much in ecological research. The landscape perspective sees all the world, and all parts of it, as the product of processes in which humans and nonhuman entities interact constantly. It is an enormous relief to be able to think like the Maya, and see the world as the product of human-environment interactions that go back indefinitely far in history or prehistory.

Landscape theorists usually have a historical perspective. Landscape is a dynamic process. Forces that create and shape a landscape are always operating, and are themselves always changing. A description of a landscape can never be anything more than a freeze-frame. Moreover, the history in question is contingent; it cannot be predicted from broad parameters. However, it does follow rules: water always flows downhill. It also has to live within limits: one cannot live by hunting seals on the ice in the Yucatan Peninsula.

The landscape approach is very close to traditional cultural ecology as originally envisioned by Julian Steward (1955). This is not surprising, as Steward was part of the same intellectual climate as Sauer and was his contemporary. Sauer, however, gave more place to phenomenological considerations of how people experience landscapes. More recently, environmental history and political ecology began a process of moving back to the human agent after a period of flirtation with biological determinism and heavy emphasis on functionalism. The return to landscape studies continues this progress. However, much of the recent work on landscapes in anthropology and history is done with too little awareness of and discussion of the actual biology and ecology of the landscapes in question. (This is *not* true of the better literature, especially solidly Sauerian work such as the "cultivated landscapes" trilogy [Denevan 2001; Doolittle 2000; Whitmore and Turner 2001].) Escobar has charged some of this literature with essentialism (Escobar 1999). Vayda and Walters (1999) have accused political ecology of being about politics, with too little attention paid to ecology. It is true that many authors of recent works in environmental anthropology and geography seem to be so captivated with philosophy or politics that they forget that people have to eat. Yet people do have to eat, and no system can become "traditional" if it does not provide food, clothing, and shelter. Foucault was certainly right in arguing that knowledge is inseparably involved with power, but recent writers sometimes forget that knowledge involves making a living and feeding the family as well.

Similarly, advocates of agent-based political approaches sometimes forget that systemic emergent effects are real. Political ecologists are also prone to forget that "nature always bats last" (an old watchword of ecologists). And humanists forget that deforestation, soil erosion, water contamination, and game depletion are not just social constructions; they are horribly real and horribly fatal. Conservation and sustainable management are not colonial or neocolonial impositions but survival needs for indigenous societies, and for all humanity. However, some conservation methods, especially those that move people off the land to create "wild" reserves, are often colonial or at least hegemonic, and are often pernicious. Work with Maya communities can and does succeed in saving resources. The Maya have a better knowledge of the forests than do outside biologists or ecologists. Landscape knowledge has real value and real applications.

In the meantime, Quintana Roo's Maya communities struggle on, sustained by their powerful system of local responsibility, local control, and local negotiation of resource management and other important matters. Modernization and change are taking their toll on this system among many younger people, but others continue the tradition, and it will not soon die out.

References

Abram, David. 1996. *The Spell of the Sensuous*. New York: Pantheon.

Alvard, Michael. 1995. "Intraspecific Prey Choice by Amazonian Hunters." *Current Anthropology* 36: 789–818.

Anderson, E. N. 1987. "A Malaysian Tragedy of the Commons." In *The Question of the Commons*, ed. Bonnie McCay and James Acheson. Tucson: University of Arizona Press.

———. 1995. "The Wodewose." Paper presented at American Anthropological Association, annual conference, Washington, D.C.

———. 1996. *Ecologies of the Heart*. New York: Oxford University Press.

———. 2003. *Those Who Bring the Flowers*. Chetumal: ECOSUR.

———. 2005. *The Lords of the Forest*. Tucson: University of Arizona Press.

Anderson, E. N., and M. L. Anderson. 1978. *Fishing in Troubled Waters*. Taiwan: Orient Cultural Service.

Anderson, E. N., and Felix Medina Tzuc. 2005. *Animals and the Maya in Southeast Mexico*. Tucson: University of Arizona Press.

Atran, Scott, Douglas Medin, Norbert Ross, Elizabeth Lynch, John Coley, Edilberto Ucan Ek, and Valentina Vapnarsky. 1999. "Folkecology and Commons Management in the Maya Lowlands." *Proc Natl Acad Sci* 96: 7598–7603.

Becker, Gary. 1996. *Accounting for Tastes*. Cambridge: Harvard University Press.

Burton, Sharon. 1992. "Managing the Commons of Tourist Dollars." Unpublished MS.

Damasio, Antonio. 1994. *Descartes' Error*. New York: G. P. Putnam's Sons.

Demarest, Arthur, Prudence Rice, and Donald Rice, eds. 2004. *The Terminal Classic in the Maya Lowlands: Collapse, Transition, and Transformation*. Boulder: University Press of Colorado.

Denevan, William. 2001. *Cultivated Landscapes of Native North America and the Andes*. New York: Oxford University Press.

Diamond, Jared. 2005. *Collapse: How Societies Choose to Fail or Succeed*. New York: Viking.

Doolittle, William. 2000. *Cultivated Landscapes of Native North America*. New York: Oxford University Press.

Dzib, Alfonso, et al. 1992. *Relatos del Centro del Mundo / U Tsikbalo'obi Chuumuk Lu'um* (collection of texts that accompanies Terán and Rasmussen 1994). 3 vols., ed. by Silvia Terán and Christian Rasmussen. Mérida: Govt. of Yucatán.

Escobar, Arturo. 1999. "After Nature: Steps toward an Antiessentialist Political Ecology." *Current Anthropology* 40: 1–30.

Faust, Betty. 1998. *Mexican Rural Development and the Plumed Serpent*. Westport, CT: Greenwood Press (subdivision of Bergin and Garvey).

Faust, Betty B., E. N. Anderson, and John G. Frazier, eds. 2004. *Rights, Resources, Culture, and Conservation in the Land of the Maya*. Westport, CT: Praeger.

Faust, Betty, and Richard Bilsborrow. 2000. "Maya Culture, Population, and the Environment on the Yucatán Peninsula." In *Population, Development, and Environment on the Yucatán Peninsula: From Ancient Maya to 2030,* ed. Wolfgang Lutz, Leonel Prieto, and Warren Sanderson. Laxenburg, Austria: International Institute for Applied Systems Analysis.

Fedick, Scott, ed. 1996. *The Managed Mosaic.* Salt Lake City: University of Utah Press.

Flachsenberg, Henning, and Hugo A. Galletti. 1998. "Forest Management in Quintana Roo, Mexico." In *Timber, Tourists, and Temples: Conservation and Development in the Maya Forest of Belize, Guatemala, and Mexico,* ed. Richard B. Primack, David Bray, Hugo A. Galletti, and Ismael Ponciano. Washington, D.C., and Covelo, CA: Island Press.

Folan, William J., Betty Faust, Wolfgang Lutz, and Joel D. Gunn. 2000. "Social and Environmental Factors in the Classic Maya Collapse." In *Population, Development, and Environment on the Yucatán Peninsula: From Ancient Maya to 2030,* ed. Wolfgang Lutz, Leonel Prieto, and Warren Sanderson. Laxenburg, Austria: International Institute for Applied Systems Analysis.

Frankl, Viktor. 1988. *The Will to Meaning.* 2nd edn. New York: Penguin Books.

Frazier, John. 2004. "The 'Yucatan Syndrome': Its Relevance to Biological Conservation and Anthropological Activities." In *Rights, Resources, Culture, and Conservation in the Land of the Maya,* ed. Betty B. Faust, E. N. Anderson, and John Frazier. Westport, CT: Praeger.

Galletti, Hugo A. 1998. "The Maya Forest of Quintana Roo, Mexico: Thirteen Years of Conservation and Community Development." In *Timber, Tourists, and Temples: Conservation and Development in the Maya Forest of Belize, Guatemala, and Mexico,* ed. Richard B. Primack, David Bray, Hugo A. Galletti, and Ismael Ponciano. Washington, D.C., and Covelo, CA: Island Press.

Gerth, Hans, and C. Wright Mills, eds. 1958. *From Max Weber.* New York: Oxford University Press.

Gill, Richardson B. 2000. *The Great Maya Droughts.* Albuquerque: University of New Mexico Press.

Gomez-Pompa, A., Michael Allen, Scott Fedick, and J. Jimenez-Osornio, eds. 2003. *The Lowland Maya Area: Three Millennia at the Human-Wildland Interface.* New York: Haworth.

Green, Donald P., and Ian Shapiro. 1994. *Pathologies of Rational Choice Theory.* New Haven, CT: Yale University Press.

Haenn, Nora. 2005. *Fields of Power, Forests of Discontent: Culture, Conservation and the State in Mexico.* Tucson: University of Arizona Press.

Hardin, Garrett. 1968. "The Tragedy of the Commons." *Science* 162: 1243–1248.

Hill, Jane. 2001. "Proto-Uto-Aztecan: A Community of Cultivators in Central Mexico?" *American Anthropologist* 103: 913–934.

INEGI (Instituto nacional de Estadística Geografia e Informatica, Govt. of Mexico). 1986. *Anuario Estadistico del Estado de Quintana Roo.* Mexico City: INEGI.

Janzen, Daniel. 1998. "Gardenification of Wildland Nature and the Human Footprint." *Science* 279: 1312–1313.

Jorgensen, Jeffrey. 1993. "Gardens, Wildlife Densities, and Subsistence Hunting by Maya Indians in Quintana Roo, Mexico." PhD diss., University of Florida, Dept. of Forest Resources and Conservation.

———. 1994. "La cacería de subsistencia practicada por la gente Maya en Quintana Roo." In *Madera, chicle, caza y milpa: Contribuciones al manejo integral de las selvas de Quintana Roo, Mexico,* ed. Laura K. Snook and Amanda Barrera de Jorgenson. Mérida, Yucatán: INIFAP.

Kant, Immanuel. 1970. "Idea for a Universal History with a Cosmopolitan Purpose." In *Kant's Political Writings*, ed. and trans. by Hans Reiss. Cambridge: Cambridge University Press.

———. 1978. *Anthropology from a Pragmatic Point of View*. Trans. Victor Lyle Dowdell (German orig. 1798). Carbondale: Southern Illinois University Press.

Kiernan, Michael J. and Curtis H. Freese. 1997. "Mexico's Plan Piloto Forestal: The Search for Balance between Socioeconomic and Ecological Sustainability." In *Harvesting Wild Species: Implications for Biodiversity Conservation*, ed. Curtis H. Freese. Baltimore: Johns Hopkins University Press.

Kintz, Ellen. 1990. *Life Under the Tropical Canopy*. New York: Holt, Rinehart, and Winston.

Llanes Pasos, Eleuterio. 1993. *Cuentos de cazadores*. Chetumal: Govt. of Quintana Roo.

Low, Bobbi. 1993. "Behavioral Ecology of Conservation in Traditional Societies." Paper, American Anthropological Association, Annual Conference, Washington, D.C.

Merleau-Ponty, Maurice. 2003. *Nature: Course Notes from the Collège de France*. Ed. Dominique Ségard; trans. Robert Vallier (French orig. 1995). Evanston, IL: Northwestern University Press.

Milton, Kay. 2002. *Loving Nature, Towards and Ecology of Emotion*. London, New York: Routdledge.

Murphy, Julia. 1990. *Indigenous Forest Use and Development in the "Maya Zone" of Quintana Roo, Mexico*. MA diss., York University, Ontario, Canada.

North, Douglass. 1990. *Institutions, Institutional Change and Economic Performance*. Cambridge: Cambridge University Press.

Ostrom, Elinor. 1990. *Governing the Commons*. New York: Cambridge University Press.

Primack, Richard B., David Bray, Hugo A. Galletti, and Ismael Ponciano, eds. 1998. *Timber, Tourists, and Temples: Conservation and Development in the Maya Forest of Belize, Guatemala, and Mexico*. Washington, D.C., and Covelo, CA: Island Press.

Putnam, Robert D. 1993. *Making Democracy Work*. Princeton, NJ: Princeton University Press.

Redfield, Robert, and Alfonso Villa Rojas. 1934. *Chan Kom: A Maya Village*. Washington, D.C.: Carnegie Institute of Washington

Redman, Charles. 1999. *Human Impact on Ancient Environments*. Tucson: University of Arizona Press.

Robbins, Paul. 2004. *Political Ecology*. Oxford: Blackwell.

Sauer, Carl. 1925. *The Morphology of Landscape*. University of California Publications in Geology 2:2. Berkeley: University of California Press.

———. 1963. *Land and Life: A Selection from the Writings of Carl Ortwin Sauer*. Ed. John Leighly. Berkeley: University of California Press.

———. 1971. *Sixteenth Century North America*. Berkeley: University of California Press.

Schwartz, Norman. 1990. *Forest Society*. Philadelphia: University of Pennsylvania Press.

Scott, James. 1985. *Weapons of the Weak*. New Haven, CT: Yale University Press.

Sillitoe, Paul. 2004. *Investigating Local Knowledge: New Directions, New Approaches*. Aldershot: Ashgate.

———. 2007. *Local Science vs Global Science: Approaches To Indigenous Knowledge In International Development*. Oxford: Berghahn.

Steward, Julian. 1955. *Theory of Culture Change: The Methodology of Multilinear Evolution*. Urbana: University of Illinois.

Sullivan, Paul. 1989. *Unfinished Conversations.* New Haven, CT: Yale University Press.

Taube, Karl. 2004. "Flower Mountain: Concepts of Life, Beauty and Paradise among the Classic Maya." *Res: Anthropology and Aesthetics* 45:69–98.

Terán, Silvia, and Christian Rasmussen. 1994. *La milpa entre los Maya.* Mérida: Govt. of Yucatan.

Tuan, Yi-Fu. 1977. *Space and Place: The Perspective of Experience.* Minneapolis: University of Minnesota Press.

———. 1979. *Landscapes of Fear.* New York: Pantheon Books.

———. 1990. *Topophilia: A Study of Environmental Perceptions, Attitudes, and Values.* Albuquerque: University of New Mexico Press.

Vayda, Andrew P., and Bradley Walters. 1999. "Against Political Ecology." *Human Ecology* 27: 167–179.

Villa Rojas, Alfonso. 1945. *The Maya of East Central Quintana Roo.* Washington, D.C.: Carnegie Institute of Washington.

Webster, David. 2002. *The Fall of the Ancient Maya.* London: Thames and Hudson.

Whitmore, T., and B. L. Turner III. 2001. *Cultivated Landscapes of Middle America on the Eve of Conquest.* New York: Oxford University Press.

Zajonc, Robert. 1980. "Feeling and Thinking: Preferences Need No Inferences." *American Psychologist* 35: 151–175.

PART 4

Conclusions

Chapter 13

Landscape Ethnoecology
Reflections

Leslie Main Johnson and Eugene S. Hunn

Landscape ethnoecology is grounded in the relationship between people and land—particular people and particular tracts of land. Moreover, landscape ethnoecology implies a *homeland*. The landscape is not mere substrate, nor a bundle of (actual or potential) resources, but instead is invested with a framework of deep meaning. In landscape ethnoecology, relevant understandings range from the very particular grounded ecotopic knowledge of geomorphology, biogeography, and hydrology to overarching cosmological formulations, and levels are mutually interactive. In this inclusive and multileveled perspective, landscape ethnoecology is compatible with the original formulation of cultural ecology by Julian Steward in the 1950s (Steward 1955).

In our treatment we privilege the grounded local knowledge of place and the domain of landscape, rather than extending our consideration to ways of making a living and the whole cultural edifice. Neither do we regard landscape ethnoecology simply as a social or cultural construction, though of course there is deep social and cultural patterning in ethnoecological perception of and interaction with landscapes. We believe that landscape ethnoecology responds to biophysical reality as well as human understanding. Landscape ethnoecology ties pragmatic knowledge of how to make a living on the land to self-definition, cosmology, meaning, and morality. It encompasses interactions among the multiple dimensions of landscape, place, and homeland. In discussing what we mean by landscape ethnoecology, we reflect on the common root of ecology and economy: *oikos,* the Greek word for house. Landscape ethnoecology, then, is the sum of what one needs to know to live in a place.[1]

A range of methodological and theoretical paradigms have been used by the contributors to this volume, reflecting our convergence on the realm of landscape from a number of disciplinary starting points and the multidisciplinary synthesis which landscape ethnoecology represents. It is evident from the theoretical expositions of Hunn and Meilleur, and Mark and his coauthors, that landscape ethnoecology differs from ethnobotany and ethnozoology in that the natural discontinuities that constrain ethnobiological classification are lacking among the geomorphological, climatological, and biogeographical dimensions that help define landscapes. Relations among elements of a landscape are thus more often partonomic than taxonomic, and classifications of landscape features typically involve continuously and independently varying patterns of soil, water, and vegetation, variously modified though long bouts of human history and imbued with spiritual or sacred meanings. Virtually every case study in this volume affirms this. In this sense we substantiate the early intuitions of Sauer (1925), who included physical, biological, hydrological, and cultural features in complex combination as constituents of landscape, and some of the more contemporary formulations of landscape ecology in the European tradition (cf. Pedroli, Pinto-Correia, and Cornish 2006).

The degree to which people's recognition of environmental variation is lexicalized appears to vary between the cases presented here, as, indeed, do the richness and variability of local environments. There is also some disagreement among our authors with regard to how they name culturally recognized landscape variability. Ellen questions why the Nuaulu with whom he has worked for forty years systematically name just a fraction of the number of tropical forest ecotopes that are reported for the Baniwa of the Amazon by Abraão and her coauthors. Is this a real sociocultural difference, a difference grounded in ecological contrasts between Amazonian forests and those of insular Southeast Asia, or perhaps more a consequence of variable methodological standards for recognizing valid "names"? In any case, it seems clear that counting names is not the best measure of indigenous knowledge of landscape. In a similar vein, Anderson finds that people of Chunhuhub recognize, but do not name, many kinds of forest and name forest types by regrowth stage in a non-Amazonian New World environment. However, Chunhuhub has but recently been resettled after having been abandoned for more than a century, which may in this case account for not only the limited number of landscape features named but also the puzzling lack of place names. This reinforces what several of our authors have stressed, the importance of the historical dimension for understanding local landscape ethnoecology.

In her exposition of place and Paiute song, Fowler (p. 243) reminds us of the significant perspectives of Sapir and of Boas, which prefigure interests we pursue here. She writes that "[p]art of Sapir's interest in this case [Paiute delineation of topographic features] was in *the interactive role of language in a group's environmental adaptation,* and thus the utility (and necessity) of its study as a reflection of a group's culture; but in addition, he was also pointing to the role that *the grammar of a language might play* in setting at least some of the parameters for naming" (emphasis added).

Given the increasing importance of geographic information systems (GIS) and remote sensing in rendering landscape information and serving as the methodological foundation for the contemporary management of resources and landscapes, it is useful to consider how the understandings of landscape elaborated by the authors of this volume reflect on the assumptions underlying GIS. Geographic information systems are based on a system of bounded landscape units (polygons) and vectors (linear features like roads) on a set of gridded cells of fixed size that are permanent and geo-referenced. These units are not inherently relational, though geographic information systems have layers that may reveal spatial overlap of different kinds of units based on their geographic coordinates (Haines-Young, Green, and Cousins 1993). Underlying a geographic information system is a defined set of feature types, a "geographic ontology" (Mark and Turk n.d., 2003). Geographic ontologies—the list of "kinds of place" that different cultures and languages allow, or modern polities or government agencies define—are not givens, as Mark, Turk, and Stea eloquently demonstrate for seemingly innocent and obvious terms like "hill" and "stream" (this volume and Mark and Turk n.d., 2003). One needs to keep in mind that the defining features and entailments of these spaces may vary in unexpected ways from one culture to another, and thus that vigilance is warranted against inadvertent distortion of local knowledge in the process of translation.

Two broad formulations of the cultural knowledge of landscape have been presented in this volume: "ethnophysiography" and "landscape ethnoecology." Both of these approaches derive from ethnoscience traditions (especially ethnobiology) and cognitive anthropology (cf. Berlin 1992; Medin and Atran 1999). In contrast to ethnobotany and ethnozoology, landscape elements do not fall into hierarchical taxonomies, but are intrinsically more partonomic as noted above. Partonomy was originally elaborated by Cecil Brown (1976) to describe anatomical terminology, and indeed, landscape terminology is often explicitly anatomic in its phrasing, as in the "head" of the lake, the "mouth" of the river, or in other cultures such as the San Juan Mixtepec Zapotec, the "belly," "mouth," "nose," or "back" incorporated in place names to characterize particular localities. The Wola refer to plant anatomy in their partonomy of rivers, using "sprout" to indicate upstream, "base" to indicate downstream, and "fork" to refer to a confluence (Sillitoe 1996: 111 and footnote 2). Hunn and Meilleur point out that relations of contiguity as opposed to those of similarity are preeminent in landscape ethnoecology. Furthermore, boundaries are not always sharply drawn but may be fuzzy or elusive.

Landscape ethnoecology appears to be characterized by intergrading and/or overlapping classes; ecotopes may interpenetrate as well as form neat tilings.[2] Scale continues to be a challenge. Treatment of fine mosaics or co-occurring small areas of distinctive ecological characteristics is difficult. Similarly, as discussed in the introduction and implicit in several of the chapters in this volume, it is difficult to define where "substrate" leaves off and landform, landscape unit, or ecotope begins. There are also questions, as with plant ecology (cf. McCune and

Antos 1981), as to whether different "strata" of landscape co-vary, or possibly may show discordant patterns. Krohmer partly addresses this with her musings on what seem obvious vegetation-based units to her but are classified by Fulani on the basis of the soils instead, with vegetation becoming a kind of implicit character that may not always be present, as it is possible for similar soils to support different plants.

There are, as our authors (especially Mark, Turk, and Stea) emphasize, different ways of slicing the pie: there can be relative "underdifferentiation" and "over-differentiation" of systems relative to each other. There are contrasting choices in features to attend to in the array of possible attributes of the same pieces of terrain. These choices, and questions of boundedness between and of contrast among categories influence questions of mappability of environmental schemata. Fuzzy boundaries, shifting temporalities, and fine-grained interpenetration of attributes, along with "points" of little areal extent, can all render the mapping conventions of cosmopolitan science impotent to portray in two-dimensional graphic form the embodied and dynamic perspectives of local understandings of landscapes. However, with the atemporal, fixed, and bounded territoriality of contemporary nation states and their subdivisions, such mapping is often necessary. One must, as the old adage implies, avoid mistaking the map for the territory; the representation is not equivalent to, nor more real than, the underlying referent.

Northern environments display shifting potentials, extreme seasonal contrasts, and differing entailments of sites in different seasons, forcing a dynamic, fluid, and variably bounded vision of landscape, inclusive of what may seem ephemeral, such as icescapes (Aporta, this volume). Arid lands may also show strong contrasts and different uses of environments in the wet and dry seasons, as Krohmer shows in her discussion of the use of various units of the Sahel for pasturage in different seasons. The extremely unpredictable occurrence of precipitation and water is highlighted by Mark, Turk, and Stea for their study area in the Pilbara region of northwestern Australia.

Relationships among landscape elements can be in some sense analogous to multiple successional pathways in vegetation ecology: in vegetation ecology, the same seral community may later yield different, more mature phases. Similarly, the array of characteristics of relevance in a place of sharp seasonal contrasts in one season may not be well aligned with the pertinent set of "kinds of place" of a different season, and to characterize any given site one must indicate the set of ecotopes most relevant to local inhabitants throughout the seasons. For example, in the north, wetlands in the open season may be difficult or impossible to travel through, or may require use of a canoe if they are passable at all. However, in the winter, wetlands would be similar to any very open area (including frozen lakes and rivers) that can be traversed by dogsled, snowshoe, ski, or snowmobile and would in those seasons become preferentially chosen travel routes, as they are relatively flat and lack impeding woody vegetation (Nelson 1986). Strong seasonal contrasts in biological affordances and physical traits can in a sense cause a succession of ecotopes to occupy the "same" site through the year.

It is difficult to know how to treat kinds of place defined solely by human activity, such as "trail," "camp," "village," "graveyard," or "grave site." It is similarly difficult to delimit focal place kinds defined by the behavior of animals, such as mineral licks, which are reported in environmental classifications of hunting peoples (e.g. Shepard, Yu, and Nelson 2004; Johnson, this volume). Although many of these sites may have some biophysical or geographic correlates, predicting which sites will be village sites, trails, or mineral licks cannot be accomplished solely from visible biophysical correlates such as landforms or vegetation. The mineral lick example is discussed in Johnson's chapter on the Kaska, and the question of how to define a "berry patch" has been explored by Trusler and Johnson (2008).

One significant question, given the holistic and interactive nature of ethnoecology, is where to stop. This is a rather practical question: what is the domain under consideration, and how shall we recognize the boundaries of what we propose to study? Do we include as "ecotopes" such urban features as malls and intersections? Pilgrimage sites and trails? How do these purported "kinds of place" defined by human activity fit with landscape units or ecotopes that are defined by more biophysical parameters? The editors are of two minds as to whether these sites should best be considered "special-purpose" categories not grounded in objective characteristics, or whether they are simply an aspect of the landscape, that is, something one must know about to live effectively as a member of a culture in a given environment. Johnson tends to think in terms of layers rather than a tiling of biophysical landscape units, and sees categories defined primarily by human use as a "layer" of landscape. She conceives of the question of the boundary of the landscape domain as the problem of the slippery slope of the "human geography layer," whereas Hunn is more inclined to look to a natural basis for defining cultural ecotopes in the objectively defined biophysical characteristics of sites, and to view locations such as spiritual sites as being "special-purpose classifications" analogous to Brent Berlin's differentiation of general-purpose ethnobiological classes versus special-purpose classifications of plants or animals, such as "medicinal plant" or "edible root" (Berlin 1992). Taking their cue from Berlin, Hunn and Meilleur (this volume) propose that we examine landscape ethnoecology comparatively to illuminate fundamental cognitive and adaptive characteristics of this domain. Johnson argues that separating such "place kinds" from a general-purpose ecotopic classification downplays the ubiquity of linkage to and interaction with the "social-ecological system" (sensu Berkes and Folke 1998; Davidson-Hunt and Berkes 2003). In the final analysis, there may be no "best" way to approach landscape ethnoecological classification, but rather a choice of perspectives more or less useful for particular intellectual projects.

Davidson-Hunt and Berkes demonstrate the necessity of integrating the human dimension with the biophysical to appreciate holistic emic representations of local landscape ecological knowledge. They discuss the interactive process that required them to redraw their landscapes to include the "dwelling" perspective. Sacred sites, which could be seen as an aspect of the "human" layer, are represented in the emically valid landscape diagrams presented by Davidson-Hunt and Berkes. One might argue that their "use," if we must frame the significance of such places in these terms, is strongly dependent on a particular cosmology and perspective on utility as understood within a cultural context. Arguably this is an "etic" distinction, and may not reflect the lived experience of people in landscape. It is also possible, as various analyses have indicated, that there are latent ecological functions for such sites and perceptions. For example, many sites visited by that Native American transformer Coyote are also places where people expect to catch fish, harvest roots and berries, and encounter an abundance of game, because Coyote decreed that it should be so (cf. Jacobs 1934, 1937).

Anomalous topographic or vegetation features may be explained by supernatural causation. Krohmer's chapter describes a Sahelian feature, the decayed (ablated) remnants of termite mounds, considered by Fulani to be abodes of dangerous spirits, *jinnaaji,* and avoided lest misfortune or illness befall the trespasser. The Maijuna "devil's swidden," or *mañaco taco,* is a peculiar category of Amazonian forest, likewise the creation of insect activity and similarly avoided (Gilmore 2005 and this volume). The significance of anomaly recalls Mary Douglas's classic work *Purity and Danger* (1966).

Implications of a landscape ethnoecology perspective for sustainable development are taken up by several contributors to this volume. Anderson's ruminations on the conservation of landscapes by indigenous peoples in the Zona Maya, past and present, underscores the significance of resilience (cf. Vayda and McCay 1975; Berkes and Folke 1998; Davidson-Hunt and Berkes 2003, Folke, Colding, and Berkes 2003), that is, the dynamic response of traditions to changing circumstances, emphasized by Davidson-Hunt and Berkes in their account of the Shoal Lake Anishinaabe. The relationship of local knowledge of Amazonian forest types to satellite data in mapping and inventorying the landscape is touched on by Abraão and her co-authors, and has been addressed in other publications by Shepard, Abraão and their colleagues (Shepard et al. 2004; Abraão et al. 2008). The examination of the correlation of indigenous understandings of habitat in the extensive and forested Amazon Basin with those discerned through remote sensing is a particularly interesting area, and one of potential practical and theoretical importance.

Reflections: Landscape Studies, Science, and Management

Landscape studies are one of a number of tools that can be brought to bear upon issues of impact assessment, co-management, land claims, development, and land use planning where local perspectives and more general regional or national perspectives must both inform decisions about the future of the land. Here they may complement heritage and other forms of traditional knowledge studies, as well as more conventional ecological and spatial analysis.

In the past and still to a great degree, development initiatives have privileged Western scientific ecology and geography as bases for examining resource management and use in traditional cultures (see Sillitoe 2006; Fairhead and Leach 1996; Nadasdy 2003; Mulrennan and Scott 2005; Scott 1996; Fienup-Riordan 1999; Tsing 2005; Moore 1999; McDermott Hughes 2005; Anderson this volume). This bias has prevented appreciation of what often are highly effective strategies employed by local and indigenous people in managing their land, water, and biotic resources (cf. Robbins 2004).

The value of traditional landscape knowledge to inform scientific understanding of land and of human relationships to and management of land, and to articulate particularized approaches to the understanding and management of particular landscapes, cannot be overestimated. In Europe, recent formulations of landscape ecology explicitly include human influences and values, and have informed planning efforts and strategies for conservation and environmental preservation (e.g., Pedroli et al. 2006). Emic understandings of landscape have not been strongly represented in the interdisciplinary field of landscape ecology. We believe that our approach, that is, an explicitly ethnographic approach to landscape *ethno*ecology, broadens and strengthens conventional landscape ecology and landscape planning. Careful exposition of a landscape ethnoecology perspective might, for example, help to alleviate the frequent conflicts, seen in many regions of the world, that pit global conservation movements against local and indigenous peoples whose practices and knowledge are dismissed in the effort to save species and environments (see Dwyer 1994; Igoe 2006; Fairhead and Leach 1996; West, Igoe, and Brockington 2006). In too many cases, indigenous and local practices that are seen as contributing to degradation of local environments, rather than as sustaining them, often prove in hindsight to have contributed instead to the very values outsiders wish to preserve (Robbins 2004; Fairhead and Leach 1996). Alternatively, indigenous communities may have degraded their local environments as a consequence of colonial influences destructive of aboriginal social life and livelihood (Anderson, this volume).

Equally pernicious is to essentialize indigenous and local knowledge and practice as an edenic "green" vision, equally unrelated to the complexity of actual local ethnoecology. When people fail to live up to these simplistic and idealistic outside images, they fall from grace, so to speak, and their actual knowledge, however sophisticated, may be completely discounted. Sadly, people themselves may fail to recognize their traditional knowledges and practices as valid or real knowledge, as Baviskar shows for people using the Great Himalaya National Park in India (Baviskar 2000). In these instances landscape ethnoecological work could help present some of the complexity of local knowledge and perhaps validate it both to local people and to outside interests.

Ethnoecological knowledge of landscape is localized, nuanced, and tightly integrated with other aspects of the social and cultural context. Such knowledge is significant for social/cultural resilience, which increases in importance as change accelerates in both biophysical and social realms around the globe. In applying ethnoecological knowledge, it is crucial to keep in mind social, cultural, and ecological contexts, often difficult in face of the policy and administrative pressures of state and international institutions. We cannot afford to treat the "etic" (that is, the cosmopolitan scientific approach to landscape and ecological process) as objectively true and the local "emic" ethnoecological understandings as conditional and partial, and by implication quaint or inferior. They have different goals and purposes. Yet traditional perspectives on landscape and cosmopolitan scientific perspectives can be complementary lenses that can enhance our overall understanding by examining their points of congruence and their differences and seeking to understand both. Local communities pursue different goals in their understanding of landscape and express contrasting, if not complementary, strengths and weaknesses compared to cosmopolitan scientific perspectives. Traditional and modern scientific perspectives on landscape can provide "binocular vision," which can help us understand both local and global perspectives.

Thus the relationship between cosmopolitan science of landscape and traditional ethnoecological knowledge of particular landscapes should be seen as mutually informing. Nancy Turner and Carla Burton (personal communication) suggest that traditional knowledges and cosmopolitan scientific knowledge of landscape and ecological processes can best be seen as enabling a symbiotic relationship that enriches human ability to live sustainably on the planet. A symbiotic approach to landscape enables the joining and complementary use of different systems of environmental knowledge, bringing both to bear on a fuller understanding of land and of the human relationship to it.[3]

"Data mining", the processing and insertion of "TEK bites" into gaps in the scientific edifice, will not yield new insights and understanding but may distort local understandings to the point of unrecognizability (cf. Stevenson 1998; Nadasdy 1999, 2003). Unfortunately, as these authors and others have argued, much of the "integration" of traditional understandings into planning and management structures in places like the Canadian North has in fact taken the cosmopolitan scientific perspective and the bureaucratic policy environment as a given, resulting in the dismissive cooptation of traditional ecological knowledge. Power sharing and a real willingness to learn from one another are needed to ensure that lip service and appropriation and distortion of knowledge are avoided in practice. Respect (in the sense articulated by Nadasdy 2003 and Scientific Panel for Sustainable Forest Practices in Clayoquot Sound 1995) for knowledge and knowledge holders is important, and it is fundamental to the challenging business of negotiating understanding across cultural, disciplinary, and epistemological divides.

Johnson's attention was first drawn to the difficulties of expressing local understandings of habitat and ecological relationships in ways compatible with typical land use planning paradigms when she watched wildlife researchers at the Gwich'in Renewable Resource Board attempt to delineate moose habitat "polygons" from the input of elders and hunters. The wildlife manager, following his training protocol, sought to match Gwich'in information about moose occurrence to the generalized biophysical traits on his habitat map, in order to devise a sampling program to assess the viability of the moose population. On another occasion she observed a community researcher interviewing an elder to elicit information about bear sightings using a standardized questionnaire. The usually voluble elder responded in terse monosyllables. Johnson concluded that the young researcher could have learned much more of significance about bears by letting the elder speak freely, expressing what he knew in his own way.

The admirable goals of resource co-management and community consultation have very often been defeated by the implicit and unexamined assumptions of government scientists and bureaucrats, informed by disciplines such as scientific wildlife management. These assumptions effectively preclude communication of real differences in indigenous understanding of ecological dynamics and landscape knowledge, silencing perspectives that do not fit preexisting structures of knowledge and regulation (cf. Spak 2005; Nadasdy 2003; Morrow and Hensel 1992). For example, "swamps" were long considered of no value, even hostile.[4] They were places to be redeemed, made "good" by reclamation for agriculture, a complete type conversion to another kind of place that *is* valued (see Cronon 1983; Ogden 2008). By contrast, for the Kaska a "swamp," loosely translating the Kaska term *tútsel,* is seen as a site rich in potential food, a "grocery store" valuable in itself. These attitudinal differences are often covert, but they play out strongly in the biases of systems and managers, and may cause undesired and maladaptive responses. Direct attention to local landscape categories and their entailments has the potential to bring such tacit assumptions into the open, where they can be considered and acted upon.

The accelerating globalization of economies and increasing pressure on world resources lends urgency to the need to examine perceptions and strategies for land and resource management from the perspective of traditional and local communities. Around the world, the remaining indigenous and local peoples and their landscapes are threatened by these global forces. They must make their concerns heard in an arena where modern scientific land management goals and economic development strategies are taken for granted. These issues have been particularly highlighted by recent large-scale resource development in the Canadian Arctic and sub-Arctic (e.g., the development of huge open-pit diamond mines in the tundra of the Canadian Shield in the 1990s and 2000s, and the renewal of the proposals for gas development in the Mackenzie Delta and nearby areas, with associated pipeline infrastructure to bring the gas south) as well as by ongoing concerns about development in indigenous homelands of the Amazon Basin and elsewhere in the world. Where local people are permitted a voice in speaking to the environmental and sociocultural impacts of such projects, it is difficult to integrate concerns about the environment expressed from indigenous and local perceptions of landscape with scientific assessments of ecosystems, endangered species, and local environments, and harder still to give these perceptions weight in the economistic calculus of costs and benefits as framed by developers and governments (Nadasdy 2003; Spak 2005; Hornborg 1998; Baviskar 2000; Mulrennan and Scott 2005; Spaeder 2005; see also Tsing 2005).

Reflections: Human Relationship to Landscape in the Context of Mobility and Global Linkages

In thinking about landscape ecology and human relationships to landscapes, there are several key concepts that must be considered. There is a comforting desire for the simple "pure case" of a small-scale community in long-term relationship with a discrete landscape, changing over time, but still with detectable continuity in adaptation and change. However, in the last half-century and indeed, throughout human history, the phenomena of displacement, delocalization, and relocalization have been pervasive and important, and migration of individuals and populations, voluntary or not, is a hallmark of our times. Thus, we must consider the effects and consequences of removal from Place, of becoming delocalized, and of (perhaps), relocalization as migrant populations settle, put down roots, so to speak, and develop a new relationship to place. Such a process certainly took place, for example, with the European and African peoples transplanted to North America historically, and with the reestablishment of strong relations to (transformed) cultural landscapes in new places, repeated again and again over the three centuries of westward expansion. Johnson's partner's family, for example, has lived in and around the communities of Shell Lake and Hawkeye, Saskatchewan, for 100 years now, since the early 1900s, comprising five generations on the land and still farming in the same area. Although their forebears had diverse origins in Ireland and the United States, the family now has (relatively) deep roots in one place, their own oral histories, and a deep knowledge of that landscape.

Much is now being written about persistance and change in traditional knowledge in migrant populations, including maintenance of local cultivars in new places and barriers to maintaining this knowledge (e.g. Peña 2006 on local Mexican cultivars in Los Angeles), and more locally in Quintana Roo (the creation of gardens in new communities that service Cancún; see Greenberg 2003). The persistence of Mixtec ethnozoological terms among U.S. migrants has recently been explored (Aldasoro Maya 2007), and the impact of immigrant Southeast Asian mushroom and floral gatherers in the U.S. and Canada on land and shared resources, a by-product of translation of knowledge into new contexts, has also been noted (e.g., Ballard and Huntsinger 2006).

It is also important to think about the layering or interpenetration of landscapes. Pastoralists and hunters, who traditionally had low population densities and shifting locales, are particularly susceptible to superposition of someone else's (sedentary, bounded) notion of landscape over their own. Examples include Hekkinen (2006) regarding Saami reindeer herders in modern Finland, Thunder-Hawk (2000) on formerly nomadic buffalo-hunting people now confined to the Rosebud Sioux Reservation in the United States, and Vitebsky (2005) on the relationship between the Eveny landscape, the Soviet North, and post-Soviet Siberia. William Cronon's classic study of ecological change in the northeastern United States touches on this for historic data (Cronon 1983).

The establishment of national parks on the American "Yellowstone" model or the creation of other protected areas in the homelands of local or tribal peoples also implies interpenetrating or layered landscapes. Such cases have had enormous repercussions in South Asia and in Africa. Both gathering of subsistence resources and pastoralism are implicated in the South Asian examples, as described by Sundar (2000) for an Indian Joint Forest Management case, and in an Indian National Park example discussed by Baviskar (2000). In East Africa pastoral lifeways and recent sedentarization intersect with notions of wilderness protection and wildlife conservation with complex and problematic results (West et al. 2006; Igoe 2006). Other examples include Tlingit gull egg harvests in Glacier Bay, Alaska (Hunn et al. 2003), where park management practices based on wildlife management principles excluded Tlingit from a traditional and sustainable food harvest, and Kluane National Park in the Canadian North (Nadasdy 2003), where members of the Burwash First Nation had great difficulty framing their perspectives on animals and the land in ways that could inform sheep co-management plans, despite goodwill and sincere effort on all sides. Similarly complex negotiations have been described in Mexico, concerning governance and sustainability in the Calakmul Biosphere Reserve (Haenn 2006) and competing values of indigeneity and fisheries conservation at the mouth of the Colorado River (Muehlmann 2006).

The forced shifts of populations in various countries such as Zimbabwe (e.g., Kinsey 1982; Moore 1999; McDermott Hughes 2005) and migrations for economic reasons, as in the Cancún area in Mexico, are part of the intranational processes of dis-Place-ment, which have significant ramifications for nuanced knowledge of the land and attachment to land, particularly where they intersect with the aftermath of conflict or the imposition of government land use and management policy. The promotion of internal colonization and landscape transformation by transmigrants in southern Kalimantan, Indonesia, discussed by Tsing (2005), also serves to disrupt both preexisting landscape ecologies, and to inject into local settings large populations whose interest in the land is opportunistic, and who lack both knowledge of and attachment to the lands in which they settle. Similar conditions obtain in resource frontiers in Amazonia. The imposition of reserves and alienation of land in British Columbia is also part of this process; Palmer's (2006) detailed examination of the displacement of the Secwepemc of Alkalai Lake provides a fine-scale Canadian example. The historic displacement and dispossession of Native Americans in the U.S. and Aborigines in Australia have similarly created disconnections between people and land, alluded to above in the Rosebud Sioux example (ThunderHawk 2000).

There are also shifts in cultural landscapes in Eastern Europe, detailed in Palang et al. (2006), where the landscape is "layered" and the present reconfigures the former pastoral working landscape in the Kras region of Slovenia as a recreational "pseudomorph" of the older cultural landscape, creating the form without the original function, for recreational enjoyment of more affluent populations from urban areas in Italy and Slovenia (Palang et al. 2006). The landscape shifts to become "scenery" for the consumption of elites escaping from stressful urbanized lives.

There is a strong association of traditional and local peoples, especially "indigenous" or "tribal" peoples, with protected areas (as, for example, the work of Shepard et al. [2004] with Matsigenka who inhabit the Manu Biosphere Reserve and National Park, as well as many other examples on all continents). Luisa Maffi (2001) and others have focused on this with the goal of saving biocultural diversity. Maffi champions the notion that indigenous peoples are in some sense responsible for the high biological diversity of their homelands. It is only recently that government and international policies and NGOs have been able to conceive of protected areas as including their human populations, and much debate remains about the degree of autonomy of reserve-dwelling people who continue their traditional relationships with their homelands (and even what this might mean in terms of the debate about "frozen" traditions, versus resilience and adaptation in the face of changing contexts and possibilities).

Other people's landscapes become mythic places in the imaginaries of wealthy Europeans, North Americans, and Japanese, to be consumed as "tourist destinations," a special kind of imposition of one people's vision of landscape on another's. The tourist destinations may combine natural and cultural diversity as a package, commoditizing both for those who have the money to enjoy them. This is not to imply that those who visit designated natural areas, historic sites, or ethnically interesting regions are necessarily unsophisticated consumers, and they may in fact have strong feelings for landscapes and interest in the ways of life of the peoples whose countries or homelands they visit. However, these impositions of one people's recreational landscape on another's homeland shift meanings of landscapes in complex ways. Sometimes local people can leverage the interest of outsiders to improve their own situations, whereas at other times the voices or economic pressure created by outsiders can serve to further displace or constrain local peoples (e.g., the separation of local people in Costa Rica from national parks, which are configured to cater to international ecotourists and are largely inaccessible to adjacent communities [Ramirez-Sosa pers. comm. 2007]).

Conclusions

In the complex relationships among peoples that characterize the contemporary world, landscape studies are important for conservation of the heritage and languages of traditional peoples. Their languages and relationships to the land, formative to both land and people, are richly implicated in narrative, place names, and specific linguistically coded perceptions of place kinds. This vocabulary and associated knowledge is quick to disappear with a people's removal from the land, and changes in lifeways and the insights contained therein are likewise vulnerable, as Catherine Fowler (pers. comm.) has pointed out. Landscape terminology is significant for understanding linguistic structures and the lexicon but also for understanding the meaning of landscapes for local peoples. At a time when the majority of the world's languages are endangered (Maffi 2001; Terralingua 2006; Krauss 1992), it is important to document such knowledge while it is still possible, and to record as much as possible of its social, cultural, and environmental context.

Kat Anderson has pointed out (pers. comm.) that landscape ethnoecology is significant in ecological and ecocultural restoration efforts, and such knowledge is of strong interest to tribes in the United States. As the land and relationship to it is fundamental in the relationships of many indigenous peoples around the world, landscape ethnoecology also has application in the bolstering of cultural identity and group pride, apparent in settings as diverse as the United States, the Canadian North, and Australia.

Although we have alluded above to difficulties in the relationships of indigenous and local peoples to governments, bureaucracies, scientists, and the international conservation movement, landscape ethnoecology still holds real promise of playing an important role in co-management and conservation. It can help to articulate the distinct syntheses of knowledge and bases of understanding of homelands and local environments possessed by indigenous and local peoples, and in the sense suggested by Nancy Turner, it can contribute to a respectfully negotiated mutualistic approach to co-management. The models for this are few in the literature we have surveyed, but the co-management process arrived at through extensive negotiation in Clayoquot Sound does give some hope (see Scientific Panel 1995; Goetze 2005). As articulated by contemporary expositors of political ecology (see Robbins 2004 and many specific studies), issues of power are real. Despite these structural factors, people of goodwill can explore ways to find common ground. One must not be naive about obstacles, but we cannot move forward without articulating and examining the richness of local landscape knowledge and seeking ways to apply these insights for a sustainable future.

The landscape ethnoecology approach also allows us to query human/environmental relations cross-culturally and illuminate the nature of human understanding of the natural world.

It is important not to essentialize or blur local nuance in this effort, but neither should we simply rest with a collection of particulars. We need to look for similarities or "universals" as part of understanding ourselves and the relationship of our species to its varied environments in general. We all occupy one world, and its common future is of concern to us all. Learning from the many rich and subtle ways that the world's peoples know their environments is important. Through such research we can hope to gain insights into sustainability that may give all of us options for the present and future.

Notes

1. This is similar to Ward Goodenough's classic definition of culture. He wrote: "As I see it, a society's culture consists of whatever it is one has to know or believe in order to operate in a manner acceptable to its members, and do so in any role that they accept for any one of themselves" (1957: 167).

2. One of us (Hunn) suggests that apparently overlapping or interpenetrating ecotopes may indicate that multiple logically independent classificatory schemes are in play, as in the case of special-purpose ethnobiological classifications.

3. Anna Tsing has productively examined this fraught area of communication and articulation of local understandings of land with those of governments, industrial developers, and environmentalists in her recent monograph *Friction, An Ethnography of Global Connection* (2005), which is focused on the forested Meratus Dayak homeland in Indonesia.

4. In cartoon culture, for example, swamps are full of mosquitos, alligators, and bottomless mires, and are places where characters may become waylaid, hopelessly lost, or endangered, encapsulating a long-standing prejudice against places that cannot be put to the plow, which have been associated with diseases such as malaria, and which impede travel both by land and water. This is eloquently expressed in Ogden's historical account of Everglades management and restoration. She writes: "Without reclamation, the Everglades was considered miasmic and dangerous, uncivilized, and certainly worthless" (2008: 22). Only recently has scientific understanding of "swamps" (reconfigured as "wetlands") come to appreciate the many ecosystem services these places perform, as well as their role in maintenance of biodiversity in their capacity as critical habitat for many kinds of fishes, birds, mammals, vascular and non-vascular plants, and insects.

References

Abraão, Marcia Barbosa, Bruce W. Nelson, João Claudio Baniwa, Douglas W. Yu, and Glenn H. Shepard, Jr. 2008. "Ethnobotanical Ground-Truthing: Indigenous Knowledge, Floristic

Inventories and Satellite Imagery in the Upper Rio Negro, Brazil." *Journal of Biogeography* 35(12): 2237–2248.

Aldasoro Maya, Elda Miriam. 2007. "The Ñuu Savi (Mixtec) Ethnozoological Knowledge in a Transnational Community." Paper presented at Society of Ethnobiology 30th Annual Conference, 28–31 March, Berkeley, CA.

Aporta, Claudio. 2002. "Life on the Ice: Understanding the Codes of a Changing Environment." *Polar Record* 38(207): 341–354.

Ballard, Heidi L. and Lynn Huntsinger. 2006. "Salal Harvester Local Ecological Knowledge, Harvest Practices and Understory Management on the Olympic Peninsula, Washington." *Human Ecology* 34: 529–547.

Baviskar, Amita, 2000. "Claims to Knowledge, Claims to Control: Environmental Conflict in the Great Himilayan National Park, India." In *Indigenous Environmental Knowledge and Its Transformations: Critical Anthropological Perspectives,* ed. Roy Ellen, Peter Parkes, and Alan Bicker. Australia: Harwood Academic Publishers.

Berkes, Fikret, and Carl Folke. 1998. *Linking Social and Ecological Systems: Management Practices and Social Mechanisms for Building Resilence.* Cambridge: Cambridge University Press.

Berlin, Brent. 1992. *Ethnobiological Classification: Principles of Categorization of Plants and Animals in Traditional Societies.* Princeton, NJ: Princeton University Press.

Brown, Cecil. 1976. "General Principles of Human Anatomical Partonomy and Speculations on the Growth of Partonomic Nomenclature." *American Ethnologist* 3: 400–424.

Cronon, William. 1983. *Changes in the Land: Indians, Colonists and the Ecology of New England.* New York: Hill and Wang.

Cruikshank, Julie. 2005. *Do Glaciers Listen? Local Knowledge, Colonial Encounters and Social Imagination.* Vancouver: UBC Press.

Davidson-Hunt, Iain J., and Fikret Berkes. 2003. "Nature and Society through the Lens of Resilience: Toward a Human-in-Ecosystem Perspective." In *Navigating Social-Ecological Systems: Building Resilience for Complexity and Change,* ed. F. Berkes, J. Colding, and C. Folke. Cambridge: Cambridge University Press.

Deur, Douglas, and Nancy J. Turner, eds. 2005. *Keeping It Living: Traditions of Plant Use and Cultivation on the Northwest Coast of North America.* Seattle: University of Washington Press and Vancouver: UBC Press.

Douglas, Mary. 1966. *Purity and Danger: An Analysis of the Concepts of Pollution and Taboo.* New York: Praeger.

Dwyer, Peter D. 1994. "Modern Conservation and Indigenous Peoples: In Search of Wisdom." *Pacific Conservation Biology* 1: 91–97.

———. 1996. "The Invention of Nature." In *Redefining Nature: Ecology, Culture and Domestication,* ed. Roy Ellen and Katsuyoshi Fukui. Oxford and Herendon, VA: Berg.

Fairhead, James, and Mellissa Leach. 1996. *Misreading the African Landscape: Society and Ecology in a Forest-Savanna Mosaic.* Cambridge: Cambridge University Press.

Fienup-Riordan, Ann. 1999. "*Yaqulget Qaillun Pilartat* (What the Birds Do): Yup'ik Eskimo Understanding of Geese and Those Who Study Them." *Arctic* 52(1): 1–22.

Folke, C., J. Colding, and F. Berkes. 2003. "Synthesis: Building Resilience and Adaptive Capacity in Social-Ecological Systems." In *Navigating Social-Ecological Systems: Building Resilience for Complexity and Change,* ed. F. Berkes, J. Colding, and C. Folke. Cambridge: Cambridge University Press.

Gilmore, Michael. 2005. "The Cultural Significance of the Habitat *mañaco taco* to the Maijuna of Sucusari." Chapter 3 in *An Ethnoecological and Ethnobotanical Study of the Maijuna Indians of the Peruvian Amazon.* PhD diss., Miami University, Oxford, OH.

Goetze, Tara. 2005. "Empowered Co-management: Towards Power-Sharing and Indigenous Rights in Clayoquot Sound, BC." *Anthropologica* 47(2): 247–265.

Goodenough, Ward. 1957. "Cultural Anthropology and Linguistics." In *Report of the Seventh Annual Round Table Meeting in Linguistics and Language Study,* ed. P. Garvin. Monograph Series on Language and Linguistics, No. 9. Washington, D.C.: Georgetown University.

Greenberg, Laurie S. Z. 2003. "Women in the Garden and Kitchen: The Role of Cuisine in the Conservation of Traditional House Lot Crops among Yucatec Mayan Immigrants." In *Women and Plants: Gender Relations in Biodiversity Management and Conservation,* ed. P. Howard. London and New York: Zed Books.

Haenn, Nora M. 2006. "Changing Governance Models and Green Democracy." Paper presented at American Anthropological Association Annual Meeting, 15–19 November, San Jose, CA.

Haines-Young, Roy, David R. Green, and Steven Cousins. 1993. *Landscape Ecology and Geographic Information Systems.* London: Taylor and Francis.

Hekkinen, Hannu I. 2006. "Neo-entrepreneurship as an Adaptation Model for Sustainable Reindeer Herding." Paper presented at American Anthropological Association Annual Meeting, 15–19 November, San Jose, CA.

Hornborg, Alf. 1998. "The Mi'kmaq of Nova Scotia." In Voices of the Land: Identity and Ecology in the Margins/Lund Studies in Human Ecology 1. ed. A. Hornborg, and Mikael Kurkiala. Lund: Lund University Press

Hunn, Eugene S., Darryll R. Johnson, Priscilla N. Russell, and Thomas F. Thornton. 2003. "Huna Tlingit Traditional Enviornmental Knowledge, Conservation, and the Management of a 'Wilderness' Park." *Current Anthropology* 44(suppl.): S79–S103.

Igoe, Jim. 2006. "Ecosystem Dynamics and Institutional Inertia: A Discussion of Landscape Conservation in Northern Tanzania Systems Approaches." In *Savannas and Dry Forests: Linking People with Nature,* ed. Jayalaxshmi Mistry. Ashgate: Aldershot, Hants, England, and Burlington, VT.

Jacobs, Melville. 1934. "Northwest Sahaptin Texts, Part 1." *Columbia University Contributions to Anthropology* No. 19 (English).

———. 1937. "Northwest Sahaptin Texts, Part 2." *Columbia University Contributions to Anthropology* No. 19 (Sahaptin).

Johannes, R. E. 1981. *Words of the Lagoon.* Berkeley: University of California Press.

Johnson, Leslie Main. 1998. "Trails of Story: Gitksan Understanding of Land and Place, Northwest British Columbia." Paper presented at American Anthropological Meeting November 1998.

———. 2000a. "'A Place that's Good': Gitksan Landscape Perception and Ethnoecology." *Human Ecology* 28(2): 301–325.

———. 2000b. "Envisioning Ethnoecology: Fieldwork with the Mackenzie Delta Gwich'in." Paper presented at CASCA 2000, 27th Congress of the Canadian Anthropology Society, 4–7 May 2000, University of Calgary, Calgary.

Johnson, Leslie Main, and Sharon Hargus. 2007. "Witsuwit'en Words for the Land: A Preliminary Examination of Witsuwit'en Ethnogeography." In *ANLC Working Papers in Athabas-*

kan Linguistics Volume #6, ed. Siri Tuttle, Leslie Saxon, Suzanne Gessner, and Andrea Berez. Fairbanks: Alaska Native Language Center.

Kari, James, and James Fall. 1987. *Shem Pete's Alaska: The Territory of the Upper Cook Inlet Dena'ina.* Fairbanks: Alaska Native Language Center, University of Alaska.

Kinsey, B. H. 1982. "Forever Gained: Resettlement and Land Policy in the Context of National Development in Zimbabwe." *Africa: Journal of the International African Institute* 52(3): 92–113.

Krauss, Michael. 1992. "The World's Languagues in Crisis." *Language* 68(1): 4–10.

Maffi, Luisa. 2001. *On Biocultural Diversity: Linking Language, Knowledge and the Environment.* Washington and London: Smithsonian Institution Press.

Mark, David, and Andrew G. Turk. n.d. "Ethnophysiography." Pre-conference paper for Workshop on Spatial and Geographic Ontologies, 23 September 2003 (prior to COSIT03; e-document).

———. 2003. "Landscape Categories in Yindjibarndi: Ontology, Environment, and Language." In *Spatial Information Theory: Foundations of Geographic Information Science*, ed. W. Kuhn, M. Worboys, and S. Timpf. Springer-Verlag, Lecture Notes in Computer Science (e-document).

McCune, Bruce, and Joseph A. Antos. 1981. "Correlations between Forest Layers in the Swan Valley, Montana." *Ecology* 62(5): 1196–1204.

McDermott Hughes, David. 2005. "Third Nature: Making Space and Time in the Great Limpopo Conservation Area." *Cultural Anthropology* 20(2): 157–184.

McDonald, Miriam, Lucassie Arragutainaq, and Zack Novalinga, compilers. 1997. *Voices From the Bay: Traditional Ecological Knowledge of Inuit and Cree in the Hudson Bay Bioregion.* Ottawa: Canadian Arctic Resources Committee; Sanikiluaq, NWT: Muncipality of Sanikiluaq.

Medin, D.L., and S. Atran, eds. 1999. *Folkbiology.* Cambridge, MA: MIT Press.

Moore, Donald S. 1999. "The Crucible of Cultural Politics: Reworking 'Development' in Zimbabwe's Eastern Highlands." *American Ethnologist* 26(3): 654–689.

Morrow, Phyllis, and Chase Hensel. 1992. "Hidden Dissension: Minority-Majority Relationships and the Use of Contested Terminology." *Arctic Anthropology* 29(2): 38–53.

Muehlmann, Shaylih. 2006. "Indigeneity and Endangerment at the End of the Colorado River." Paper presented at American Anthropological Association Annual Meeting, 15–19 November, San Jose, CA.

Mulrennan, M. E., and C. H. Scott. 2005. "Co-management: An Attainable Partnership? Two Cases from James Bay, Northern Quebec and Torres Strait, Northern Queensland." *Anthropologica* 47(2): 197–213.

Nadasdy, Paul. 1999. "The Politics of TEK: Power and the 'Integration' of Knowledge." *Arctic Anthropology* 36(1–2): 1–18.

———. 2003. *Hunters and Bureaucrats: Power, Knowledge, and Aboriginal-State Relations in the Southwest Yukon.* Vancouver: UBC Press.

Nelson, Richard K. 1969. *Hunters of the Northern Ice.* Chicago and London: University of Chicago Press.

———. 1986. *Hunters of the Northern Forest: Designs for Survival of the Alaskan Kutchin.* 2nd ed. Chicago and London: University of Chicago Press.

Ogden, Laura. 2008. "The Everglades Ecosystem and the Politics of Nature." *American Anthropologist* 110: 21–32.

Palang, Hannes, Anu Printsmann, Éva Konkoly Gyuró, Mimi Urbanc, Ewa Skowronek, and Witold Woloszyn. 2006. "The Forgotten Rural Landscapes of Central and Eastern Europe." *Landscape Ecology* 21: 347–357.

Palmer, Andie Diane. 2006. *Maps of Experience, the Anchoring of Land to Story in Secwepemc Discourse.* Toronto: University of Toronto Press.

Pedroli, Bas, Teresa Pinto-Correia, and Peter Cornish. 2006. "Landscape – What's in It? Trends in European Landscape Science and Priority Themes for Concerted Research." *Landscape Ecology* 21: 421–430.

Peña, Devon G. 2006. "Urban Agriculture and Environmental Justice: Indigenous Diaspora Farmers in South Central Los Angeles." Paper presented at American Anthropological Association Annual Meeting, November 15–19, San Jose, CA.

Robbins, Paul. 2004. *Political Ecology: A Critical Introduction.* Oxford: Blackwell Publishing.

Sauer, Carl. 1925. *The Morphology of Landscape.* University of California Publications in Geology 2(2). Berkeley: University of California Press.

Scientific Panel for Sustainable Forest Practices in Clayoquot Sound. 1995. First Nations Perspectives on Sustainable Forest Practices Standards in Clayoquot Sound. Victoria, B.C.: Ministry of Sustainable Resource Management, British Columbia.

Scott, Colin. 1996. "Science for the West, Myth for the Rest: The Case of James Bay Cree Knowledge Construction." In *Naked Science: Anthropological Inquiry into Boundaries, Power, and Knowledge,* ed. Laura Nader. New York: Routledge.

Shepard, Glenn H. Jr., Douglas W. Yu, and Bruce Nelson. 2004. "Ethnobotanical Ground-Truthing and Forest Diversity in the Western Amazon." In *Ethnobotany and Conservation of Biocultural Diversity,* ed. Thomas J. S. Carlson and Luisa Maffi. Advances in Economic Botany 15. Bronx: New York Botanical Garden Press.

Sillitoe, Paul. 1996. *A Place Against Time: Land and Environment in the Papua New Guinea Highlands.* Amsterdam: Harwood Academic Publications.

———. 2006. "Ethnobiology and Applied Anthropology: *Rapprochement* of the Academic with the Practical." *Journal of the Royal Anthropological Institute* (N.S.): S119–S142.

Spaeder, Joseph J. 2005. "Co-management in a Landscape of Resistance: The Political Ecology of Wildlife Management in Western Alaska." *Anthropologica* 47(2): 165–178.

Spak, Stella. 2005. "The Position of Indigenous Knowledge in Canadian Co-management Organizations." *Anthropologica* 47(2): 223.

Stevenson, Marc. 1998. *Traditional Knowledge in Environmental Management: From Commodity to Process.* Edmonton: Sustainable Forest Network.

Steward, Julian. 1955. *Theory of Culture Change: The Methodology of Multilinear Evolution.* Urbana: University of Illinois.

Sundar, Nandini. 2005. "The Construction and Destruction of "Indigenous" Knowledge in India's Joint Forest Management Programme." In *Indigenous Environmental Knowledge and its Transformations, Critical Anthropological Perspectives.* ed. Roy Ellen, Peter Parkes and Alan Bicker. Australia: Harwood Academic Pubishers.

Terralingua. 2006. "About Terralingua." http://www.terralingua.org/AboutTL.htm. Accessed 19 June.

ThunderHawk, Regina. 2000. "Indigenous Commons and Advocacy to Promote Empowerment: An Examination of Reservation Commons on the Central High Plains." Paper presented at Constituting the Commons: Crafting Sustainable Commons in the New Millennium, 8th Biennial Conference of the International Association for the Study of Common Property, 31 May-4 June, Indiana University, Bloomington, IN.

Trusler, Scott, and Leslie Main Johnson. 2008. "'Berry Patch' as a Kind of Place: The Ethnoecology of Black Huckleberry in Northwestern Canada." *Human Ecology* 36(4): 553–568.

Tsing, Anna Lowenhupt. 2005. *Friction: An Ethnography of Global Connection.* Princeton, NJ, and Oxford: Princeton University Press.

Vayda, A. P., and B. J. McCay. 1975. "New Directions in Ecology and Ecological Anthropology." *Annual Review of Anthropology* 4: 293–306.

West, Paige, James Igoe, and Dan Brockington. 2006. "Parks and Peoples: The Social Impact of Protected Areas." *Annual Review of Anthropology* 35: 251–77.

Vitebsky, Piers. 2005. *The Reindeer People, Living with Animals and Spirits in Siberia.* Boston, New York: Houghton-Mifflin

Notes on Contributors

Marcia Abraão, of Rio Grande do Sul, Brazil, studied geography at The Federal University of Rio Grande do Sul and in 2005 completed her master's degree in ecology from Instituto Nacional de Pesquisas da Amazônia (INPA), Manaus. Her research on Baniwa habitat classification and *campinarana* forest structure on the Rio Içana, Upper Rio Negro, is the basis for her chapter. Her research interests include remote sensing of tropical forests, ethnoecology, and participatory management of land and resources by indigenous Amazonian peoples. She worked from 2005 to 2008 as a consultant for Instituto Socioambiental in a participatory assessment of biodiversity on the Içana. She is currently an environmental analyst for the Chico Mendes Institute for Biodiversity and Conservation (ICMBio) of Brazil's Environment Ministry at the Mapía-Inquini Forest in Boca do Acre, AC.

E. N. Anderson is Emeritus Professor of Anthropology, University of California, Riverside. He has done research in Hong Kong, Malaysia, British Columbia, and other areas, as well as in southeastern Mexico. The major focus of research has been traditional knowledge, use, and management of plant and animal resources. His chief concern has been to find general principles of management, for the purpose of saving and sustainably using plants and animals now and in future. He has also done research on traditional medicine and its relation to public health, often with his wife Barbara Anderson (Seattle University). His books include *Ecologies of the Heart* (Oxford University Press, 1996), *Animals and the Maya in Southeast Mexico* (co-authored with Felix Medina Tzuc, Maya woodsman; University of Arizona Press, 2005), *Those Who Bring the Flowers* (ECOSUR, Chetumal, Quintana Roo, 2003), and *Everyone Eats* (New York University Press, 2005).

Geraldo Andrello holds a PhD in anthropology from University of Campinas, Brazil. He has worked with diverse indigenous groups of the Upper Rio Negro studying mythology, ethnohistory, territorial occupation, migration, and urbanization. During a series of socioeconomic, territorial, and resource-use studies that led up to the demarcation of 15 million hectares of Upper Rio Negro indigenous lands, Andrello carried out pioneering early work on Baniwa habitat classification. From 2005 he worked with Insituto Socioambiental coordinating participatory studies of socio-biodiversity in the Upper Rio Negro funded by the Moore Foundation. He is now professor in the Social Sciences Department at the Federal University of São Carlos (UFSCar) São Paulo, and maintains a permanent collaboration with Instituto Socioambiental's Rio Negro Program.

Claudio Aporta received his PhD in anthropology from the University of Alberta in 2003. He is presently Associate Professor in the Department of Sociology and Anthropology at Carleton University. His current research is documenting Inuit knowledge and use of sea ice across the Canadian Arctic, and he is principal investigator of ISIUOP (Inuit Sea Ice Use and Occupancy Project), funded by International Polar Year, Canada. As a member of the Geomatics and Cartographic Research Centre at Carleton University, he is also highly involved in several research projects, including the creation of the Nunavut Interactive Atlas. He has published a number of important recent articles on Inuit wayfinding, mapping, and geographic knowledge, including adaptation to and use of GPS devices, in *Current Anthropology, Human Ecology, Études Inuit Studies, Polar Record,* and *Arctic*.

João Cláudio Baniwa, from the Baniwa community of Vista Alegre on the Cuiari River in the Upper Rio Negro Indigenous Lands of Brazil, graduated from the Baniwa-Curripaco Indigenous High School of Pamáali is now a teacher at the newly inaugurated Máadzero Indigenous Municipal School. In 2004–2005 he held a research stipend from the "Young Amazonian Scientist" program of the Amazonas state research foundation (FAPEAM), and he worked with Marcia Abraão throughout all phases of her thesis field research.

Fikret Berkes is Professor of Natural Resources at the University of Manitoba in Winnipeg and the Canada Research Chair in Community-Based Resource Management. An applied ecologist by background, he works at the interface of natural and social sciences. Dr. Berkes has devoted most of his professional life to investigating the relations between societies and their resources, and to examining the conditions under which the "tragedy of the commons" may be avoided. He teaches in the area of social and ecological aspects of resource management systems, contributes to theory, and applies his experience both nationally and internationally in a range of areas. His publications include the books *Navigating Social Ecological Systems* (Cambridge University Press, 2003), *Managing Small-Scale Fisheries* (International Development Research Centre of Canada, 2001), and *Sacred Ecology* (Taylor and Francis, 1999).

Iain Davidson-Hunt is Assistant Professor at the Centre for Community-Based Resource Management, Natural Resources Institute, Faculty of Environment, University of Manitoba. His current interests are in the areas of traditional eco-logical knowledge (ethnobotany/ethnoecology), forests and land-use planning, non-timber forest products, rural development, common property resources, and co-management and political ecology. Prior to joining the Natural Resources Institute Davidson-Hunt worked as an applied ethnoecologist in Latin America and Northern Canada, focusing on collaborative research on systems of plant production, forest management, community-based land-use planning, and com-munity economic development. His overall research orientation is that of eth-noecology, which he defines as the study of complex, adaptive relationships of people, resources, and places. An emerging focus of his research is exploring how place-based values and meanings of landscapes can be translated so that they be-come visible to other scales and processes of resource management. The concept of cultural landscapes is an integrating research paradigm that brings together people, plants, and places. Recent publications include the co-edited book *Forest Communities in the Third Millennium: Linking Research, Business, and Policy toward a Sustainable Non-timber Forest Product Sector* (USDA Forest Service, 2001), and several papers on Anishinaabe ethnobotany, sharing knowledge, and indigenous land management.

Roy Ellen was educated at the London School of Economics and Political Sci-ence, where he obtained a PhD in social anthropology in 1972. He has under-taken extensive fieldwork in island Southeast Asia, particularly in Maluku. He has been Professor of Anthropology and Human Ecology at the University of Kent at Canterbury since 1988, and was elected to a fellowship of the British Academy in 2003. Among his numerous publications, he has most recently published *The Cultural Relations of Classification: An Analysis of Nuaulu Animal Categories from Central Seram* (1992) and *On the Edge of the Banda Zone: Past and Present in the Social Organization of a Moluccan Trading Network* (2003).

Samuel Ríos Flores was born along the headwaters of the Sucusari River in the northeastern Peruvian Amazon and continues living along this river to this day as a member of the Maijuna community of Sucusari. He is a master storyteller and singer of Maijuna traditional stories and songs. He is recognized within the Sucusari community as a bastion of traditional knowledge, as he is one of the last people within the community with extensive knowledge and expertise in Maijuna traditional folklore.

Catherine S. Fowler is Foundation Professor of Anthropology, University of Ne-vada, Reno. She earned her PhD in anthropology at the University of Pittsburgh, 1972. Dr. Fowler is a cultural anthropologist who has worked with Great Basin indigenous peoples since 1962 (Northern Paiute, Southern Paiute, Shoshone) on cultural studies, including ethnobiology, material culture, indigenous land management and land restoration, languages, and language retention/restoration programs. She served as secretary/treasurer and president of the Society of Ethnobiology from 1990 to 1997 and on the Board of Trustees, National Museum of the American Indian, Smithsonian Institution, 1996–2001, 2007–present. She is a research associate, member of the U.S. National Museum of Natural History, and is a tribal consultant on land restoration and language projects. Dr. Fowler has more than 120 publications, including *Tule Technology: Northern Paiute Uses of Marsh Resources in Western Nevada* (Smithsonian Folklife Series #6, Smithsonian Institution 1990), *In the Shadow of Fox Peak: Ethnography of the Cattail-Eater Northern Paiute of Stillwater Marsh* (USDA, 1992), and numerous others.

Michael Gilmore is currently Assistant Professor of Life Sciences/Integrative Studies at New Century College, George Mason University. He completed his PhD in 2005 at Miami University in Oxford, Ohio, and his dissertation research project focused on the ethnoecological and ethnobiological knowledge of the Maijuna of the Peruvian Amazon. His dissertation research was funded by a variety of sources, including the National Science Foundation, Phipps Conservatory and Botanical Gardens (Botany in Action), the Elizabeth Wakeman Henderson Charitable Foundation, and Miami University, among others. In 2002, he was awarded a Richard Evans Schultes Research Award from the Society for Economic Botany that also financially supported his dissertation research. Currently, he is working on a variety of community-based and applied ethnobiological research projects with the Maijuna, helping to conserve their biological and cultural resources.

Eugene S. Hunn is Professor Emeritus, Department of Anthropology, University of Washington, Seattle. He received his PhD in anthropology from the University of California, Berkeley, in 1973. His primary research interests are ethnobiology, ethnoecology, and cognitive anthropology. He has conducted fieldwork in Mexico and with Native North American communities. His books include *Tzeltal Folk Zoology: The Classification of Discontinuities in Nature* (Academic Press, 1977), *Resource Managers: North American and Australian Hunter-Gatherers*, co-edited with N. M. Williams (Westview, 1981), *Nch'i-Wána, 'The Big River': Mid-Columbia Indians and Their Land* (University of Washington Press, 1990), and *A Zapotec Natural History* (University of Arizona Press, 2008).

Leslie Main Johnson is Associate Professor in the Centre for Work and Community Studies and the Centre for Integrated Studies, Athabasca University, Alberta, Canada. She lived in northwestern British Columbia for twelve years before returning to graduate school at the University of Alberta in the 1990s, where she

earned her MA and PhD in anthropology. Her research interests include ethnoecology, traditional knowledge, ethnobiology, subsistence, and concepts of health
and healing among northwestern Canadian First Nations, and she has conducted
fieldwork with groups in northwestern Canada. Her publications include *Trails of
Story, Traveller's Path: Reflections on Ethnoecology and Landscape* (Athabasca University Press), and a number of articles on ethnoecology and ethnobiology in *Economic
Botany, Human Ecology, Journal of Ethnobiology, Ecology of Food and Nutrition,
Journal of Ethnobotany and Ethnomedicine,* and *Botany* as well as book chapters on
aboriginal burning, evidence for past plant uses, and landscape perception.

Julia Krohmer was born in 1967 at Überlingen/Lake of Constance, Germany.
She began her university studies in the field of French literature and worked
from 1986 until 1991 in Paris. Subsequently she studied geoecology in Bayreuth,
Germany, where she completed her diploma thesis in 1997 on land use changes
in the Serra de Monchique, in the hinterland of the Algarve Coast of southern
Portugal. She received her PhD in 2004 from the University of Frankfurt for
her work on the environmental classification system, ethnobotany, and land use
system of different Fulani groups in three ecological zones in Western Africa,
some of which is reported here. She has also been involved in the creation and
supervision of an ethnobotanical garden in Papatia, Northern Benin. After living
with husband and son in northern Japan from 2005 to 2007, she is now back in
Germany, working in the transfer office of the Biodiversity and Climate Research
Centre in Frankfurt.

David M. Mark is a SUNY Distinguished Professor in the Department of Geography at the University at Buffalo, State University of New York, where he is the
director of the Buffalo site of the National Center for Geographic Information
and Analysis (NCGIA). Mark also is Project Director of the Integrative Graduate
Education and Research Traineeship (IGERT) in Geographic Information Science, and is a member of UB's Center for Cognitive Science and the National
Center for Ontological Research. Mark completed his PhD in geography at Simon Fraser University (Burnaby, Canada) in 1977, and joined the University
at Buffalo in 1981. He has written or co-authored more than 230 publications,
including numerous articles, book chapters, and technical reports, and has edited five books. His research interests include ontology of the geospatial domain,
geographic cognition, cultural differences in geographic concepts, geographic information science, and digital elevation models.

Brien Meilleur is an affiliate of the CNRS research group Eco-anthropologie et
Ethnobiologie at the Muséum national d'Histoire naturelle in Paris, and is an
ethnobiology and eco-anthropology consultant living near Seattle, Washington.
He previously held positions as director of the Center for Plant Conservation,

Missouri Botanical Garden, St. Louis, and as director of the Amy B.H. Greenwell Ethnobotanical Garden, Bishop Museum, in Hawai`i. He also holds an affiliate appointment in the Department of Anthropology at the University of Washington in Seattle, where he received his PhD in 1986. His books include *Terres de Vanoise: systèmes et productions « plein champs » dans la haute montagne savoyarde* (Le Monde Alpin et Rhodanien, Musée Dauphinois, 2008), *Challenges in Managing Forest Genetic Resources for Livelihoods: Examples from Argentina and Brazil* (IPGRI/Bioversity International, 2004 [co-editor]), *Hawaiian Breadfruit: Ethnobotany, Nutrition and Human Ecology* (College of Tropical Agriculture and Human Resources, University of Hawai'i, 2004), *Hala and Wauke in Hawai`i* (Bishop Museum Press, 1997), and *Gens de Montagne: plantes et saisons, Termingon en Vanoise* (Le Monde Alpin et Rhodanien, Musée Dauphinois, 1985).

Bruce Nelson holds his PhD in botany from the National Institute for Amazon Research (INPA) in Manaus. He is a researcher in the Ecology Department at INPA. His interests include phytogeography, ethnobotany, remote sensing, and natural forest disturbance, for example by wind and *Guadua* bamboo. Nelson worked previously with Shepard and Yu (see below) analyzing satellite images for an ethnobotanical ground-truthing study among the Matsigenka of Peru.

Sebastián Ríos Ochoa is a member of the Maijuna community of Sucusari located along the Sucusari River in the northeastern Peruvian Amazon. He is recognized as one of the most knowledgeable people regarding Maijuna traditional knowledge within the Sucusari community, and he is currently the only Maijuna individual who knows how to read and write the Maijuna language. He has held several important positions within the community government of Sucusari and is a founding member of the Federación de Comunidades Nativas Maijunas (FECONAMAI). FECONAMAI is a Maijuna indigenous organization representing all of the Maijuna communities within the Peruvian Amazon. Its three main objectives are to: (1) conserve the Maijuna culture, (2) conserve the environment, and (3) to better organize all four of the Maijuna communities.

Glenn Shepard holds a PhD in anthropology from the University of California at Berkeley. He has worked with diverse indigenous groups of Peru, Brazil, and Mexico and published widely on ethnobiology, medical anthropology, human ecology, indigenous environmental knowledge and sustainable development in journals including *American Anthropologist, Medical Anthropology Quarterly, Journal of Ethnobiology, Economic Botany, Conservation Biology, Journal of Biogeography, Science, Nature,* and others. He first collaborated with Yu and Nelson on a study of habitat classification and ethnobotanical ground-truthing among the Matsigenka of Peru. Shepard began ethnobotanical research among the Baniwa in 2001 and served as Abraão's co-advisor (with Bruce Nelson) on the 2004–

2005 thesis research presented here. During this time he held a Leverhulme Trust research fellowship at the University of East Anglia, U.K., and is currently a researcher and curator of indigenous ethnology at the Museu Paraense Emilio Goeldi in Belém, Brazil, at the mouth of the Amazon.

David Stea received a BS (hons) in mechanical/aeronautical engineering from (now) Carnegie-Mellon University (1957), an MS in psychology from the University of New Mexico (1961), and a PhD in psychology from Stanford University (1964). He is Professor Emeritus of Geography and International Studies, Texas State University, San Marcos, and Research Associate, Center for Global Justice (Mexico). A co-founder of the field of environmental psychology, his research interests have included spatial cognition, map learning in young children, and developing techniques of effective participatory planning with indigenous peoples around the world. His books include *Image and Environment, Maps in Minds, Environmental Mapping,* and *Placemaking.*

Andrew Turk received the following degrees from Australian universities: B. Surveying, University of Queensland (1971); B. Applied Science in cartography, R.M.I.T. Melbourne (1980); B. Arts (honours in psychology), The University of Melbourne (1992); PhD, The University of Melbourne (1992). His research interests have concerned surveying, cartography, geographic information systems, human factors in IT, ethical aspects of IT, indigenous land rights, usability evaluation of interfaces and websites, bioinformatics, interactive television, digital divide remediation, indigenous community development, conceptions of landscape, and phenomenology. He has published many conference papers, journal articles, and book chapters regarding these research activities.

Douglas Yu holds a PhD in biology from Harvard University. He conducts research in Peru on spatial ecology and evolution of mutualisms. Since 1996, he has worked with Shepard on studies of indigenous habitat knowledge and resource use in Manu National Park, Peru. He is Lecturer in conservation biology at the University of East Anglia, U.K., and Professor of ecology at the Kunming Institute of Zoology, China.

Index

CPSIA information can be obtained at www.ICGtesting.com
Printed in the USA
BVOW071234050212

282170BV00005B/1/P